INTRODUCTION TO DEVICE MODELING AND CIRCUIT SIMULATION

INTRODUCTION TO DEVICE MODELING AND CIRCUIT SIMULATION

Tor A. Fjeldly, Trond Ytterdal, and Michael Shur

A Wiley-Interscience Publication

JOHN WILEY & SONS, INC.

New York / Chichester / Weinheim / Brisbane / Singapore / Toronto

This text is printed on acid-free paper ∞.

Copyright © 1998 by John Wiley & Sons, Inc.

All rights reserved. Published simultaneously in Canada.

Library of Congress Cataloging in Publication Data:
Fjeldly, Tor A.
    Introduction to device modeling and circuit simulation / Tor A.
Fjeldly, Trond Ytterdal, Michael Shur.
        p.   cm.
    "A Wiley-Interscience publication."
    Includes bibliographical references and index.
    ISBN 0-471-15778-3 (cloth : alk. paper)
    1. Semiconductors—Computer simulation.   2. Integrated circuits—
Design and construction—Data processing.   3. Computer-aided
design.   I. Ytterdal, Trond.   II. Shur, Michael.   III. Title.
TK7871.85.F593   1997
621.3815'2'01135369—dc21

                                           96-29929
                                              CIP

10 9 8 7 6 5 4 3 2 1

CONTENTS

APPENDICES

PREFACE

An engineer should not only acquire knowledge but also make it practical and use it. Thirty years ago, the tools that made the engineering knowledge practical included the drawing board and the slide rule. Nowadays, the tool chest includes computers and engineering software. Of the great variety of different computer programs, the electronic circuit simulator SPICE is probably the most important for electrical engineers. The SPICE program, originated and developed at the University of California at Berkeley, is truly a wonderful present to the electrical engineering community worldwide from one of the best American public universities. It allows us to simulate both individual devices and electronic circuits, performing a large number of different analyses needed for tasks such as verification of circuit designs and prediction of circuit performance. It is so flexible and usually so reliable that many engineers use it as a "software oscilloscope." Several excellent books explain how to use SPICE for circuit simulation and analysis, including those by Vladimirescu, Tuinenga, and Sedra, often used as textbooks or as supplementary texts for circuit courses.

However, the results of a SPICE simulation are only as good as the device models and the device parameters used in the simulation. Device technologies change so fast and device characteristics are so different that just using default parameters is almost never justified. If wrong device parameters or models are used in a SPICE simulation, all this computer power will be wasted—true to the old adage "Garbage in, garbage out." Accordingly, it is only natural to link a course in electronic device modeling to SPICE, which is exactly the purpose of the present book: to provide a text for an introductory junior-, senior-, or even graduate-level course on SPICE-oriented semiconductor device modeling. (We also envision this book as a useful reference for practicing electrical engineers.) The reader is expected to possess a basic familiarity with semiconductor device physics on the level of, for example, the text by Streetman (1995).

Less than a decade ago, a typical circuit simulator would run only on mainframe computers. However, the rapid progress in microcomputers has enabled the development of SPICE versions that can run on inexpensive machines, making advanced circuit simulators readily available practically to every electrical engineer and electrical engineering student. Hence, we are now in the privileged situation that we can teach a course on the basics of semiconductor device physics and device modeling using the same computer-aided design (CAD) tool that electrical engineering students will almost certainly use as practicing engineers. This fortunate circumstance allows students to try actual circuit design, bringing semiconductor device physics and modeling (which are often taught as a fairly theoretical subject) down to a very practical level. Two educational versions of SPICE run on microcomputers—PSpice, from Microsim Corporation, and AIM-Spice, developed by the authors of this book together with Kwyro Lee. Most examples and problems discussed in this book can be run on either version of SPICE.

This book relies primarily on the circuit simulator AIM-Spice (for IBM PCs and compatibles) and includes basic information on SPICE along with a detailed AIM-Spice manual (AIM-Spice also comes with a comprehensive on-line help support). This makes the book very useful, not only as an introductory text on semiconductor device modeling, but also as a supplementary text for junior-level courses on electronics and circuit theory. The student version of AIM-Spice can be downloaded from the AIM-Spice home page on the Internet. The address is

http://www.aimspice.com

AIM-Spice is a version of SPICE with standard SPICE parameters that is familiar to many electrical engineers and electrical engineering students. Running under the Microsoft Windows family of operating systems, it takes full advantage of the available graphics user interface. The program will run on PCs with 80386 or compatible microprocessors with at least 4 MB RAM. However, we recommend using more powerful computers with a 80486 or a Pentium microprocessor and with 8 MB RAM. Even this recommended hardware configuration already looks fairly modest at the moment of writing.

In addition to all the models included in Berkeley SPICE (Version 3e.1), AIM-Spice incorporates many new models for silicon and compound semiconductor devices, such as metal–oxide–semiconductor field-effect transistors (MOSFETs), amorphous silicon and poly–silicon thin-film transistors (TFTs), metal–semiconductor field-effect transistors (MESFETs), heterostructure field-effect transistors (HFETs), and heterostructure bipolar transistors (HBTs). The student version of AIM-Spice was used for the device and circuit simulation examples included in the book. A complete professional version of AIM-Spice is available from the authors.

Lee et al. (1993) gave a detailed description of most of the device models used in this new circuit simulator. This book presents a combination of background device physics and technology, a review of existing device models, and, more importantly, a set of new and improved models compatible with the most advanced technology, along with new device characterization techniques for extraction of device parameters. The book was intended to serve as a reference for professional engineers and as a text primarily for a graduate-level course on semiconductor device modeling. Instructors adopting the present book for an introductory course may consider using the advanced book by Lee et al. as their personal reference and, perhaps, as an additional source of problems to be assigned to the students.

Much of the knowledge of software packages is fleeting: You keep up on a particular software package only if you use it on a regular basis. However, once you understand the capabilities and limitations of a program and are provided with a good user's manual, you can get up to speed fairly quickly. Perhaps, the easiest way to get ahead is to use examples that come close to your particular task at hand and modify these examples to suit your needs. Keeping this in mind, we put most of the computer-related information into appendices and provided a large number of examples.

The book is organized as follows: Chapter 1 (Introduction) starts with a simple, motivating example explaining the capabilities and limitations of SPICE simulations. It continues with a historical overview of SPICE and ends with a circuit simulation tutorial intended to quickly enable the reader to understand and operate AIM-Spice. Chapter 2 includes basic semiconductor equations and describes basic material properties of important semiconductors. Chapter 3 is devoted to the modeling of two-terminal devices: p–n junction diodes, Schottky barrier diodes, heterojunctions, and metal–insulator–semiconductor (MIS) capacitors. In Chapter 4, we consider the modeling of bipolar junction transistors, including the HBT. Chapter 5 deals with the operation of and basic model issues of FETs, including the MOSFET, the TFT, and compound semiconductor FETs. The appendices include basic information about SPICE and AIM-Spice.

SPICE (the AIM-Spice package, in particular) allows students to solve engineering problems, bridging the gap between theory and practice. Many problems included in the book are design problems. These problems do not have unique solutions and involve tradeable parameter values.

Most of the problems in the book can be solved (or their analytical solutions can be checked) using AIM-Spice. We believe that students should be encouraged to do just that since AIM-Spice is a state-of-the art VLSI design tool and since real-life problems often do not have analytical solutions.

For a course based on this book, we recommend that two-thirds of the credit hours be allocated to engineering design and one-third to engineering science.

A detailed solution manual and sets of transparencies for lectures based on this book will be made available to instructors.

REFERENCES

K. Lee, M. Shur, T. A. Fjeldly, and T. Ytterdal, *Semiconductor Device Modeling for VLSI*, Prentice-Hall, Englewood Cliffs, NJ (1993).

B. Streetman, *Solid State Electronic Devices*, 4th ed. Prentice-Hall, Englewood Cliffs, NJ (1995).

INTRODUCTION TO DEVICE MODELING AND CIRCUIT SIMULATION

CHAPTER 1

INTRODUCTION TO SPICE

1.1. INTRODUCTION

The SPICE program—originated and developed at the University of California at Berkeley—is possibly the most important computer program package for electrical engineers. Although SPICE is a general-purpose analog simulator, it comes with models for most circuit elements, making it an outstanding tool for performing precise simulations of complex, nonlinear circuits.

With the SPICE simulator, we can calculate direct-current (dc) operating points; perform transient analysis; locate poles and zeros for different kinds of transfer functions; find the small-signal frequency response, small-signal transfer functions, and small-signal sensitivities; and perform Fourier, noise, and distortion analyses. For this reason, it is at the heart of most contemporary computer-aided design (CAD) tools used for circuit analysis and design, which is also why SPICE has such a unique position among computer programs for electrical engineers.

However, the results of a SPICE simulation are only as good as the device models and the device parameters used in the simulation. Device technologies change so fast and device characteristics are so different that just using default parameters is almost never justified. If wrong device parameters or models are used in a SPICE simulation, this wonderful computer power will be wasted, true to the old adage "Garbage in, garbage out."

To illustrate the capabilities and limitations of SPICE for simulation of electronic devices and circuits, we start by considering the simple electronic circuit shown in Fig. 1.1.1, consisting of a p–n junction diode, a resistor, and a voltage supply.

An ideal p–n junction diode is an electronic device with a current–voltage characteristic given by the equation

1

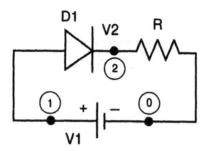

FIG. 1.1.1. Electronic circuit consisting of a *p–n* junction diode, a resistor, and a voltage supply with nodes marked for the circuit analysis using SPICE.

$$I = I_s \left[\exp\left(\frac{V}{\eta V_{th}}\right) - 1 \right] \tag{1.1.1}$$

where I_s is the saturation current, V is the bias voltage across the diode, $V_{th} = k_B T/q$ is the thermal voltage, $k_B = 1.38 \times 10^{-23}$ J/K is the Boltzmann constant, T is the

and η is the ideality factor. This current–voltage characteristic can easily be calculated for any given temperature if the parameters I_s and η for this particular temperature are known. However, if we account for the series resistance R, Eq. (1.1.1) becomes more complex:

$$I = I_s \left[\exp\left(\frac{V - IR}{\eta V_{th}}\right) - 1 \right] \tag{1.1.2}$$

Equation (1.1.2) does not have an exact analytical solution for I [even though an approximate solution can be obtained; see Lee et al. (1993)]. However, we have no difficulty solving this problem using SPICE. All we have to do is to prepare an input file that we open using PSpice or AIM-Spice (or any other version of SPICE for that matter). If we use PSpice, this file may look like this:

```
Diode Circuit
D1 1 2 Diode
V1 1 0 1V
R 2 0 5k
.model Diode D
.DC V1 -2 1.5 0.1
.probe
.end
```

The first line in this file is a title of our choice. The next three lines describe the elements of the circuit. The information about the notation used in SPICE for different circuit elements is given in Appendix A1. Here we simply say that these lines specify to which nodes the elements are connected (as marked in Fig. 1.1.1) and the values of the elements (1 V for the voltage supply V_1, 5 kΩ for the resistance R). The model to be used by SPICE to describe the diode D_1 (called model D) is specified in line 5, followed by a line saying that we want to calculate currents and voltages in this circuit for values of the voltage V_1 varying from –2 V to +1.5 V in steps of 0.1 V. The line .probe is the only statement specific to PSpice, and it tells the program to prepare the computed voltages and currents for plotting using a special plotting program PROBE that comes with PSpice. The last line denotes the end of the input file.

The results of a SPICE simulation using this input file can be plotted by invoking PROBE. (The use of this program is self-explanatory; you learn it by trying it.) Figures 1.1.2 and 1.1.3 show such plots for the current I and the voltage $(V_2 - V_1)$ across the diode as a function of V_1.

It all looks deceptively easy and quite elegant. By just writing a few simple lines of code, we were able to solve a problem that does not have an exact analytical solution and plot the results in a nice way. Moreover, if we are interested in a similar circuit where $R = 100$ kΩ, for example, we need only change a single value in a single line of code and run the program again.

However, before we decide that SPICE is so easy to use that we do not need to learn much about the operation of solid-state devices, we should answer a couple of questions. Semiconductor diodes come in all shapes and sizes. Some diodes are power devices used in power plants for controlling huge currents. These devices have a large cross section, have a large breakdown voltage, and can withstand very large currents. Other diodes are tiny specks of semiconductor material used in high-frequency circuits. They burn out easily and have a small cross section and a small

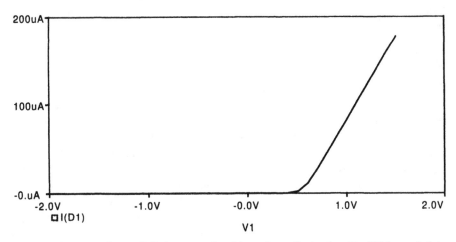

FIG. 1.1.2. Dependence of diode current I on bias voltage V_1 simulated by PSpice and plotted by PROBE.

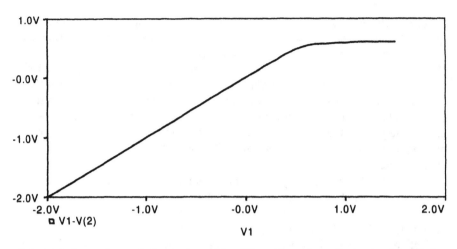

FIG. 1.1.3. Dependence of diode voltage ($V_1 - V_2$) on bias voltage V_1 simulated by PSpice and plotted by PROBE.

answer is it did not. It did not even know what temperature we had in mind (and the diode current is a very strong function of temperature). In this simulation, SPICE used so-called default parameter values given in Table 1.1.1. If these parameters were to describe our diode (but why should they?) and if the operating temperature was exactly 27°C (300 K), which is too warm for a typical room, we would be lucky and the results of the simulation would be essentially correct. If not, these results would be misleading, having little in common with our device.

This can be remedied by supplying SPICE with information about the parameters of our diode. For example, if we know that $I_s = 10^{-10}$ A, the fifth line of the input SPICE file should be replaced with

```
.model Diode D IS=1E-10
```

But how will we know the value of I_s? There are only two ways to find out: Either we have to be able to determine this value "exactly" from the measured data or we have to estimate this value from a model, given the geometric and material parameters of the diode. (Sometimes, however, there is the third, easy way: The SPICE parameters may be supplied by the manufacturer, but we should not count on this.) So, if determining the SPICE parameters is up to us, we must understand how the diode works, especially since most parameters in Table 1.1.1 look ominously strange when we compare them with Eq. (1.1.1). Should the transit time and grading coefficient be taken to be equal to their default values no matter what diode we want to simulate and what the temperature is? The answer to this question is no. In fact, it is very easy to make mistakes in SPICE simulations if we use default parameters.

TABLE 1.1.1. Parameters of *p–n* Diode SPICE Model.

SPICE Parameter	SPICE Parameter Name	Unit	SPICE Default	Chapter 1 Notation
IS	Saturation current	A	1.0×10^{-14}	I_s
RS	Series resistance	Ω	0	R_s
N	Ideality factor	—	1	η
TT	Transit time	s	0	
CJO	Zero-bias capacitance	F	0	
VJ	Built-in potential	V	1	
M	Grading coefficient	—	0.5	
EG	Energy gap	eV	1.11	
XTI	Saturation current temperature exponent	—	3.0	
KF	Flicker noise coefficient	—	0	
AF	Flicker noise exponent	—	1	
FC	Coefficient for forward-bias depletion capacitance	—	0.5	
BV	Reverse breakdown voltage	V	Infinite	
IBV	Current at breakdown	A	10^{-3}	
TNOM	Temperature at which parameters are specified	°C	27	T (K) $(T = \text{TNOM} + 273)$

An even more important question relates to device models. In our input SPICE file, we used the line

```
.model Diode D
```

What exactly is this model "diode D"? It was developed many years ago, and device technology has dramatically changed since then. Is this model still accurate and realistic? Is it correct at elevated temperatures? At cryogenic temperatures? Does it apply to diodes from new semiconductor materials and structures that did not even exist a few years ago?

In order to illustrate this problem, in Fig. 1.1.4 we again plot the current through the diode but in a different scale that allows us to see the small reverse diode current at negative bias voltages. Let us now compare the results with what we expect from Eqs. (1.1.1) and (1.1.2). A quick analysis of these expressions shows that if $-V \gg \eta k_B T/q$, the current must become independent of V_1 and approach the value $-I_s$. According to Table 1.1.1, the default value of I_s is 10^{-14} A. Instead, the simulation predicts a Linear dependence of the diode current on the reverse bias, and at $V_1 = -2$ V, the reverse current is equal to -2 pA, 200 times larger than the expected value!

If we obtain such a surprising and totally incorrect result even for a *p–n* junction diode—one of the simplest semiconductor devices—should we not display a certain

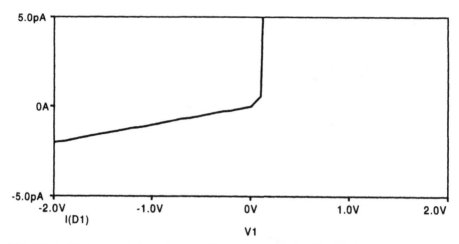

FIG. 1.1.4. Diode current dependence on bias voltage simulated by PSpice and plotted by PROBE, emphasizing the reverse diode current.

such as field-effect transistors (FETs), for example? (We should mention that some SPICE models for FETs may contain more than 100 parameters.) Also, we note that, as a rule of thumb, modern electronic device technology is ahead of our most advanced simulation capabilities. It is certainly far ahead of the relatively simple (by necessity) analytical models used in SPICE.

Therefore, the best way to understand how to use SPICE is to study the SPICE models simultaneously with or as a continuation of a course on the basics of electronics devices. This way, we can acquire an understanding of the SPICE models, their limitations, and the physical meaning of the parameters used, which is exactly the approach that we have taken in this book.

In the next chapter, we review the basic semiconductor equations with emphasis on the relationships needed for the SPICE models. Armed with this knowledge, we will return to the modeling of the p–n junction diode. Subsequently, we will also consider models for other two-terminal devices, for bipolar transistors, and for FETs.

In order to restore the faith in SPICE, let us hasten to explain that SPICE adds to every branch a parallel conductance: GMIN = 10^{-12} mho. This conductance shunts the diode, resulting in a reverse current that is much larger than I_s. If we change GMIN to 10^{-15} mho by adding the line

```
.OPTIONS GMIN=1e-15
```

to the input line, we will obtain a result more in line with what could be expected, as indicated in Fig. 1.1.5.

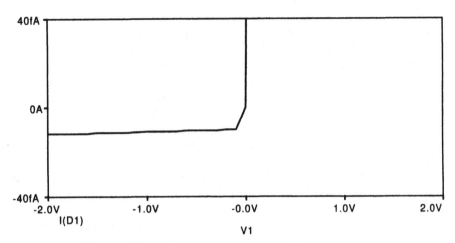

FIG. 1.1.5. Diode current dependence on bias simulated by PSpice showing the reverse diode current with GMIN = 1 × 10^{-15} mho.

1.2. SPICE AND BEYOND

1.2.1. Historical Overview and Versions of SPICE

In 1971, Nagel and Rohrer published the first paper describing the nonlinear computer circuit simulator CANCER developed at the Department of Electrical Engineering at the University of California at Berkeley. The goal of this program was to support the emerging integrated circuit technology. Clearly, a better name was in order, and the next version of this program was called SPICE—an acronym for simulation program with integrated circuit emphasis— and was released in 1972. Many scientists at Berkeley and other institutions contributed to the development and improvement of subsequent versions of SPICE. We should mention Pederson, Vladimirescu, Sangiovanni-Vincentelli, and their students, but literally hundreds of graduate students, professors, engineers, and scientists participated in the development of this enormously important software for the electronics industry. [A good overview of the history of SPICE has been given by Vladimirescu (1994).]

In 1975, Nagel described the next major release of SPICE, called SPICE2. The core of the program still remains intact, even after many improvements and additions. These included improved device models, numerical techniques, and user interfaces, but just as important was the improvement in user support. The Berkeley version of SPICE is in the public domain, but since the late 1970s a number of companies have released commercial versions of SPICE, providing further improvements in user support, in user interfaces, and, in certain cases, also in models and in simulation convergence.

The first commercial version of SPICE, I-SPICE, was developed by NCSS Timesharing in the second half of the 1970s. Other commercial versions include IG-

SPICE from A. B. Associates, HSPICE and RAD-SPICE from Meta-Software, PSpice from Microsim, DSPICE from Daisy Systems, SmartSpice from Silvaco, and AIM-Spice by Lee et al. (1993). Many useful features were introduced by these programs and include schematic capture (allowing users to display circuit diagrams on the computer screen and automatically generate SPICE input files), interactive simulation control, improved convergence, better models, the ability to implement additional user-generated models, good users manuals, and, in certain cases, verified and documented parameter extraction procedures.

Table 1.2.1 summarizes the history of SPICE along with some major historical events.

Two student versions of SPICE—PSpice from Microsim and AIM-Spice— are available for free. (AIM-Spice can easily be downloaded via the Internet; see Appendix A1.)

Many versions of SPICE exist. There have been good reports about HSPICE from fellow engineers and IC designers and we used a student version of PSpice in our courses and a professional version in our research with good results. Together with Kwyro Lee of KAIST, we developed AIM-Spice. Two features of this program that we believe will become progressively more important as device technology develops further are (i) the use of device models based on a precise description of the

also physically based. As demonstrated in Section 1.1, an appropriate choice of model parameters is a paramount condition for a successful SPICE simulation or any other circuit simulation.

In his book, Vladimirescu (1994) quotes a forecast made by Bill Joy at the Twenty-Seventh Design Automation Conference in 1990. In his keynote address, Joy predicted that, in a few years, engineering workstations will operate at 1000 MIPS (millions of instructions per second), have 1000 megabytes of memory, and have 1000 gigabytes of disk storage. At the time of this writing we are still far from this (1996). Sophisticated circuit simulations can certainly use these capabilities, especially for statistical analysis and quality assurance (QA) and quality control (QC) applications. In the meantime, even with available workstations and personal computers, we already have an awesome power at our disposal when we use SPICE. In subsequent chapters, we discuss how to utilize this power.

1.2.2. Principle of Operation

SPICE performs three main analyses: nonlinear dc analysis, transient analysis, and small-signal alternating-current (ac) analysis. All other available analyses, including transfer function analysis, distortion analysis, and noise analysis, are special cases of these three main types. In the following, we briefly describe the numerical methods utilized by SPICE in performing such analyses.

SPICE consists of four main modules, as indicated in Fig. 1.2.1. The input module reads the input file, builds a data structure representing the circuit, and performs a topology check to search for obvious user errors. After the input part is successfully completed, the setup module constructs a pointer system used by the sparse ma-

TABLE 1.2.1. History of SPICE and Some Major Historical Events.

Year	Historical Events	SPICE History
1970	Charles de Gaulle dies Rich oil reserves discovered in the North Sea	CANCER program developed at Berkeley
1971	Idi Amin becomes dictator of Uganda	First version of SPICE released for public use
1975	End of the Vietnam war	SPICE2 released
1976	Mao Zedong dies in China	Most integrated circuit manufacturers adopt SPICE2
1981–1982	Ronald Reagan and Pope John Paul II survive assassination attempts Anwar Sadat assassinated The Falklands War Yuri Andropov comes to power in the Soviet Union	SPICE2 adopted for parallel computer architectures Daisy, Mentor, and Valid (known as DMV) introduced integrated software packages for electronic design on engineering workstations
1984	Indira Gandhi assassinated Ronald Reagan reelected	PSpice introduced by Microsim for IBM PC-XT
1985	Mikhail Gorbachev comes to power in the Soviet Union Volcano erupts in Columbia	ADT introduces Analog Workbench using Graphics User Interface SPICE3 released
1986	The space shuttle *Challenger* explodes	PSpice gains wide acceptance
1988	Iran-Iraq war ends George Bush becomes U.S. president	Tuinenga's book on PSpice published by Prentice Hall
1990–1991	Collapse of the Soviet Union The Gulf war U.S. invasion of Panama Maastricht Accord signed	Analog Artist released by Cadence. Accusim released by Mentor
1993	Boris Yeltsin storms the Parliament building in Moscow GATT agreement on world trade	AIM-Spice released and modeling book by Lee et al. published by Prentice-Hall
1994	Nelson Mandela becomes president of South Africa Israeli-Palestinian agreement signed	Vladimirescu's book on SPICE published by Wiley

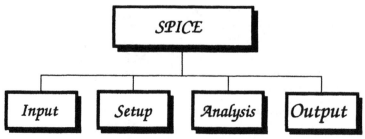

FIG. 1.2.1. Main modules of SPICE.

trix procedures of the analysis module. The analysis module is the main calculating engine of SPICE. This subprogram performs all the analyses requested from the user and passes the results to the output module for presentation.

Although the input and output modules determine the user friendliness of SPICE, the efficiency and accuracy of the program are determined by the algorithms implemented in the analysis module. The main purpose of this module is to find a numerical solution of the mathematical representation of the circuit. The transformation from the physical circuit to a mathematical system of equations is accomplished by representing each circuit element using a mathematical device model and imposing topological constraints dictated by Kirchhoff's current law (KCL) and Kirchhoff's voltage law (KVL). In general, the system of circuit equations can be written as

$$F(\mathbf{x}, \dot{\mathbf{x}}, t) = 0 \qquad (1.2.1)$$

where $\mathbf{x}$ is the unknown vector of circuit variables (currents and voltages), $\dot{\mathbf{x}}$ is the time derivative of $\mathbf{x}$, t is the time, and F is a nonlinear operator.

The circuit analysis gives the solution of Eq. (1.2.1) for a wide variety of problems relevant to electrical engineers. In a dc analysis, for example, a stationary solution is sought and the matrix $\dot{\mathbf{x}}$ is zero. Equation (1.2.1) therefore reduces to a set of nonlinear equations that can be solved by iterative techniques and sparse-matrix methods. In an ac analysis, all nonlinear circuit elements are linearized around the operating point, rendering Eq. (1.2.1) linear. In a transient analysis, however, we are interested in nonstationary phenomena involving a nonzero $\dot{\mathbf{x}}$. This analysis is more complex than the others and requires a numerical integration scheme for the temporal evolution, based on a series of dc analyses at different time points. In the rest of this section we discuss in more detail the numerical techniques used by SPICE. In the next section, we present some recent advances made in the development of so-called third-generation circuit simulators, which pose a challenge to SPICE in the commercial circuit simulator market.

Formulation of Circuit Equations. In SPICE, a technique called modified nodal analysis (MNA) is used to generate the circuit equations. To simplify our discussion, we illustrate the process by referring to the nodal analysis method, a less sophisticated version of MNA. In this method, the node voltages in the circuit are the unknown variables, and the necessary constraints are imposed to ensure that the solution satisfies KVL. In general, the system of equations of the following form are generated by writing a KCL equation for each node, excluding the ground node:

$$\mathbf{Y} \cdot \mathbf{v} = \mathbf{j} \qquad (1.2.2)$$

Here $\mathbf{Y}$ is the nodal admittance matrix, $\mathbf{v}$ is the vector of node voltages, and $\mathbf{j}$ is the so-called current excitation vector. The strength of the nodal analysis method is its simplicity. For the analysis of a linear circuit, containing only resistors and independent current sources, $\mathbf{Y}$ and $\mathbf{j}$ are constructed by the following rules: y_{kk} is the sum of all conductances connected to node k, y_{kl} is the negative sum of all conductances connected between nodes k and l, and j_k is the sum of all independent currents that flow into node k.

Circuit Simulation in the Time Domain (Transient Analysis). When the user requests a transient simulation, SPICE has to calculate the time-domain response of the circuit over a time interval specified by the user. In general, the circuit contains nonlinear elements such as diodes and transistors, and MNA transforms the circuit into a system of nonlinear differential equations [see Eq. (1.2.1)]. The solution method can be explained in terms of the following pseudocode:

For each time point in the simulation interval {
  Discretize the set of nonlinear differential equations using numerical integration
  /* The result is a set of nonlinear algebraic equations */
  Make initial guess for the unknown voltages
  converged = FALSE
  while not converged {
    Linearize the system of nonlinear algebraic equations using Newton–Raphson
    Solve the resulting system of linear equations
    if solution did not change since last iteration
      converged = TRUE
  }
}

As can be seen from this code, numerical methods are required for integrations, solutions of nonlinear algebraic equations, and solutions of linear equations.

Numerical integration is carried out by dividing the time interval into a discrete series of time points. The initial time point is arbitrarily defined as time zero, and the initial solution is either given by the user or determined by a dc operating point analysis. Then, at each time point, a numerical integration routine transforms the differential equations into equivalent nonlinear algebraic equations. In SPICE, we can choose between two numerical integration schemes: the trapezoidal method or Gear's method.

The system of nonlinear equations is solved in an iterative fashion using the Newton–Raphson algorithm. This algorithm approximates each nonlinear element in the circuit by a Taylor series that is truncated after the first-order term.

The iterative sequence starts with an initial guess at the solution. The circuit is then linearized around this presumed operating point, and a set of linear equations is generated and solved using standard techniques. In SPICE, methods equivalent to Gaussian elimination and LU factorization are used. To reduce the computational effort, the inherent sparsity of the Y matrix in Eq. (1.2.2) is utilized by implementing sparse-matrix methods in SPICE.

The circuit solution thus obtained is subsequently used as the next guess, and the process is repeated. The iteration stops when successive solutions agree to within a given tolerance.

Circuit Simulation in the Frequency Domain (Small-Signal ac Analysis).
SPICE is also capable of simulating a general circuit in the frequency domain. First, a dc operating point analysis is performed and then all nonlinear elements are lin-
niques discussed above for systems of linear equations (modified to handle complex quantities).

The ac analysis determines the small-signal solution of the circuit in one-frequency sinusoidal steady states. Hence, the phasor method can be used to transform the differential equations into the frequency domain. The admittances of resistors (R), capacitors (C), and inductors (L) are given by $Y_R = 1/R$, $Y_C = j\omega C$, and $Y_L = 1/j\omega L$, respectively, where ω is the circular input frequency.

1.2.3. Beyond SPICE

Advanced very large scale integrated (VLSI) circuits today are made up of millions of transistors. For example, the latest generation of INTEL processors (Pentium Pro) contains 5.5 million transistors and the latest version of the DEC Alpha processor has as many as 9.3 million transistor functions on a chip. In contrast, the algorithms used in SPICE were developed in the 1970s, targeting circuits containing at most a few hundred transistors. Therefore, with the VLSI revolution in the 1980s came an intense search for "third-generation" methods for circuit simulation having SPICE-type accuracy and robustness and, in addition, being suitable for simulation of large circuits containing thousands of transistors.

In the present "second-generation" SPICE-like simulators, the central processing unit (CPU) time is proportional to M^2, where M is the number of transistors. This quadratic dependence on transistor count was early recognized as a major drawback of SPICE and its close "relatives." During years of experience with second-generation circuit simulators, other problems and peculiarities have surfaced as well, including problems with numerical convergence. In fact, the Newton–Raphson technique for solving nonlinear equations has quadratic convergence if, and only if the initial guess is close enough to the solution, but even so, convergence is not guaranteed. Hence, as circuits became larger and more nonlinear, the problem of convergence was accentuated. Another important drawback of SPICE is that the entire circuit has to be simulated within each time step, even if large parts of the circuit remain inactive.

These limitations prompted a search for improved numerical techniques already by the end of the 1970s. However, it took almost 20 years before third-generation simulators were ripe for implementation in commercial CAD packages. Today, such simulators are available from several vendors, including ANACAD and Mentor Graphics. The following are some major advantages of third generation circuit simulators:

- *Circuit Decomposition.* Strongly coupled nodes are grouped into weakly coupled modules. Each module can operate with its own step size.

- *Exploit Subsystem Inactivity, or Latency.* A node voltage or circuit module will not be computed if its environment remains unchanged.

- *Efficient for Large Circuits.* CPU time is proportional to M, as illustrated in Fig. 1.2.2. For a circuit containing 1000 transistors, the simulation time of the

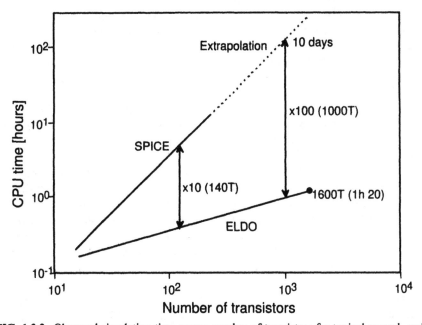

FIG. 1.2.2. Observed simulation time versus number of transistors for typical second- and third-generation circuit simulators. (After Hennion et al., 1987.)

third generation circuit simulator ELDO is reduced by a factor of 100 compared to the time required by SPICE.

- *Satisfy Contrary Requirements.* Fast analysis and large circuits.

To be able to satisfy these requirements, several new numerical techniques were explored. The waveform relaxation method (WRM) soon stood out as the best candidate (Lelarasme et al., 1982). This method is based on iterative techniques, such as Gauss–Jacobi or Gauss–Seidel, for solving sets of linear equations. The algorithm can be explained as follows:

Find an initial guess for the waveforms
For each relaxation step
  For each group of nodes
    Solve the nonlinear differential equations for the group of nodes over the entire time interval

The relaxation step is repeated until all waveforms have converged at the boundaries. The major advantage of the WRM is that a separate time step can be applied to each circuit module (multirate integration). Hence, latency is naturally exploited within the framework of the method.

In addition to the WRM algorithm itself, schemes for circuit decomposition and for finding solutions for each group of nodes were also required. One of the drawbacks of the WRM is that it has only linear convergence. Hence, the convergence can be very slow for tightly coupled nodes. Therefore, in practical implementations of the WRM, the circuit is not decomposed into single nodes. Instead, the circuit is divided into modules. Strongly coupled nodes are grouped into weakly coupled modules, and the relaxation iterations force the waveforms at the boundaries of each circuit module to converge. For the solution of each group of nodes (or circuit module), Newton-based algorithms are usually applied, and the overall simulation scheme is commonly referred to as Newton waveform relaxation or the WRN (see, e.g., Erdman and Rose, 1992).

Circuit decomposition techniques generally fall into two categories: one-level block decomposition and hierarchical circuit partitioning. One-level block decomposition has been shown to be an NP-complete problem, and heuristics have to be applied. The methods come in many flavors. Here we briefly present some of the techniques that have been reported in the literature. As mentioned, to improve the convergence of the WRM, the circuit should be partitioned into weakly coupled modules where the coupling within each module is strong. In metal–oxide–semiconductor FET (MOSFET) circuits, the coupling between the gate and the drain/source of a transistor is weak. Hence "node tearing" in MOSFET circuits is often done at the gates of the transistors [Nishigaki et al. (1992)]. Alternatively, the following procedure can be used [Kage et al. (1994)]:

- Transistors that share common source and/or drain nodes are combined into a transistor group. Linear elements are included in the transistor group to which they are connected.
- Initial partitioning of transistor groups into a predefined number of subcircuits is performed.
- The number of interconnection nodes are minimized.
- The subcircuits are balanced

However, this procedure is time consuming and may violate the wish to keep the CPU time of the circuit partitioning a small fraction of the total simulation time. In ELDO, for example, nodes are ordered according to the direction of signal propagation, and circuit blocks are limited to contain linearly coupled nodes (Hennion et al., 1987). RELAX2 simplifies further by letting the user specify the circuit decomposition (White and Sangiovanni-Vincentelli, 1983).

In hierarchical circuit partitioning, also called multilevel block decomposition, the partition philosophy is brought one step further by dividing the circuit into hierarchical levels. According to several authors [see, e.g., Nishigaki et al. (1992) and Saviz and Wing (1993)] further progress can be made by applying so-called multilevel decomposition techniques. The concept of hierarchical circuit partitioning is illustrated in Fig. 1.2.3 for a circuit containing four operational amplifiers (opamps).

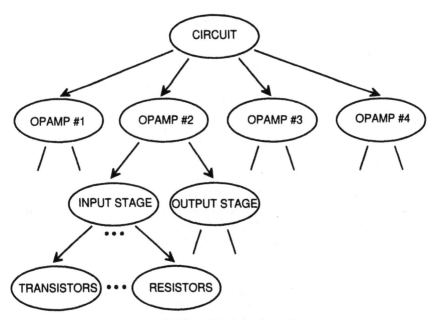

FIG. 1.2.3. Hierarchical circuit partitioning.

The topmost level is the complete circuit. In level 2, the circuit is partitioned into the four opamps. Next, each opamp is divided into an input stage and an output stage. At the bottom level, the circuit has been broken down into single elements such as resistors and transistors. The simulation of the circuit can now be looked upon as traversing this tree structure. This allows for an efficient determination of which parts of the circuit need to be solved at which time points.

At each circuit block at a specific level, the simulator checks if the block is active. If not, the block and the whole subtree of that block is skipped. The method also allows the solution at each level to occur with a different error criterion, allowing critical sections of the circuit to be analyzed more accurately than noncritical sections. Also, each module can be solved according to the strength of the coupling between the variables. If the coupling is week, waveform relaxation techniques can be applied. Otherwise, Newton iterations should be used.

Both one-level and multilevel circuit decomposition techniques can be implemented in a static or a dynamic mode. In the static mode, the decomposition is carried out before the simulation starts and the decomposition remains unaltered during the simulation. However, in a general circuit, the efficiency of the partitioning may decrease during the course of the simulation. Dynamic partitioning accounts for this circumstance by allowing the circuit to be repartitioned at any time during

cuit modules that exhibit slow convergence may be rearranged to increase the speed of the simulation.

Many modifications to the original WRM have been reported in the literature for the purpose of optimizing properties such as speed, convergence, and memory requirement. One drawback of the WRM is that the waveforms for the entire simulation interval of each circuit block have to be stored, which may require a considerable amount of memory. To avoid this problem and to improve convergence, RELAX2 uses a "windowed" waveform relaxation method, where the entire time interval is broken up into windows and the WRM is applied within each window (White and Sangiovanni-Vincentelli, 1983). In ELDO, a special case of windowed WRM where each window is assigned only one time point, called the one-step relaxation method, is utilized (Hennion et al, 1985).

Breaking the circuit up into blocks, where the coupling between the blocks is weak, makes it possible to utilize separate CPUs to compute each block simultaneously. With the introduction of new and inexpensive hardware, commercially available general-purpose concurrent computers, and third-generation simulation techniques, parallel circuit simulation is gaining increasing attention. The following is some of the literature covering this field: Kage et al. (1994), Hajj and Tejayadi (1989), Peterson and Mattisson (1993), Xia and Saleh (1992), White and Sangiovanni-Vincentelli (1985), and Odent et al. (1990).

A completely different approach to circuit simulation is to apply piecewise linear models to all nonlinear components (Hajj and Tejayadi, 1989; van Eijndhoven, 1984; Yu and Wing, 1984; van Stiphout et al., 1990). The piecewise linear segments and the corresponding boundaries (piecewise linear mapping) are stored in

TABLE 1.2.2. Simulation Time (min) (VAX 11/780) for Three Different Circuits Using RELAX2 and SPICE.

Circuit	SPICE	RELAX2
Shift cell	12.52	2.1
Two-phase clock	43.13	5.47
Memory cell	13.63	2.98

a single matrix using a compact matrix description. Very powerful and specialized algorithms are used to solve the sets of piecewise linear equations. While Newton methods have only local convergence (i.e., the initial guess has to be sufficiently close to the solution), the algorithms applied for piecewise linear analysis can have very strong global convergence. However, the most important advantage of this method is probably the ability to model a wide range of devices, such as resistors, transistors, logic gates, and opamps, in a uniform way. It becomes natural to mix low-level models with macromodels. Even devices with memory (or hysteresis) can be modeled. Another advantage is that the models are completely separated from the simulator. All device descriptions reside in a model library that can be easily modified or extended. Even measured device characteristics can be used directly in the model.

We end this section by comparing results obtained using some of the third-generation simulators and SPICE. Table 1.2.2 shows simulation times and number of iterations for three different circuits using RELAX2 (White and Sangiovanni-Vincentelli, 1983) and SPICE. We note that a speed advantage by up to a factor of 10 can be achieved with RELAX2 compared to SPICE. Table 1.2.3 shows a similar comparison between SPICE, RELAX (Lelarasme and Sangiovanni-Vincentelli, 1982), and PYRAMID (Saviz and Wing, 1993), where the latter makes use of hierarchical circuit partitioning. We note that for large circuits (e.g., circuit 4), the speed advantages of multilevel block decomposition becomes apparent.

TABLE 1.2.3. CPU time (min) (VAX 11/750) for Four Different Circuits Using SPICE, RELAX, and PYRAMID.

Circuit	Number of Devices	SPICE	RELAX	PYRAMID
1	266	31.2	1.2	1.0
2	285	67.6	5.5	2.7
3	2850	—	173	31.8
4	7300	—	1458	246

Note: Circuit 1 is a 4-bit complementary metal–oxide–semiconductor (CMOS) counter; circuit 2 is a CMOS asynchronous sequential circuit (1 stage); circuit 3 is the same as 2, but with 10 stages; circuit 4 is an emitter coupled logic (ECL) asynchronous sequential circuit (20 stages).

1.3. SPICE SIMULATION TUTORIAL

In this book, we include numerous examples and problems involving SPICE simulations. We therefore recommend that the reader acquires some elementary skills in working with SPICE. To this end, we will now run step by step through a brief tutorial session with simple simulation examples using AIM-Spice. The commands, actions, and simulation results in this tutorial are specific for AIM-Spice, but PSpice can be used as well since the similarities between the two simulators are many. Some explanatory text on the simulation procedure is given, but for a detailed discussion, we refer to the AIM-Spice Users Manual in Appendices A1–A3, the AIM-Spice on-line help function, or the PSpice Users Manual.

Our sample circuit, a bipolar junction transistor (BJT) differential pair, is shown in Fig. 1.3.1. This is a basic element in a differential amplifier and also in emitter coupled logic (ECL). Using AIM-Spice, we shall investigate the large-signal transient response, the large-signal transfer characteristic, and the small-signal frequency response of this circuit. At this point, the principle of BJT operation or of how the circuit works is not important to us, since our goal at this stage of the game is solely to acquire some experience in how to use SPICE.

The first task is to name all the nodes in the circuit diagram. A node name can be

to name nodes is to assign them numbers, as shown in the circuit diagram.

The AIM-Spice circuit description for the differential pair is shown in Fig. 1.3.2.

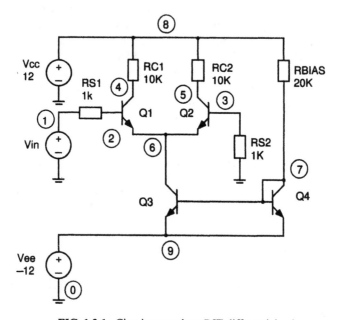

FIG. 1.3.1. Circuit example: a BJT differential pair.

```
DIFFPAIR CKT - SIMPLE DIFFERENTIAL PAIR
VIN 1 0 DC 0 PULSE(-0.5 0.5 0.1u 1n 1n 0.3u 2u) AC 1
VCC 8 0 12
VEE 9 0 -12
Q1 4 2 6 QNL
Q2 5 3 6 QNL
RS1 1 2 1K
RS2 3 0 1K
RC1 4 8 10K
RC2 5 8 10K
Q3 6 7 9 QNL
Q4 7 7 9 QNL
RBIAS 7 8 20K
.MODEL QNL NPN(BF=80 RB=100 CJS=2PF TF=0.3NS TR=6NS
+ CJE=3PF CJC=2PF VA=50)
```

FIG. 1.3.2. AIM-Spice circuit description of the circuit in Fig. 1.3.1.

The first line in the list is the title of the circuit. The real description of the circuit starts at line 2. The elements in the circuit are named, and the first letter of the name is unique to each element type. For example, the names of resistors must start with the letter R, voltage sources with V, bipolar transistors with Q, and so on. The subsequent letters in the name can be any alphanumeric string. The circuit nodes to which a given element is connected follows immediately after the element name. The last field of the element description is the set of parameter values of the circuit element. In order to continue the description of an element on a new line, a plus is placed in column 1 of the new line.

In line 2, we have a voltage source with the name VIN between nodes 1 and 0. This source is specified with three sets of parameters: DC 0 allows us to sweep the input voltage during a DC Analysis, PULSE(-0.5 0.5 0.1u 1n 1n 0.3u 2u) is used during a Transient Analysis, and AC 1 is used in an AC Analysis. The predefined function PULSE means that VIN is switched between two values. The arguments of this function have the following meaning: initial voltage –0.5 V, pulsed value 0.5 V, delay time 0.1 μs, rise time 1 ns, fall time 1 ns, pulse length 0.3 μs, and period 2 μs. Parameter AC 1 means that the source voltage is given an amplitude of 1 V during an AC Analysis, which makes the output at node 5 numerically equal to the voltage gain. (Note that the circuit is linearized in an AC Analysis, so any value of the amplitude is permitted.)

Lines 5, 6, 11, and 12 describe bipolar transistors. The last field in these lines, QNL, is a device model specification. Bipolar transistors, like other semiconductor devices, are described in terms of specific models and associated sets of model parameters. Each model is assigned a set of default parameter values that are listed in the AIM-Spice reference in Appendix A2. Parameter values different from default are specified in lines that start with .MODEL, which in our case is line 14 for model

QNL. Once model QNL is defined, it can be used for several circuit elements. In our circuit, the same model is used for all the bipolar transistors.

To simulate our example circuit using PSpice, we have to add one or more lines specifying which analyses to run and which analysis parameters to use. Appropriate PSpice specifications for our tutorial example are contained in the following lines that have to be added to those in Fig. 1.3.2:

```
.tran 10ns 700ns 0 10ns
.dc vin -0.3 0.3 0.01
.ac dec 10 10k 100meg
.probe
.end
```

The statement .probe is needed to prepare the computed voltages and currents for plotting using PROBE in PSpice, and the statement .end denotes the end of the input file. Note that three different analyses are specified in the above lines. To run one at a time, remove the two lines not needed.

Here follows a step-by-step simulation session of our example circuit using AIM-Spice. (Information on how to install AIM-Spice on your computer is given

using a pulsed input as specified in Fig. 1.3.2. Next, we perform a DC Analysis from which we obtain the large-signal transfer characteristic. Finally, we use an AC Analysis to find the frequency response of the differential pair. To get the most out of this tutorial, read it in front of your computer and follow the steps outlined below. Text lines in italic are your actions and we encourage you to try them out.

Load AIM-Spice.
Enter the circuit description shown in Fig. 1.3.2 in the untitled circuit window.

Before doing the simulation, it is wise to save the circuit description in a circuit file. For this, we use the File Save command.

Choose the Save command from the File menu (Shift+F12).

If this is the first time the circuit is filed, a Save As dialog box appears. (Otherwise the circuit is saved automatically.)

Type the filename under which you want to save the circuit. Then choose OK.

We want to run a Transient Analysis.

Choose Transient Analysis from the Analysis menu (Ctrl+T).
Specify the analysis parameters shown in Fig. 1.3.3.

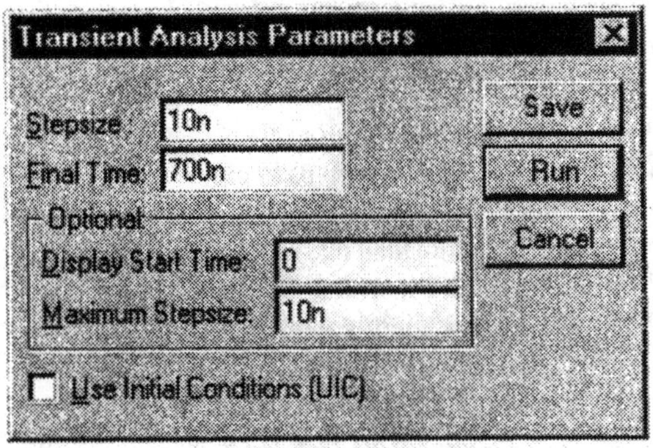

FIG. 1.3.3. Transient Analysis parameters used for the example circuit.

We are now ready to run the simulation.

Choose the button labeled Run in the dialog box.

AIM-Spice is now entering the analysis mode. The circuit window is hidden and the status bar at the bottom of the main window displays "Parsing Circuit, Please Wait." After the circuit is loaded into the AIM-Spice kernel, the status bar displays which analysis type is about to be run. In our case, it is a Transient Analysis. Before starting the simulation, we have to select circuit variables to be plotted during the simulation.

Choose Select Variables to Plot from the Control menu (F5).

A dialog box with a list of all variables in the circuit is shown. The list box is of a multiple selection type. This means that you can select more than one variable to display in a single plot window. In our case, we are interested in plotting the input voltage V(1) and the output voltage V(5) in separate windows.

Select V(1) in the list box.

To select from the list box with the keyboard, use the arrow keys to move the highlighted rectangle to the line containing V(1) and then press the space bar. To select V(1) with the mouse, just click on it with the left mouse button. Normally, we have to specify the *y*-axis limits of each plot window.

Select the button Y Axis Format in the Select Variables to Plot dialog box.

It is wise to specify fairly large intervals as a first guess, and as the simulation progresses, the plot traces on the screen can be used as a guide for resetting the limits.

Specify the y-axis format shown in Fig. 1.3.4 and choose the OK button.
Choose the OK button in the Select Variables to Plot dialog box.

An empty plot window is now created with a title reflecting which variables we chose to plot in the window. Repeat the above steps to create another plot window containing the variable V(5). Use –2 and 14 V as lower and upper limits for V(5) and 2 V as the increment. Note that with more than one plot window present, any window can be brought to the front by activating it with a click of the mouse anywhere within the window. With the keyboard, it is slightly more complicated. Use Ctrl+F6 to activate one plot window after another until the one you want to be active is in front and its caption bar is highlighted. The plot windows can also be tiled or cascaded using the commands in the Window menu. We are now ready to start the simulation.

Choose Start Simulation from the Control menu (Ctrl+S).

During the simulation, AIM-Spice displays "Doing Transient Analysis ..." in the status bar. While running a simulation, we may reset the limits in one or more of the plot windows, we can cancel the simulation with the Stop Simulation command in the Control menu, or we can switch to another application to do other work while displays the message "Simulation Done" in the status bar and a window with Simulation Statistics appears in front of the plot windows. The plot windows can be viewed by removing the Simulation Statistics window.

Choose the OK button of the Simulation Statistics window.

Figure 1.3.5 shows a screen dump of the tiled plot windows for the Transient Analysis. We can now save the simulation results.

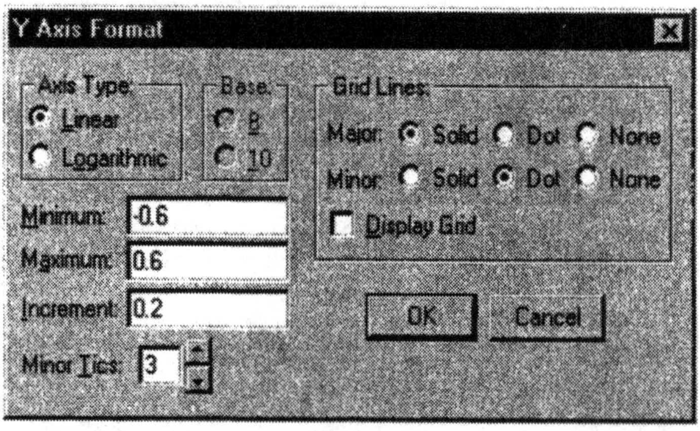

FIG. 1.3.4. The *y*-axis format used in the Transient Analysis of the example circuit.

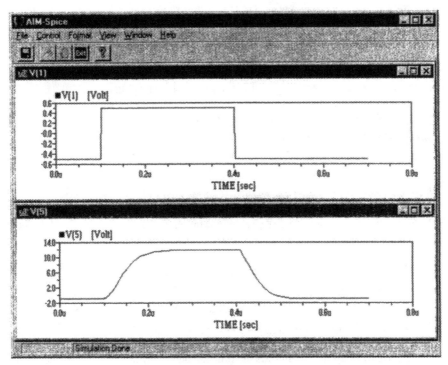

FIG. 1.3.5. Results from the Transient Analysis simulation of the differential pair. The input voltage is shown on top and the output voltage at the bottom.

Choose Save Plots from the File menu (Alt+F12).

A dialog box with all accumulated plots since the last time we saved plots or since we started AIM-Spice is displayed.

Choose the button labeled Save All Plots.

A Save As dialog box appears. The default name for the output file is the same as the name of the circuit file, but with the extension OUT. However, note that every time you save to a file, all the previous plots stored in the file will be erased. You may therefore want to either change the filename or wait until the end of the simulation session to save all plots in one file. However, if you want to save now:

Choose OK.

The next step would normally be to load the accompanying postprocessing program called AIM-Postprocessor and prepare the simulation results for printing. However, we will skip this for now and simply refer to the discussion of the postprocessor in

Appendix A1. Instead, we turn to a DC Analysis of our example circuit. But first, we have to leave the analysis mode.

Choose Exit Analysis Mode from the Control menu (Ctrl+E).

The plot windows are destroyed and the circuit description appears.

Choose DC Analysis from the Analysis menu (Ctrl+D).
Specify the analysis parameters shown in Fig. 1.3.6 and choose the Run button.
Choose Select Variables to Plot from the Control menu (F5).
Select V(5) in the list box.
Select the button Y Axis Format in the Select Variables to Plot dialog box.

As before, we use −2 V and 14 V as lower and upper limits for V(5) and 2 V as the increment.

Specify the y-axis format and choose the OK button.
Choose the OK button in the Select Variables to Plot dialog box.
Choose Start Simulation from the Control menu (Ctrl+S).
Choose the OK button of the Simulation Statistics window.

Figure 1.3.7 shows a screen dump of the plot window with the transfer characteristic obtained from the DC Analysis. Again, if you want to save these results (recall the previous comment on saving plots):

Choose Save Plots from the File menu (Alt+F12).
Choose the button labeled Save All Plots.
Choose OK.
Choose Exit Analysis Mode from the Control menu (Ctrl+E).

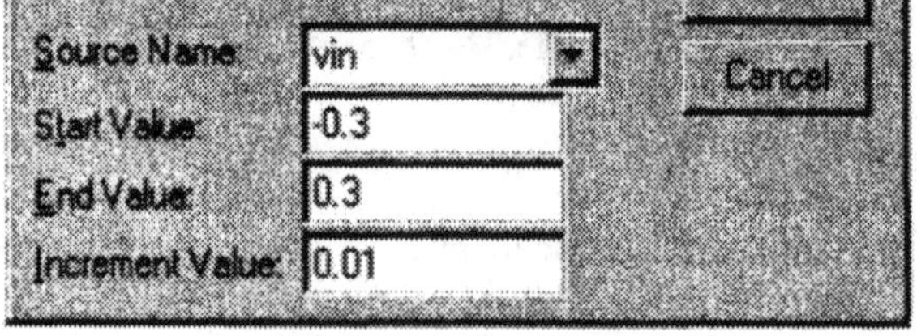

FIG. 1.3.6. Dialog box for the DC Transfer Analysis.

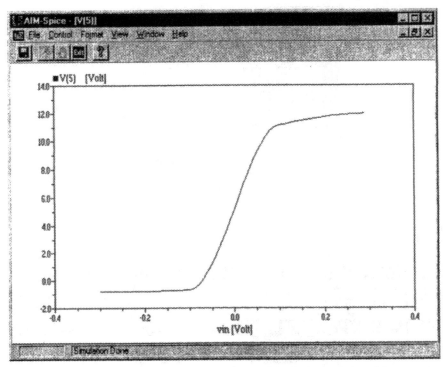

FIG. 1.3.7. Transfer characteristic obtained from the DC Analysis simulation of the example circuit.

Finally, we carry out an AC Analysis of our example circuit.

Choose AC Analysis from the Analysis menu (Ctrl+A).
Specify the analysis parameters shown in Fig. 1.3.8 and choose the Run button.
Choose Select Variables to Plot from the Control menu (F5).

We would like to plot the small-signal frequency response in terms of the gain on a decibel scale and the phase in separate windows. We start by selecting the gain.

Select V(5) and then Magnitude and DB Scale from AC Options in the list box.
Select the button Y Axis Format in the Select Variables to Plot dialog box.

Select: Minimum, –10 dB; Maximum, 50 dB; Increment, 10 dB.

Specify the y-axis format and choose the OK button.
Repeat the above steps to again select V(5) and then choose Phase from AC Options in the list box.
Select the button Y Axis Format in the Select Variables To Plot dialog box.

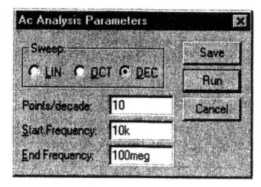

FIG. 1.3.8. Dialog box for the AC Analysis.

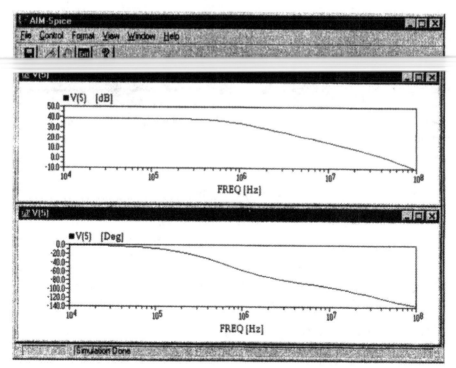

FIG. 1.3.9. Small-signal frequency response obtained from the AC Analysis simulation of the example circuit.

Select: Minimum, −140 deg; Maximum, 0 deg; Increment, 20 deg.

Specify the y-axis format and choose the OK button.
Choose the OK button in the Select Variables to Plot dialog box.
Choose Start Simulation from the Control menu (Ctrl+S).
Choose the OK button of the Simulation Statistics window.

Figure 1.3.9 shows a screen dump of the plot window with the magnitude and the phase of the small-signal frequency response of the differential pair.

Choose Save Plots from the File menu (Alt+F12).
Choose the button labeled Save All Plots.
Choose OK.
Choose Exit Analysis Mode from the Control menu (Ctrl+E).
Choose Exit from the File menu.

This concludes the tutorial.

REFERENCES

D. J. Erdman and D. J. Rose, "Newton Waveform Relaxation Techniques for Tightly Coupled Systems," *IEEE Transactions on Computer Aided Design*, vol. 11, no. 5, pp. 598–606 (1992).

I. N. Hajj and O. Tejayadi, "Parallel Solution of Piecewise-Linear Transistor Circuits," in *Proceedings of the IEEE International Symposium on Circuits and Systems*, pp. 681–684 (1989).

B. Hennion, P. Senn, and D. Coquelle, "A New Algorithm for Third Generation Circuit Simulators: The One-Step Relaxation Method," in *Proceedings of the 22nd Design Automation Conference*, pp. 137–143 (1985).

B. Hennion, Y. Paradis, and P. Senn, "ELDO: a General Purpose Third Generation Circuit Simulator Based on the O. S. R. Method," in *Proceedings of the European Conference on Circuits Theory and Design*, Paris, France, pp. 113–120 (1987).

T. Kage, F. Kawafuji, and J. Niitsuma, "A Circuit Partitioning Approach for Parallel Circuit Simulation," *IEICE Transactions on Fundamentals*, vol. E77-A, no. 3, pp. 461–466 (1994).

K. Lee, M. Shur, T. A. Fjeldly and T. Ytterdal, *Semiconductor Device Modeling for VLSI*, Prentice-Hall, Englewood Cliffs, NJ (1993).

E. Lelarasme and A. Sangiovanni-Vincentelli, "RELAX: a New Circuit Simulator for Large Scale MOS Integrated Circuits," in *Proceedings of the 1982 Design Automation Conference*, June (1982).

E. Lelarasme, A. Ruheli, and A. L. Sangiovanni-Vincentelli, "The Waveform Relaxation Method for the Time-Domain Analysis of Large Scale Integrated Circuits," *IEEE Transactions on Computer Aided Design*, vol. 1, no. 3, pp. 131–145 (1982).

L. W. Nagel, *SPICE 2: A Computer Program to Simulate Semiconductor Circuits*, Memorandum No. ERL-M520, Electronic Research Laboratory, College of Engineering, University of California, Berkeley (1975).

L. W. Nagel and R. A. Rohrer, "Computer Analysis of Nonlinear Circuits, Excluding Radiation (CANCER)," *IEEE Journal of Solid-State Circuits*, vol. SC-6, pp. 166–182 (1971).

M. Nishigaki, N. Tanaka, and H. Asai, "Hierarchical Decomposition and Latency for Circuit Simulation by Direct Method," *IEICE Transactions on Fundamentals*, vol. E75-A, no. 3, pp. 347–351 (1992).

M. Nishigaki, N. Tanaka, and H. Asai, "Mixed Mode Circuit Simulation Using Dynamic Network Separation and Selective Trace," *IEICE Transactions on Fundamentals*, vol. E77-A, no. 3, pp. 454–460 (1994).

P. Odent, L. Claesen, and H. de Man, "Acceleration of Relaxation-Based Circuit Simulation Using a Multiprocessor System," *IEEE Transactions on Computer Aided Design*, vol. 9, no. 10, pp. 1063–1072 (1990).

L. Peterson and S. Mattisson, "The Design and Implementation of a Concurrent Circuit Simulation Program for Multicomputers," *IEEE Transactions on Computer Aided Design*, vol. 12, no. 7, pp. 1004–1014 (1993).

P. Saviz and O. Wing, "Circuit Simulation by Hierarchical Waveform Relaxation," *IEEE Transactions on Computer Aided Design*, vol. 12, no. 6, pp. 845–860 (1993).

J. T. J. van Eijndhoven, "A Piecewise Linear Simulator for Large Scale Integrated Circuits," Ph.D. Thesis, Eindhoven University of Technology, Eindhoven, The Netherlands (1984).

M. T. van Stiphout, J. T. J. Eijndhoven and H. W. Buurman, "PLATO: A New Piecewise Linear Simulation Tool," in *Proceedings of the European Design Automation Conference*, Glasgow, United Kingdom, pp. 235–239 (1990).

A. Vladimirescu, *The SPICE Book*, Wiley, New York (1994).

J. White and A. Sangiovanni-Vincentelli, "RELAX2: A Modified Waveform Relaxation Approach to the Simulation of MOS Digital Circuits," in *Proceedings of the 1983 International Symposium on Circuits and Systems*, IEEE, New York, pp. 756–759 (1983).

J. White and A. Sangiovanni-Vincentelli, "Partitioning Algorithms and Parallel Implementation of Waveform Relaxation Algorithms for Circuit Simulation," in *Proceedings of the IEEE International Symposium on Circuits and Systems*, pp. 221–224 (1985).

E. Z. Xia and R. A. Saleh, "Parallel Waveform-Newton Algorithms for Circuit Simulation," *IEEE Transactions on Computer Aided Design*, vol. 11, no. 4, pp. 432–442 (1992).

Q. J. Yu and O. Wing, "PLMAP—a Piecewise Linear MOS Circuit Analysis Program," *Integration, the VLSI Journal*, pp. 27-48 (1984).

PROBLEMS

1.2.1. Prepare a SPICE input file describing a series L–C–R circuit with $L = 1$ nH, $C = 1$ pF, and $R = 1$ Ω fed by a 1-V ideal voltage source. Simulate the frequency response (i.e., the current) for frequencies between 1 MHz and 10 GHz and plot this response. (You may use PSpice with PROBE or AIM-Spice with AIM-Spice Postprocessor.)

1.2.2. Search the Internet and continue the list of SPICE development in Table 1.2.1 beyond 1994.

1.3.1. (a) Search the Internet to locate and download AIM-Spice.

(b) Review the tutorial given in Section 1.3.

1.3.2. Repeat the Transient Analysis of the tutorial with a sinusoidal instead of a pulsed voltage input. Use different frequencies and amplitudes for the input signal. (See AIM-Spice Reference, Appendix 2, under "V–Independent Voltage Source" for details on the input waveform definition.)

1.3.3. Use the sinusoidal voltage source `vin` of Problem 1.3.2 as an input to both Q_1 and Q_2. Perform a Transient Analysis of this case using different frequencies and amplitudes for the input signal. Comment on the result.

CHAPTER 2

CHARGE TRANSPORT IN
SEMICONDUCTORS

2.1. INTRODUCTION

The purpose of this chapter is to give a brief summary of basic semiconductor equations used in SPICE device models and to discuss material parameters that are relevant to modeling and simulation of modern semiconductor devices and integrated circuits. This information is needed in order to understand how device parameters used in circuit simulators depend on different factors, including temperature and material properties.

An elementary discussion of the underlying semiconductor physics may be found, for example, in textbooks by Sze (1985), Streetman (1995), and Shur (1996). A more comprehensive discussion of semiconductor device physics and semiconductor properties may be found, for example, in monographs by Sze (1981), Seeger (1985), Hess (1988), Shur (1990), and Lee et al. (1993).

2.2. BASIC SEMICONDUCTOR EQUATIONS

In this section, we briefly discuss the basic semiconductor equations that constitute the backbone of most of the semiconductor device models implemented in SPICE. These equations are based on the assumption that electron and hole velocities in semiconductor devices are instantaneous functions of the electric field. However, we should recognize that this approximation may be too crude for state-of-the-art submicrometer devices. One way of dealing with this problem is to adjust the values of the carrier velocities to account for the shortcomings of the basic equations.

In a low electric field, the current densities for electrons and holes are given by

$$\mathbf{J}_n = qn\mu_n\mathbf{F} + qD_n\,\nabla n \tag{2.2.1}$$

$$\mathbf{J}_p = qp\mu_p\mathbf{F} + qD_p\,\nabla p \tag{2.2.2}$$

where n and p are the mobile electron and hole densities, respectively, $\mathbf{F}$ is the electric field, μ_n and μ_p are the low-field electron and hole mobilities, D_n and D_p are the electron and hole diffusion coefficients, and q is the unit charge.

The first term on the right-hand side of Eqs. (2.2.1) and (2.2.2) corresponds to a drift contribution to the current density that is proportional to the electric field and to the carrier density. Here, $-\mu_n\mathbf{F}$ and $\mu_p\mathbf{F}$ represent low-field drift velocities for the electrons and holes, respectively. Taking into account that electrons and holes have opposite charges, this always gives rise to a current density contribution in the direction of the electric field. The second term in each of the above expressions represents a diffusion current density. This term is independent of the electric field but is proportional to the gradient of the charge carrier density. Hence, gradients in n and p of the same sign result in opposite contributions to the current density. The directions of carrier flow and current densities versus the directions of the electric field and the concentration gradients are illustrated in Table 2.2.1.

In a semiconductor device, the relative magnitude of the drift and the diffusion component may vary widely with position within the device, with bias conditions, and with the type of device considered.

In low electric fields, the diffusion coefficients are related to the mobilities via the Einstein relationships, valid for nondegenerate semiconductors:

$$D_n = \frac{\mu_n k_B T}{q} \tag{2.2.3}$$

TABLE 2.2.1. Directions of Electron and Hole Velocities and Current Densities versus Directions of Electric Field and Concentration Gradient.

		Electrons		Holes	
		$\mathbf{v_n}$	$\mathbf{J_n}$	$\mathbf{v_p}$	$\mathbf{J_p}$
$\mathbf{F}$	$\rightarrow$	◄⊖	⇒	⊕►	⇒
	$\leftarrow$	⊖►	⇐	◄⊕	⇐
$\nabla n, \nabla p$		◄⊖	⇒	◄⊕	⇐
		⊖►	⇐	⊕►	⇒

$$D_p = \frac{\mu_p k_B T}{q} \tag{2.2.4}$$

Here, T is the absolute temperature and k_B is the Boltzmann constant. At high carrier concentrations, these relationships have to be modified to account for degeneracy.

The low-field mobilities are determined by the effective mass of the carriers and by their rate of collisions with lattice vibrations (phonons) and with impurities in the semiconductor, according to

$$\mu_n = \frac{q \tau_n}{m_n} \tag{2.2.5}$$

$$\mu_p = \frac{q \tau_p}{m_p} \tag{2.2.6}$$

Here, τ_n and τ_p are so-called momentum relaxation times, a measure of how fast the carrier momentum changes on the average owing to collisions, and m_n and m_p are the effective masses of electrons and holes, respectively.

ductor caused by the atoms of the crystal lattice, as reflected in the electronic energy band structure. Since the charge carriers normally collect near the band extrema (electrons at the bottom of the lowest conduction band and holes at the top of the highest valence bands), they will experience a parabolic variation in kinetic energy with (crystal) momentum. We note that the parabolicity of the bands near the band extrema is analogous to the quadratic-energy-versus-momentum relationship of free particles. This allows us to apply classical laws of motion for electrons and holes in crystals. Their effective masses can be identified with the inverse curvature of the energy bands in question.

Important scattering mechanisms encountered by the charge carriers are ionized impurity scattering, neutral impurity scattering, acoustic deformation potential scattering, nonpolar optical scattering, polar optical scattering, and piezoelectric scattering. The last two mechanisms are present only in partially heteropolar crystals, such as GaAs. Typically, only one or two scattering processes are dominant for a given value of temperature and impurity concentration. In nonpolar semiconductors such as silicon and germanium, scattering by acoustic phonons and ionized impurities determine the electron and hole mobilities in low electric fields. Scattering by nonpolar optical phonons becomes dominant in these materials in high electric fields, leading to a saturation of the electron and hole velocities. Scattering by polar optical phonons and ionized impurities is normally important for electrons and holes in GaAs, while scattering by acoustic phonons (deformation potential scattering) and scattering caused by piezoelectric properties (piezoelectric scattering) become important in pure GaAs samples at low temperatures. At high carrier concentrations, collisions between charge carriers, so-called carrier–carrier scattering, also have to be taken into account.

In a high electric field, the charge carriers are heated by the field and the carrier kinetic energy may become larger than the average thermal energy, $3k_B T_l/2$, per particle in the crystal lattice determined by the lattice temperature T_l. This changes the conditions for scattering. Scattering by ionized impurities, for example, becomes less important since the carriers will be traveling at a higher average speed, thus spending less time in the vicinity of scattering centers. On the other hand, the scattering involving emission of phonons becomes more important since the probability of a carrier having enough energy to emit a phonon increases. In addition, large carrier energies cause a redistribution of electrons between the conduction band valleys and a redistribution of holes between the light and heavy hole valence bands, thereby changing the transport conditions. As a consequence, the electron and hole velocities are no longer proportional to the electric field when the field is high. The diffusion coefficients are also dependent on the electric field.

Clearly, a very detailed description of the various processes contributing to charge transport is unsuitable in the context of device modeling for circuit simulation. Instead, we wish to emphasize the use of simplified relationships where the contributing mechanisms are taken into account in an average, phenomenological way. As an example, the following phenomenological equations, which are generalized versions of Eqs. (2.2.1) and (2.2.2), are frequently used in device modeling in order to describe electron and hole transport in a wide range of electric fields:

$$\mathbf{j}_n = q[-n\mathbf{v}_n(\mathbf{F}) + D_n(\mathbf{F})\nabla n] \tag{2.2.7}$$

$$\mathbf{j}_p = q[p\mathbf{v}_p(\mathbf{F}) - D_p(\mathbf{F})\nabla p] \tag{2.2.8}$$

Here, the carrier velocities [$\mathbf{v}_n(\mathbf{F})$ and $\mathbf{v}_p(\mathbf{F})$] and the diffusion coefficients [$D_n(\mathbf{F})$ and $D_p(\mathbf{F})$] are assumed to be more general functions of the electric field, derived from computations or measured for a uniform sample under steady-state conditions. In a low electric field, these expressions reduce to Eqs. (2.2.1) and (2.2.2). However, we should be aware that even the modified basic semiconductor equations may lead to considerable errors in describing hot electron behavior. Difficulties are also encountered in describing near-equilibrium transport in simple systems with large built-in electric fields, such as p–n junctions.

Under high-frequency operation, comparable to the inverse energy relaxation time (a measure of how fast the carrier energy changes on the average owing to collisions), the velocity and diffusion do not instantaneously follow the variations of the electric field. In such cases, the effective mobility, for example, becomes frequency dependent.

The advantage of using Eqs. (2.2.7) and (2.2.8), or similar equations with field-independent diffusion coefficients, is the relative simplicity of the analysis, an analysis that nonetheless allows us to achieve some insight into the device physics.

In order to obtain a self-consistent solution of the basic semiconductor equations in a nonhomogeneous semiconductor, space charge has to be taken into account.

This means that Eqs. (2.2.7) and (2.2.8) [or Eqs. (2.2.1) and (2.2.2) in the low-field case] should be solved together with Poisson's equation,

$$-\nabla^2\phi = \nabla \cdot \mathbf{F} = \frac{\rho}{\epsilon_s} \tag{2.2.9}$$

where ϕ is the electric potential, and the continuity equations are

$$\frac{\partial n_t}{\partial t} + \frac{\partial n}{\partial t} = \frac{1}{q}\nabla \cdot \mathbf{j}_n + G - R \tag{2.2.10}$$

$$\frac{\partial p_t}{\partial t} + \frac{\partial p}{\partial t} = -\frac{1}{q}\nabla \cdot \mathbf{j}_p + G - R \tag{2.2.11}$$

Here ϵ_s is the dielectric permittivity of the semiconductor,

$$\rho = q(N_d - N_a - n + p) + q(p_t - n_t) \tag{2.2.12}$$

is the space charge density, p_t and n_t are the densities of trapped holes and electrons.

spectively, G is the generation rate of electron–hole pairs (caused by radiation or impact ionization), and R is the electron–hole recombination rate.

As an example, let us now consider the application of the semiconductor equations to a quasi-neutral n-type region that contains electrons and holes at concentrations n_n and p_n that are different from the equilibrium concentrations n_{no} and p_{no}. We assume that there are no traps, that is, that $p_t = 0$ and $n_t = 0$. This would be the situation in a good-quality piece of n-type semiconductor where extra carriers are homogeneously generated by light. For simplicity, we consider a one-dimensional steady-state situation where $\partial j_n/\partial t = 0$ and $\partial j_p/\partial t = 0$. Then, in the low-field case, the one-dimensional continuity equations (2.2.10) and (2.2.11) can be combined with Eqs. (2.2.1) and (2.2.2) to give

$$\frac{D_n\partial^2 n_n}{\partial x^2} + \mu_n F\frac{\partial n_n}{\partial x} + \mu_n n_n\frac{\partial F}{\partial x} + G - R = 0 \tag{2.2.13}$$

$$\frac{D_p\partial^2 p_n}{\partial x^2} - \mu_p F\frac{\partial p_n}{\partial x} - \mu_p p_n\frac{\partial F}{\partial x} + G - R = 0 \tag{2.2.14}$$

With the assumption that the semiconductor is nearly neutral ($\rho \approx 0$), we have

$$n_n - n_{no} = p_n - p_{no} \tag{2.2.15}$$

By eliminating $\partial F/\partial x$ from Eqs. (2.2.13) and (2.2.14), we obtain the following equation:

$$\mu_p p_n D_n \frac{\partial^2 n_n}{\partial x^2} + \mu_n n_n D_p \frac{\partial^2 p_n}{\partial x^2} + \mu_n \mu_p p_n F \frac{\partial n_n}{\partial x}$$

$$+ \mu_n \mu_p n_n F \frac{\partial p_n}{\partial x} + (G - R)(\mu_n n_n + \mu_p p_n) = 0 \qquad (2.2.16)$$

Combining this expression with the quasi-neutrality condition of of Eq. (2.2.15), we derive the so-called ambipolar transport equation

$$D_a \frac{\partial^2 p_n}{\partial x^2} - \mu_a F \frac{\partial p_n}{\partial x} + G - R = 0 \qquad (2.2.17)$$

where

$$\mu_a = \frac{\mu_n \mu_p (n_n - p_n)}{\mu_n n_n + \mu_p p_n} \qquad (2.2.18)$$

is called the ambipolar mobility and

$$D_a = \frac{\mu_p p_n D_n + \mu_n n_n D_p}{\mu_n n_n + \mu_p p_n} \qquad (2.2.19)$$

is called the ambipolar diffusion coefficient.

When $n_n \gg p_n$, which would be the case in an n-type semiconductor, we obtain $D_a \approx D_p$ and $\mu_a \approx \mu_p$ and Eq. (2.2.17) reduces to

$$D_p \frac{\partial^2 p_n}{\partial x^2} - \mu_p F \frac{\partial p_n}{\partial x} + G - R = 0 \qquad (2.2.20)$$

Equation (2.2.20) is the continuity equation for minority carriers (holes) in an n-type sample under steady-state conditions. This equation is extremely useful for the analysis of different semiconductor devices, such as p–n junctions, bipolar junction transistors, and solar cells.

Another useful semiconductor equation can be derived under the condition that the difference in densities of trapped electrons and holes is time independent or can be neglected. Using this condition and subtracting Eq. (2.2.10) from Eq. (2.2.11) yield

$$q \frac{\partial}{\partial t}(p - n) + \nabla \cdot (\mathbf{j}_n + \mathbf{j}_p) = 0 \qquad (2.2.21)$$

From Eq. (2.2.12), we find

$$q \frac{\partial}{\partial t}(p - n) = \frac{\partial \rho}{\partial t} \qquad (2.2.22)$$

Taking the time derivative of Poisson's equation [Eq. (2.2.9)], the result can be used to eliminate $\partial \rho / \partial t$ and $\partial (p - n)/\partial t$ from Eqs. (2.2.21) and (2.2.22) to give

$$\epsilon_s \frac{\partial}{\partial t} \nabla \cdot \mathbf{F} + \nabla \cdot (\mathbf{j}_n + \mathbf{j}_p) = 0 \qquad (2.2.23)$$

Integrating this expression over the space coordinates yields the total current density, including the displacement current density, $\epsilon_s \, \partial \mathbf{F}/\partial t$:

$$\mathbf{j}(t) = \mathbf{j}_n + \mathbf{j}_p + \epsilon_s \frac{\partial}{\partial t} \mathbf{F} \qquad (2.2.24)$$

Furthermore, by integrating the current density over the sample cross section, we obtain the total current

$$I = \int_S \mathbf{j} \, ds \qquad (2.2.25)$$

For a sample with a constant cross section S and a uniform current density, we have

$$J \quad S \qquad (2.2.26)$$

In low electric fields, the Einstein relationship is valid and the electron and hole current densities of Eqs. (2.2.1) and (2.2.2) can be rewritten as

$$\mathbf{j}_n = q\mu_n n\mathbf{F} + qD_n \, \nabla n$$
$$= \mu_n (n \, \nabla E_c + k_B T \, \nabla n)$$
$$= \mu_n n \, \nabla E_{Fn} \qquad (2.2.27)$$

$$\mathbf{j}_p = \mu_p p \, \nabla E_{Fp} \qquad (2.2.28)$$

Here, we used the relationship $\mathbf{F} = \nabla E_c / q$, where E_c is the electron energy of the conduction band edge. Furthermore, we assumed that the semiconductor is nondegenerate and that the carrier density can be described in terms of near-equilibrium particle statistics by means of the quasi-Fermi levels, E_{Fn} and E_{Fp}. For a discussion of carrier statistics and the quasi-Fermi level concept, we refer to any elementary text on semiconductor device physics, for example, Sze (1985) or Streetman (1995).

2.3. MATERIAL PROPERTIES OF IMPORTANT SEMICONDUCTORS

In this section, we discuss semiconductor material parameters of special importance for device and circuit modeling. Fundamentally, this involves semiconductor properties such as the lattice constant, the dielectric constant, the density, and details about band structure (including band gap and effective masses), and scattering processes. However, of more direct interest for device modeling are combined manifestations of these parameters on the transport properties of charge carriers, conveniently expressed in terms of the carrier mobilities and drift velocities. Relevant data for some important semiconductors are shown in Table 2.3.1.

In a low electric field, the electron drift velocity $v_n = -\mu_n F$, where $\mu_n = q\tau_n/m_n$ is the low-field mobility, F is the electric field, q is the unit charge, τ_n is the electron momentum relaxation time, and m_n is the electron effective mass. As was discussed in Section 2.2, the momentum relaxation time and the mobility are strongly influenced by the scattering processes. Mathiessen's rule states that the net momentum relaxation rate can be expressed as follows:

$$\frac{1}{\tau_n} = \sum_i \frac{1}{\tau_{ni}} \tag{2.3.1}$$

where the right-hand side is a sum over the relaxation rates for individual scattering mechanisms.

The linear dependence of the drift velocity on the electric field does not hold at high fields where electrons gain considerable energy from the field. At the same time, the exchange of energy and momentum between the carriers and the crystal lattice increases, and the net effect is a drift velocity that, in most semiconductors, becomes nearly independent of the electric field, as indicated in Fig. 2.3.1. In many semiconductors such as GaAs, InP, and InGaAs, the electron velocity in a certain range of electric fields may actually decrease with an increase of the electric field due to transfer of hot electrons from the central valley (Γ) of the conduction band into the satellite valleys (X and L) of higher energy. As can clearly be seen from Fig. 2.3.1, compound semiconductors have the potential for a higher speed of operation than silicon since the electrons in these materials move faster at relevant values of the electric field.

The electron and hole drift velocities in high electric fields depend on tempera-

TABLE 2.3.1. Properties of Important Semiconductors.

	a (Å)	ϵ_r (rel.)	ρ (g/cm³)	E_g (eV)	m_n	m_p	μ_n (cm²/V s)	μ_p (cm²/V s)
Si	5.43	11.8	2.33	1.12	1.08	0.56	1350	480
Ge	5.66	16.0	5.32	0.67	0.55	0.37	3900	1900
GaAs	5.65	13.2	5.31	1.42	0.067	0.48	8500	400
InP	5.87	12.1	4.79	1.35	0.080	—	4000	100

Note: a = lattice constant, ϵ_r = dielectric constant, ρ = density, E_g = energy band gap, m_n = electron effective mass (density of states effective mass for Si and Ge and central valley effective mass for GaAs and InP), m_p = hole effective mass (density of states effective mass), μ_n = electron mobility, μ_p = hole mobility. All at 300 K.

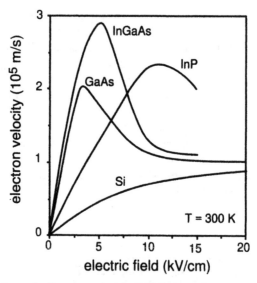

FIG. 2.3.1. Electron velocity versus electric field for some important semiconductors.

ture and on the total concentration of charged impurities, $N_T = N_a + N_d$. The stationary dependence of the electron velocity on the electric field in bulk silicon can be approximated by

$$v_n = \frac{\mu_n F}{\sqrt{1 + (\mu_n F/v_s)^2}} \qquad (2.3.2)$$

where

$$\mu_n(N_T, T) = \mu_{mn} + \frac{\mu_{on}}{1 + (N_T/N_{cn})^\nu} \qquad (2.3.3)$$

is the dependence of electron mobility on the impurity concentration (see Caugley and Thomas, 1967; Thornber, 1982; Sze, 1981; Arora et al., 1982; and Yu and Dutton, 1985). The variation of the mobility parameters μ_{mn}, μ_{on}, N_{cn}, and ν with temperature are given in Appendix A3. Figure 2.3.2a shows the dependence of the electron mobility on temperature for different impurity concentrations.

Correspondingly, for the hole velocity and mobility in silicon, we have

$$v_p = \frac{\mu_p F}{1 + (\mu_p F/v_s)} \qquad (2.3.4)$$

$$\mu_p(N_T, T) = \mu_{mp} + \frac{\mu_{op}}{1 + (N_T/N_{cp})^\nu} \qquad (2.3.5)$$

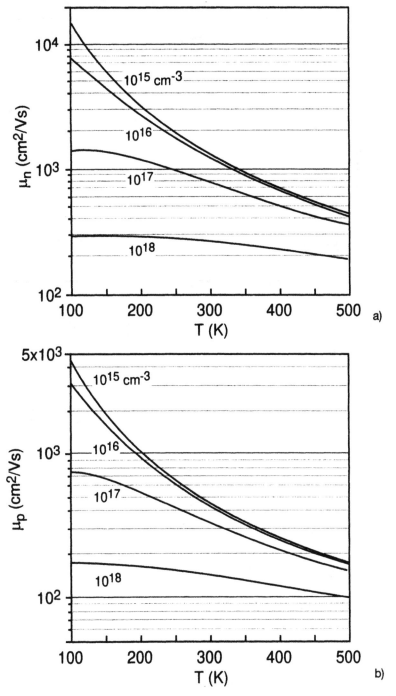

FIG. 2.3.2. (*a*) Electron and (*b*) hole mobilities in *n*-type and *p*-type silicon, respectively, versus temperature for different impurity concentrations.

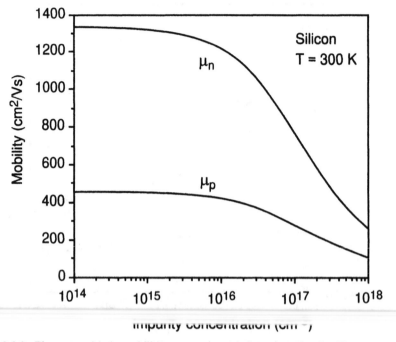

FIG. 2.3.3. Electron and hole mobilities versus impurity concentration in silicon at room temperature.

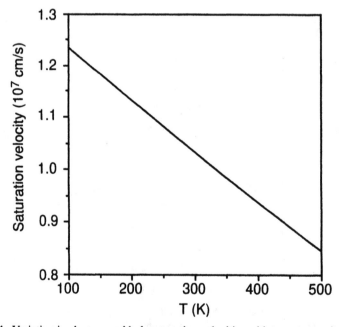

FIG. 2.3.4. Variation in electron and hole saturation velocities with temperature in silicon.

where, again, the temperature dependencies of the mobility parameters are given in Appendix A3. Figure 2.3.2*b* shows the dependence of the hole mobility on temperature for different impurity concentrations.

In Fig. 2.3.3, we compare the variation of the electron and hole mobilities with impurity concentration in silicon at room temperature.

The saturation velocities for electrons and holes in silicon are very similar and can be approximated as follows:

$$v_s = \frac{2.4 \times 10^7}{1 + 0.8 \exp(T/600)} \quad \text{(cm/s)} \tag{2.3.6}$$

This relationship is shown in Fig. 2.3.4.

Note that the above expressions are valid for majority carriers in bulk silicon. The mobility and velocity of minority carriers may be quite different because of electron–hole scattering. Nevertheless, the same equations are frequently used for minority carriers as well because of a lack of sufficient experimental data.

REFERENCES

N. D. Arora, J. R. Hauser, and D. J. Rulson, *IEEE Transactions on Electron Devices*, vol. ED-29, p. 292 (1982).

D. M. Caugley and R. E. Thomas, *Proceedings of the IEEE*, vol. 55, pp. 2192–2193 (1967).

K. Hess, *Advanced Theory of Semiconductor Devices*, Prentice-Hall, Englewood Cliffs, NJ (1988).

K. Lee, M. Shur, T. A. Fjeldly, and T. Ytterdal, *Semiconductor Device Modeling for VLSI*, Prentice-Hall, Englewood Cliffs, NJ (1993).

K. Seeger, *Semiconductor Physics, An Introduction*, 3rd ed., Series on Solid-State Sciences, Vol. 40, Springer-Verlag, Berlin (1985).

M. Shur, *Introduction to Electronic Devices*, Wiley, New York (1996).

M. Shur, *Physics of Semiconductor Devices*, Prentice-Hall, Englewood Cliffs, NJ (1990).

B. Streetman, *Solid State Electronic Devices*, 4th ed., Prentice-Hall, Englewood Cliffs, NJ (1995).

S. M. Sze, *Physics of Semiconductor Devices*, 2nd ed., Wiley, New York (1981).

S. M. Sze, *Semiconductor Devices. Physics and Technology*, Wiley, New York (1985).

K. K. Thornber, *IEEE Electron Device Letters*, vol. EDL-3, no. 3, pp. 69–71 (1982).

Z. Yu and R. W. Dutton, *Sedan III—A General Electronic Material Device Analysis Program*, Program manual, Stanford University (July 1985).

PROBLEMS

2.2.1. (a) Find the resistivity of intrinsic Si and Ge at room temperature.

(b) Add shallow donor impurities in the amount of 1 atom per 10^8 atoms of the host material, and find out how the resistivities change. Use the data in Table 2.3.1, intrinsic carrier densities n_i of 1.5×10^{10} cm^{-3} for Si and 2.5×10^{13} cm^{-3} for Ge, and the atomic weights 28 a.u. for Si and 72 a.u. for Ge (1 a.u. $\approx 1.66 \times 10^{-24}$ g).

2.2.2. For a constant uniform generation rate, $G = 10^{20}$ cm^{-3}/s, and assuming a recombination rate of the form $R = (n - n_o)/\tau$, where $n_o = 10^{15}$ cm^{-3} and $\tau = 10^{-7}$ s, calculate and plot $n(t)$ (neglect the effects of traps).

2.2.3. Express the ambipolar diffusion coefficient D$_a$ in terms of the ambipolar mobility μ_a, V_{th}, p_n, and n_n. Compare the resulting equation with the Einstein relationship.

2.2.4. Assume that electrons move with saturation velocity. Calculate and plot the transit time as a function of the sample length in the range 0.05 μm $< L < 1$ μm for silicon at 77, 300, and 400 K. (This gives a crude idea of the minimum delay in silicon devices with a given minimum feature size.)

2.2.5. Use the basic semiconductor equations to derive Ohm's law for a uniform semiconductor sample. Express the sample resistivity in terms of electron and hole concentrations and mobilities.

2.3.1 Plot and compare the velocity-field curves for electrons and holes in silicon

for the temperatures 100, 300, and 400 K. (Use the relationships in Appendix A3.)

CHAPTER 3

TWO-TERMINAL DEVICES

3.1. INTRODUCTION

Junctions between semiconductors of different doping and material composition and between semiconductors and metals or insulators are basic building blocks of all semiconductor devices.

An insulator separating a semiconductor from a metal forms a metal–insulator–semiconductor (MIS) capacitor, as shown in Fig. 3.1.1. Certain metals make low-resistive contacts with a semiconductor (see Fig. 3.1.2). Other metals, or even the same metals with a different technological treatment, may form rectifying contacts. These so-called Schottky junctions are much more conductive for one polarity of the applied voltage (forward bias) than for the other (reverse bias), as indicated in Fig. 3.1.3. A p–n junction is another example of a rectifying contact. Its current–voltage characteristic is similar to that of the Schottky junction and the same circuit symbol is used. However, the turn-on voltage for a p–n diode is typically larger than that for a Schottky diode, and the current at reverse bias is usually much smaller.

Semiconductor devices typically include a combination of these building blocks. It is well known that p–n junctions are crucial elements in p–n diodes and bipolar transistors, but they are also essential in FETs. Normally, we think of the p–n junction as two adjoining regions of opposite type doping in the same semiconductor material. However, junctions can also be made by joining different semiconductor materials, creating a different class of structures with new properties and additional opportunities for device design. The fabrication of a wide variety of these so-called heterostructures has become possible in recent years owing to significant advances in epitaxial crystal growth.

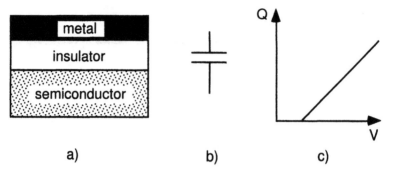

a)   b)   c)

FIG. 3.1.1. (*a*) Metal–insulator–semiconductor structure, (*b*) its equivalent circuit (a capacitor), and (*c*) the simplified charge–voltage characteristic. (Q and $-Q$ are the charges induced in the metal and in the semiconductor, respectively, and V is the bias voltage).

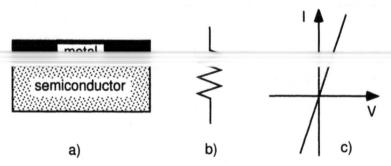

a)   b)   c)

FIG. 3.1.2. (*a*) Metal–semiconductor ohmic contact, (*b*) its equivalent circuit (a resistor, hopefully small compared to the semiconductor resistance), and (*c*) the current–voltage characteristic. (I is the current through the contact and V is the bias voltage).

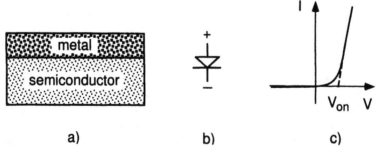

a)   b)   c)

FIG. 3.1.3. (*a*) Metal–semiconductor Schottky barrier contact (a different type of metal compared to that of Fig. 3.1.2), (*b*) its circuit symbol (a diode), and (*c*) the current–voltage characteristic (I is the current through the Schottky contact and V is the bias voltage; V_{on} is the turn-on voltage).

Metal–semiconductor junctions are obviously needed for connecting semiconductors to the surrounding circuitry. The ohmic contacts provide low-resistance, nonrectifying junctions, while the Schottky barriers can be used either as high-speed diodes or as high-impedance gate structures in certain types of FETs.

Finally, high quality MIS structures are essential as gate elements of MOSFETs (metal–oxide–semiconductor field-effect transistors).

In this chapter, we review the basic properties of the various types of two-terminal devices, present models for these devices implemented in SPICE, and give SPICE simulation examples.

3.2. p–n JUNCTIONS

3.2.1. p–n Junction at Equilibrium

Figure 3.2.1 shows the qualitative energy band diagram of a p–n junction at thermal equilibrium (no applied bias voltage, no thermal gradients, etc.). As indicated, the Fermi level is constant throughout the structure. This is consistent with zero net electron and hole currents across the junction [see Eqs. (2.2.27) and (2.2.28)]. Also, the charge carrier densities obey equilibrium carrier statistics everywhere, and for parabolic bands and nondegenerate doping, the electron and hole densities are given by

$$n = n_i \exp\left(\frac{E_F - E_i}{k_B T}\right) \tag{3.2.1}$$

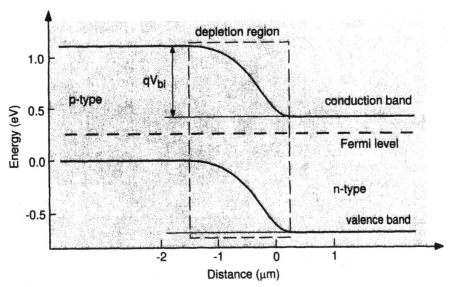

FIG. 3.2.1. Energy band diagram of a silicon p–n junction with doping concentrations $N_a = 3 \times 10^{14}$ (p-side) and $N_d = 8 \times 10^{15}$ (n-side).

$$p = n_i \exp\left(\frac{E_i - E_F}{k_B T}\right) \tag{3.2.2}$$

where n_i is the intrinsic electron density, E_F is the Fermi level, and E_i is the intrinsic Fermi level (usually located close to the middle of the forbidden gap). The *mass action law*—a consequence of the above equations—states that $np = n_i^2$ at equilibrium.

As shown in Fig. 3.2.1, the energy bands are bent in a transition region near the junction (within the dashed box of the figure), resulting in a total energy shift qV_{bi} across the junction. Here, V_{bi} is called the built-in potential and is a fundamental property of the junction.

Let us suppose that the concentration of completely ionized donors N_d is dominant and constant inside the n-type region and that the corresponding density of completely ionized acceptors N_a is constant inside the p-type region. Well away from the junction, we can assume that the semiconductor is electrically neutral such that $n \approx N_d$ in the n-type region and $p \approx N_a$ in the p-type region, and the energy bands are flat.

From Eqs. (3.2.1) and (3.2.2), we observe that the equilibrium electron and hole concentrations change by a factor of e when the bands (including the intrinsic Fermi band bending qV_{bi} is typically several tenths of an electron volt or more (depending on the doping levels, the energy gap, and the effective densities of states in the conduction and valence bands), that is, $qV_{bi} \gg qV_{th}$. As a consequence, we have $n \ll N_d$ and $p \ll N_a$ in almost the entire transition region. Hence, the charge density in the transition region can safely be approximated by the density of fixed charges associated with the dopants, that is,

$$\rho = \begin{cases} qN_d & \text{on the } n\text{-type side} \\ -qN_a & \text{on the } p\text{-type side} \end{cases} \tag{3.2.3}$$

This corresponds to the so-called depletion approximation. The transition region is frequently called the depletion region or the space charge region. The space charge density distribution in a p–n junction according to the depletion approximation is illustrated in Fig. 3.2.2.

The depletion approximation implies that the charge carrier concentrations in the depletion region are such that $N_d \gg n \gg n_{po}$ and $N_a \gg p \gg p_{no}$, where, according to the *mass action law*, $n_{po} \approx n_i^2/N_a$ and $p_{no} \approx n_i^2/N_d$ are the equilibrium minority carrier concentrations in the p- and n-type regions, respectively. The depletion approximation is not valid in the boundary layers between the neutral parts and the depletion region. However, these layers are relatively thin as long as the total band bending, that is, the difference between the bottom of the conduction band in the p- and n-regions, is much greater than the thermal energy qV_{th}.

The built-in voltage is determined by the doping densities of the p–n junction and can be derived by using the expressions for equilibrium carrier statistics in Eqs.

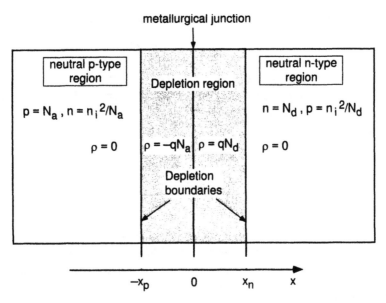

FIG. 3.2.2. Space charge densities in a *p–n* junction according to the depletion approximation.

(3.2.1) and (3.2.2). According to these expressions, the positions of the intrinsic level E_{in} deep inside the *n*-region and E_{ip} deep inside the *p*-region are determined by the respective carrier concentrations. In practice, we have, for the majority carriers, $n_n = N_d$ (*n*-region) and $p_p = N_a$ (*p*-region), which gives

$$E_{in} = E_F - k_B T \ln\left(\frac{N_d}{n_i}\right)$$

$$E_{ip} = E_F + k_B T \ln\left(\frac{N_a}{n_i}\right)$$

(3.2.4)

This leads to the following expression for the built-in potential:

$$V_{bi} = \frac{E_{ip} - E_{in}}{q} = V_{th} \ln\left(\frac{N_d N_a}{n_i^2}\right)$$

(3.2.5)

The distribution of the electric potential ϕ across the junction is determined from Poisson's equation

$$\frac{d^2\phi}{dx^2} = -\frac{\rho}{\epsilon_s}$$

(3.2.6)

where ϵ_s is the dielectric permittivity of the semiconductor and

$$\rho(x) = q[p(x) - n(x) + N(x)] \tag{3.2.7}$$

is the space charge density. The term $N(x)$ is the net doping density (donor density minus acceptor density) at position x. Using the depletion approximation, we can rewrite Poisson's equation as

$$\frac{d^2\phi}{dx^2} = \begin{cases} \dfrac{qN_a}{\epsilon_s} & \text{for } -x_p < x < 0 \\[3mm] -\dfrac{qN_d}{\epsilon_s} & \text{for } 0 < x < x_n \end{cases} \tag{3.2.8}$$

Here, $x = 0$ corresponds to the metallurgical junction between the p- and n-type regions, and x_p and x_n are the depletion widths on the two sides of the junction, as indicated in Fig. 3.2.2. A typical charge distribution for a silicon p–n junction according to the depletion approximation is shown in Fig. 3.2.3a.

The electric field

$$F = -\frac{d\psi}{dx} = \frac{1}{q}\frac{dE_c}{dx} \tag{3.2.9}$$

is obtained by integrating Eq. (3.2.8) once. Using the conditions $F(x = x_n) = 0$ and $F(x = -x_p) = 0$, we find

$$F(x) \begin{cases} -F_m\left(1 + \dfrac{x}{x_p}\right) & \text{for } -x_p < x < 0 \\[3mm] -F_m\left(1 - \dfrac{x}{x_n}\right) & \text{for } 0 < x < x_n \end{cases} \tag{3.2.10}$$

The maximum magnitude of the electric field F_m in the junction, reached at the p–n interface ($x = 0$), is obtained from elementary electrostatics:

$$F_m = \frac{qN_d x_n}{\epsilon_s} = \frac{qN_a x_p}{\epsilon_s} \tag{3.2.11}$$

The electric field profile for our example is shown in Fig. 3.2.3b.

The potential distribution is found by integration of Eq. (3.2.10):

$$\phi(x) = \begin{cases} \dfrac{qN_a(x + x_p)^2}{2\epsilon_s} & \text{for } -x_p < x < 0 \\[3mm] V_{bi} - \dfrac{qN_d(x - x_n)^2}{2\epsilon_s} & \text{for } 0 < x < x_n \end{cases} \tag{3.2.12}$$

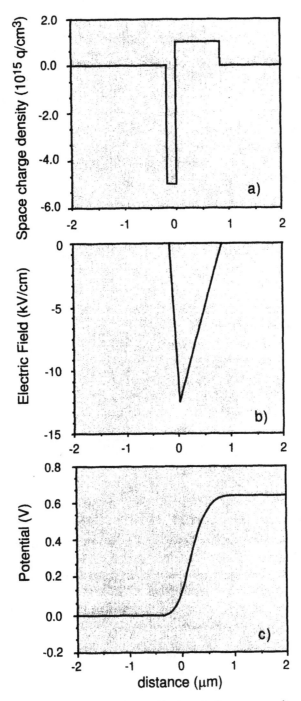

FIG. 3.2.3. (*a*) Example of charge density profile (in unit charges per cm³) and (*b*) associated electric field and (*c*) potential in a silicon *p–n* junction at equilibrium. (After Shur, 1990.)

where we used the (arbitrary) reference potential $\phi(-x_p) = 0$ and where $\phi(x_n) = V_{bi}$. The potential distribution for our example p–n junction at equilibrium is shown in Fig. 3.2.3c.

We note that the electron potential energy at any position x corresponds to the bottom of the conduction band, that is, $\phi(x) = -[E_c(x) - E_c(-x_p)]/q$. Hence, the energy band diagram, such as the one shown in Fig. 3.2.1, can be directly obtained from the potential distribution and vice versa.

Since the semiconductor junction has overall neutrality, the total amount of negative charge of ionized acceptors $qN_a x_p$ on the p-side of the depletion region must be equal to the total amount of positive charge of ionized donors $qN_d x_n$ on the n-side of the depletion region. Hence

$$\frac{x_n}{x_p} = \frac{N_a}{N_d} \tag{3.2.13}$$

The total voltage drop across the depletion region is equal to the built-in voltage. Using Eq. (3.2.12), we find

$$\frac{qN_d x_n^2}{} + \frac{qN_a x_p^2}{} = V_{..} \tag{3.2.14}$$

Solving Eqs. (3.2.13) and (3.2.14) with respect to x_n and x_p, we obtain

$$x_n = \sqrt{\frac{2\epsilon_s V_{bi}}{qN_d(1 + N_d/N_a)}} \tag{3.2.17}$$

$$x_p = \sqrt{\frac{2\epsilon_s V_{bi}}{qN_a(1 + N_a/N_d)}} \tag{3.2.18}$$

Note that for a one-sided p^+–n junction where $N_a \gg N_d$, we have $x_n \gg x_p$.

3.2.2. Current–Voltage Characteristic of an Ideal p–n Diode

As seen from Fig. 3.2.1, a potential barrier separates electrons in the n-region from the p-region. The same potential barrier separates holes in the p-region from the n-region. This potential barrier can be decreased when a positive bias is applied to the p-region with respect to the n-region (forward bias) or it can be increased when a voltage of opposite polarity (reverse bias) is applied. At zero bias the height of the potential barrier is V_{bi}. Normally, under reverse or small forward bias V, the resistance of the space charge region is much greater than that of the neutral sections of the device. As a result, almost all of the applied voltage drops across the depletion region, changing the potential barrier to $V_{bi} - V$. Hence, a forward bias (V positive) decreases the barrier while a reverse bias (V negative) increases the barrier. The width of the depletion region varies accordingly, shrinking under a forward bias and increasing under a reverse bias.

   The electric field in the depletion region gives rise to drift electron and hole currents. At zero bias, these current components are exactly compensated by diffusion current components to give zero net current and a constant Fermi level [see Eqs. (2.2.27) and (2.2.28)]

   When an external voltage is applied, this balance between the drift and diffusion components of the electric current is disturbed. However, for moderate voltages, the carrier distributions are still approximately governed by equilibrium statistics, reflecting a quasi-equilibrium state within each of the populations of electrons and holes. Mathematically, this can be expressed using separate Fermi levels E_{Fn} and E_{Fp} for electrons and holes—the so-called quasi–Fermi levels. Hence, Eqs. (3.2.1) and (3.2.2) can be modified to read

$$n = n_i \exp\left(\frac{E_{Fn} - E_i}{k_B T}\right) \tag{3.2.19}$$

$$p = n_i \exp\left(\frac{E_i - E_{Fp}}{k_B T}\right) \tag{3.2.20}$$

   As a consequence of this near-equilibrium state within each of the carrier populations, both E_{Fn} and E_{Fp} remain nearly constant throughout the depletion region, as indicated in Fig. 3.2.4, but with an energy separation given by the applied bias voltage (see below). Well away from the depletion region, the electron and hole quasi–Fermi levels come together. This reflects the fact that deep inside the *n*- and *p*-regions the electron and hole populations are very close to thermal equilibrium with each other. However, near the depletion region, E_{Fp} and E_{Fn} for the minority carriers (holes in the

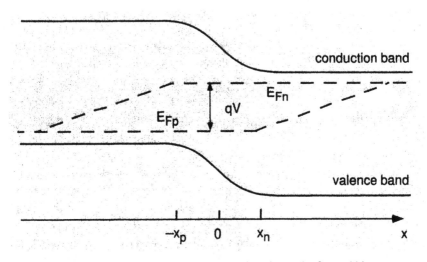

FIG. 3.2.4. Quasi–Fermi levels in a *p–n* junction under forward bias.

n-region and electrons in the p-region) vary almost linearly with distance. This corresponds to the exponential decay of the minority carrier concentrations away from the junction because of recombination. The characteristic distances of this decay are the so-called diffusion lengths for electrons and holes, L_n and L_p, which are both typically much larger than the width of the depletion region. This description is supported by numerical simulations of p–n junctions at reverse and low forward bias.

Let us now consider the consequences of an applied voltage in terms of band structure and carrier transport across the junction. We assume for simplicity that the applied voltage V falls entirely across the space charge region, with negligible voltage drops across the neutral regions and the contacts. Then the potential barrier between the p-type and n-type regions will obviously change from V_{bi} to $V_{bi} - V$ and the difference between the quasi–Fermi levels in the depletion region will be (see Fig. 3.2.4)

$$qV = E_{Fn} - E_{Fp} \qquad (3.2.21)$$

From the above equations, we obtain the *law of the junction:*

$$\ldots = n^2 \ldots \left(\frac{E_{Fn} - E_{Fp}}{\ldots} \right) = n^2 \ldots \left(\frac{V}{\ldots} \right) \qquad (3.2.22)$$

Numerical calculations show that this relationship is valid even at relatively high injection levels, where the minority carrier densities close to the depletion region boundaries become comparable to those of the majority carriers.

The charge transport in the p–n junction is conveniently discussed by focusing on the neutral n- and p-regions adjacent to the depletion region. In stationary condition, the current through all cross sections of the junction will be the same and we could, in principle, have based our discussion on any region of the junction. However, the transport within the depletion region is a complicated mixture of drift and diffusion current. On the other hand, deep inside the neutral regions the current converts to a pure majority carrier drift transport that cannot be discussed independently of what takes place at the junction. By concentrating on the electrically neutral regions in the vicinity of the depletion region, the current can be evaluated based on the following strategy and assumptions (see Fig. 3.2.5):

(a) The minority carrier concentrations at the depletion region boundaries are obtained from the law of the junction. We assume that the injection level is sufficiently low so that the majority carrier densities (dictated by the neutrality condition) do not deviate significantly from the doping densities.

(b) The variation in the minority carrier densities inside the neutral regions are obtained by solving the continuity equations [see Eqs. (2.2.13) and (2.2.14)] using the densities from (a) and the equilibrium densities deep inside the n- and p-regions as boundary conditions. We assume that the electric field in the neutral regions is negligible and that the net recombination rates can be expressed in terms of minority carrier lifetimes.

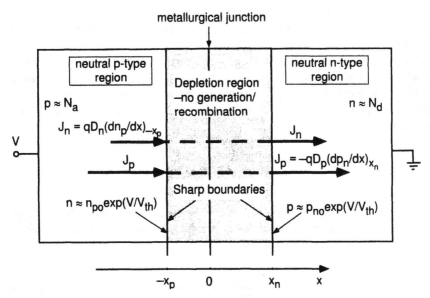

FIG. 3.2.5. Some of the assumptions and relationships used in determining the current–voltage characteristic of a p–n junction.

(c) The minority carrier diffusion current inside each of the neutral regions can be calculated based on the variation of the minority carrier densities from (b) using the transport equations (see Section 2.2). The drift component of minority carrier current in the neutral regions can be neglected since the electric field in these regions is assumed to be very small.

(d) The total current is taken as the sum of the minority carrier diffusion currents evaluated at the two boundaries of the depletion region. This is based on the assumption of continuity of both electron and hole currents through the depletion region, which is true if recombinations in this region can be neglected.

Following this logic, we start by considering the carrier densities at the boundaries between the depletion region and the neutral regions at the n- and p-sides of the junction. Assuming low-level injection and using the law of the junction, we have

$$n \approx N_d \qquad p \approx p_{no} \exp\left(\frac{V}{V_{th}}\right) \qquad \text{at } x = x_n \qquad (3.2.23)$$

$$p \approx N_a \qquad n \approx n_{no} \exp\left(\frac{V}{V_{th}}\right) \qquad \text{at } x = -x_p \qquad (3.2.24)$$

We note that x_n and x_p have to be adjusted for the applied voltage. For an abrupt junction, this is accomplished by replacing V_{bi} by $V_{bi} - V$ in Eqs. (3.2.17) and (3.2.18).

The distribution of minority carriers in the neutral regions are determined by solving the continuity equations (see Section 2.2). The net recombination rate for minority carriers may normally be expressed in terms of a relaxation time approximation. We consider, for example, the hole distribution in the neutral n-type region where we have

$$R - G = \frac{p_n - p_{no}}{\tau_p} \tag{3.2.25}$$

and where τ_p is called the hole lifetime. In the continuity equation for holes [Eq. (2.2.14)], we neglect the terms involving the electric field and traps and obtain, for stationary conditions, the so-called diffusion equation

$$D_p \frac{\partial^2 p_n}{\partial x^2} - \frac{p_n - p_{no}}{\tau_p} = 0 \tag{3.2.26}$$

For a long neutral region, this equation can be solved using the boundary conditions

$$p_n(x_n) = p_{no} \exp\left(\frac{V}{V_{th}}\right) \quad p_n(x \to \infty) = p_{no} \tag{3.2.27}$$

resulting in the exponential distribution

$$p_n(x) - p_{no} = p_{no}\left[\exp\left(\frac{V}{V_{th}}\right) - 1\right]\exp\left(-\frac{x - x_n}{L_p}\right) \tag{3.2.28}$$

where

$$L_p = \sqrt{D_p \tau_p} \tag{3.2.29}$$

is the hole diffusion length—a characteristic length for the diffusion of injected holes into the neutral n-type region.

The hole current density at the boundary between the depletion region and the neutral n-type region is primarily caused by diffusion (since the electric field is nearly zero in this region) and can be expressed as the spatial derivative of the hole distribution at $x = x_n$, that is,

$$J_p = -qD_p \frac{\partial p_n}{\partial x}\bigg|_{x=x_n} = \frac{qD_p p_{no}}{L_p}\left[\exp\left(\frac{V}{V_{th}}\right) - 1\right] \tag{3.2.30}$$

Similarly, we find, for electrons in the p-type region,

$$J_n = qD_n \frac{\partial n_p}{\partial x}\bigg|_{x=-x_p} = \frac{qD_n n_{po}}{L_n}\left[\exp\left(\frac{V}{V_{th}}\right) - 1\right] \tag{3.2.31}$$

where

$$L_n = \sqrt{D_n \tau_n} \tag{3.2.32}$$

is the diffusion length for electrons.

Typically, we have

$$L_n \gg x_n + x_p \qquad L_p \gg x_n + x_p \tag{3.2.33}$$

in which case relatively little recombination occurs in the depletion region and the hole and electron current densities remain nearly constant throughout this region. Therefore, the total current density through the p–n junction can be written as $J = J_n + J_p$, or

$$J = J_s\left[\exp\left(\frac{V}{V_{th}}\right) - 1\right] \tag{3.2.34}$$

where

$$J_s = \frac{qD_p p_{no}}{L_p} + \frac{qD_n n_{no}}{L_n} \tag{3.2.35}$$

Equation (3.2.34) is the well-known *diode equation*, also called the Shockley equation, and J_s is the junction saturation current density.

Very often, the diode is constructed as a one-sided abrupt p^+–n or n^+–p junction. For example, in a p^+–n diode, $p_{no} \gg n_{po}$ and the second term in Eq. (3.2.35) can be neglected.

3.2.3. Nonideal Effects

The ideal diode equation described by Eqs. (3.2.34) and (3.2.35) is in reasonable agreement with experimental data for germanium diodes, less so for silicon diodes, and does not describe very well the current–voltage characteristics of gallium arsenide diodes. The deviation from ideality is illustrated for a silicon diode in the AIM-Spice simulation of Fig. 3.2.6, where the effects of several nonideal mechanisms are indicated. Some of these mechanisms, such as carrier generation/recombination in the depletion region, series resistance, and high-level injection, were explicitly neglected in the above derivation of the diode equation. Other effects are related to junction breakdown, thermal runaway, and short diodes. Here follows a brief discussion of some important nonideal mechanisms.

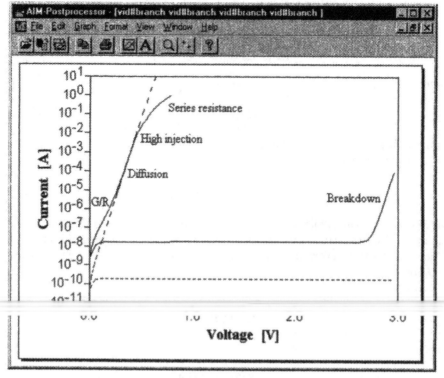

FIG. 3.2.6. Ideal (broken curves) and nonideal (solid curves) I–V characteristics of a Si p–n diode simulated with AIM-Spice and plotted in a semilog scale. The two nearly horizontal curves correspond to reverse bias. Various nonideal effects are indicated. The AIM-Spice model parameters used in the calculation are is=192.1p, n=1, xti=3, eg=1.1, cjo=893.8f, m=98.29m, vj=.75, fc=.5, bv=5, ibv=10u, nr=2, isr=16.91n, ikf=1e-2, rs=0.1 (similar to the example in Section 3.2.6; see also Appendix A1 for an explanation of the AIM-Spice parameter notation).

Series Resistance. In real semiconductor diodes, the parasitic series resistance R_s of the device contacts and of the semiconductor neutral regions may play an important role, especially at high forward bias. Then the diode equation for the total current has to be modified according to (see Section 1.1)

$$I = I_s \left[\exp\left(\frac{V - IR_s}{V_{\text{th}}} \right) - 1 \right] \tag{3.2.36}$$

where, for a uniform junction with a cross section S, we have written the total current as $I = SJ$. An approximate analytical solution of I versus V for this case was discussed by Fjeldly et al. (1991) (see also Lee et al., 1993). We note that at sufficiently large forward bias, most of the applied voltage will fall across the series resistance and the I–V characteristic approaches a linear form with a slope determined by R_s ($V \approx V_{on} + IR_s$).

High Injection. At large forward bias, when the voltage V_{di} across the intrinsic diode approaches the built-in potential V_{bi}, the junction barrier is severely reduced and the assumption of a low-level injection (where the densities of minority carriers are small compared to those of the majority carriers) is no longer valid. Instead, the injected minority carrier concentrations may become comparable to those of the majority carriers, and we enter the so-called high-injection regime. Since the law of the junction is still approximately valid, we can write [compare with Eqs. (3.2.23) and (3.2.24)]

$$pn = p(N_d + p) = n_i^2 \exp\left(\frac{V_{di}}{V_{th}}\right) \quad \text{at } x = x_n \tag{3.2.37}$$

$$pn = n(N_a + n) = n_i^2 \exp\left(\frac{V_{di}}{V_{th}}\right) \quad \text{at } x = -x_p \tag{3.2.38}$$

This shows that the minority carrier concentrations at the boundaries of the depletion region approach the value $n_i \exp(V_{di}/2V_{th})$ in the extreme case, indicating that the growth of the diode current becomes roughly proportional to

$$I \propto \exp\left(\frac{V_{di}}{2V_{th}}\right) \tag{3.2.39}$$

which means a reduced rate of increase.

Generation/Recombination in the Depletion Region. In certain regimes of operation, generation and recombination processes taking place in the depletion region (neglected in the ideal diode) give rise to current contributions that may play an important or even dominant role. These contributions depend on the concentration, distribution, and energy levels of traps in the depletion region. Traps associated with various impurities and defects are always present in any semiconductor material. The simplest model describing the generation and recombination currents is based on the assumption that we have only one type of dominant trap: uniformly distributed in the device. In reality, traps may be nonuniformly distributed, and more than one type of trap may be involved.

Practically no recombinations take place in the depletion region in reverse bias ($V < 0$) since few carriers are available. The generation process seeks to restore equilibrium ($pn \to n_i^2$), but the generated carriers are swept away leading to a generation current. This current can be found by integrating the net recombination rate over the depletion region, which leads to (see, e.g., Lee et al., 1993)

$$J_{gen} = q \int_{-x_p}^{x_n} |G - R| \, dx \approx \frac{qn_i x_d}{\tau_{gen}} \tag{3.2.40}$$

where τ_{gen} is called the generation lifetime and $x_d = x_n + x_p$ is the width of the depletion region. For an abrupt junction, x_d and, hence, J_{gen} are proportional to $(V_{bi} - V)^{1/2}$.

In reverse bias, the total current density J_R is given by

$$J_R = J_s + J_{gen} \tag{3.2.41}$$

where the diffusion component J_s is equal to the saturation current density [see Eq. (3.2.35)], which also can be written as

$$J_s = \left(\frac{qD_p}{N_d L_p} + \frac{qD_n}{N_a L_n} \right) n_i^2 \tag{3.2.42}$$

Since J_s is proportional to n_i^2 and J_{gen} is proportional to n_i, the generation current will be dominant when n_i is sufficiently small. In practice, this is always the case at reverse bias for GaAs and Si diodes at room temperature and below.

Under forward-biasing conditions ($V > 0$), we have an excess of carriers in the depletion region ($np > n_i^2$), resulting in a net recombination in this region. Again, integrating the net recombination rate over the depletion region leads to the following approximate recombination current density (see van der Ziel, 1976):

$$J_{rec} = q \int_{-x_p}^{x_n} (G - R) \, dx \approx J_{recs} \exp\left(\frac{V}{2V_{th}} \right) \tag{3.2.43}$$

where

$$J_{recs} = \frac{\pi}{2} \frac{q n_i V_{th}}{\tau_{rec} F_{max}} \tag{3.2.44}$$

Here F_{max} is the maximum electric field in the depletion layer and τ_{rec} is called the recombination lifetime. Usually, a more accurate expression for the recombination current in practical devices is obtained by replacing the factor of $\frac{1}{2}$ in the exponential function of Eq. (3.2.43) by an adjustable factor $1/m_r$.

The total forward-current density now becomes

$$J_F = J_s \exp\left(\frac{V}{V_{th}} \right) + J_{recs} \exp\left(\frac{V}{m_r V_{th}} \right) \tag{3.2.45}$$

Since J_s is proportional to n_i^2 and J_{recs} is proportional to n_i, the recombination current dominates over the diffusion current at small forward bias when n_i is sufficiently small. However, at larger forward voltages, the diffusion component will dominate since it is proportional to $\exp(V/V_{th})$, whereas the recombination current density varies only as $\exp[V/(m_r V_{th})]$, where $m_r \approx 2$.

Frequently, the following empirical diode expression is used to include the nonideal effects discussed so far in an average and simple manner:

$$J = J_{seff}\left[\exp\left(\frac{V}{\eta V_{th}}\right) - 1\right] \tag{3.2.46}$$

Here η is called the ideality factor and J_{seff} is an effective saturation current. The deviation of η from unity may be considered a measure of the importance of the non-ideal effects, such as the recombination current at small forward bias and the high-injection effects at high forward bias. This simple equation may be quite useful, but we must remember that it approximates the diode current–voltage characteristics only in a certain range of currents (which may be as large as a few decades of current variation).

Junction Breakdown. When a reverse bias applied to a *p–n* junction exceeds some critical value, the reverse current increases rapidly, as indicated for *p⁺–n* diodes in Fig. 3.2.7. One of two basic mechanisms, avalanche or tunneling breakdown, may be responsible for this rapid increase in current.

Avalanche breakdown is caused by impact ionization where an electron (or a

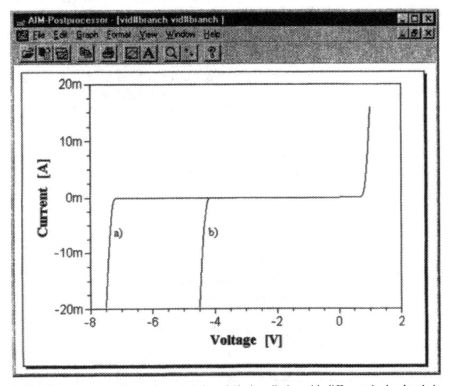

FIG. 3.2.7. Current–voltage characteristics of Si *p⁺–n* diodes with different doping levels in the *n*-region simulated with AIM-Spice. The breakdown voltages are (a) 7 V and (b) 4 V.

hole) gains so much energy from the electric field that it initiates a transition of an electron from the valence band to the conduction band, creating an additional electron–hole pair. Newly created carriers are in turn accelerated by the electric field and create new electron–hole pairs by repeating the process. If the applied voltage is sufficiently high, this leads to an uncontrolled rise of the current (until either the current is limited by an external load or the sample is destroyed).

A crude estimate of the critical voltage V_{abr} causing the avalanche breakdown may be obtained by assuming that the breakdown occurs when the electric field in the reverse biased p–n junction exceeds a certain critical field F_{br} (see, e.g., Sze, 1985). For a p^+–n junction, V_{abr} is obtained by combining Eq. (3.2.11) for the maximum field with Eq. (3.2.17) for the depletion width x_n (replacing V_{bi} by $V_{bi} - V_{abr}$ and assuming that $V_{abr} \gg V_{bi}$):

$$V_{abr} = \frac{\epsilon_s F_{br}^2}{2qN_d} \qquad (3.2.47)$$

The breakdown field at 300 K is of the order of 100 kV/cm for germanium, 300 kV/cm for silicon, 400 kV/cm for GaAs, and 2300 kV/cm for silicon carbide. Hence, V_{abr} can be quite large for diodes with low-doped n-regions. For SiC diodes

Tunneling breakdown, also called Zener breakdown, usually occurs in fairly highly doped semiconductors when the maximum electric field in the depletion layer approaches values of the order of 10^6 V/cm. Under such conditions, the distance between occupied states in the valence band of the p-type region and empty states of equal energy in the conduction band of the n-type region is so small that electron tunneling may take place across the junction. An exponential rise of the tunneling current in highly doped p–n junctions with reverse voltage leads to tunneling breakdown. The critical voltage for tunneling breakdown V_{tbr} may be estimated by defining the onset of breakdown as the reverse voltage that gives a 10-fold increase in the reverse current over the saturation current, that is,

$$I(V_{tbr}) \approx 10 S J_s \qquad (3.2.48)$$

According to Sze (1981), the critical voltage for tunneling breakdown is usually less than approximately $4E_g/q$. When the breakdown voltage is higher than $6E_g/q$, the breakdown is typically caused by avalanche effect.

Thermal Breakdown. In p–n diodes with a large power dissipation, a so-called thermal breakdown may lead to a runaway increase in the reverse current and to an S-type negative differential resistance. This mechanism is especially important in devices made from materials with a relatively narrow band gap (such as Ge) where the current density may increase appreciably even at a relatively moderate increase in temperature.

Short Diodes. In most practical semiconductor diodes, the total lengths X_n and X_p of the n-type and p-type regions are comparable to or smaller than the diffusion lengths of the minority carriers, and the continuity equation has to be solved with modified boundary conditions. If we assume that all excess carriers are extracted at $x = X_n$ in the n-region (and at $x = -X_p$ in the p-region), we can use the boundary condition

$$p_n(x = X_n) = p_{no} \tag{3.2.49}$$

and for a short p^+–n diode with $X_n \ll L_p$, the solution of the diffusion equation becomes a linear function of distance:

$$p_n(x) \approx p_{no} + p_{no}\left[\exp\left(\frac{V}{V_{th}}\right) - 1\right]\left(\frac{X_n - x}{X_n - x_n}\right) \tag{3.2.50}$$

Hence, the total current density in this case is approximately given by

$$J \approx -qD_p\frac{\partial p_n}{\partial x}\bigg|_{x=x_n} = \frac{qD_p p_{no}}{X_n - x_n}\left[\exp\left(\frac{V}{V_{th}}\right) - 1\right] \tag{3.2.51}$$

which is similar to the corresponding equation for a long p^+–n junction, except that L_p is replaced by $X_n - x_n$. This result for a short p^+–n diode is very useful when considering the carrier distributions in a bipolar junction transistor (see Chapter 4).

3.2.4. Capacitance and Small-Signal Equivalent Circuit

The *depletion layer capacitance*

$$C_d = S\left|\frac{dQ_d}{dV}\right| \tag{3.2.52}$$

is the dominant contribution to the small-signal capacitance of a p–n diode at reverse and small forward bias. Here Q_d is the charge per unit area in the depletion layer of the n-type section of the device. One can show that Eq. (3.2.52) leads to

$$C_d = \frac{\epsilon_s S}{x_d} \tag{3.2.53}$$

where $x_d = x_n + x_p$ is the total width of the depletion layer and S is the junction cross section. For an abrupt junction with an applied bias V, x_n and x_p are given by

$$x_n = \sqrt{\frac{2\epsilon_s(V_{bi} - V)}{qN_d(1 + N_d/N_a)}} \tag{3.2.54}$$

$$x_p = \sqrt{\frac{2\epsilon_s(V_{bi} - V)}{qN_a(1 + N_a/N_d)}} \qquad (3.2.55)$$

From these expressions, we obtain

$$C_d = S\sqrt{\frac{q\epsilon_s N_{eff}}{2(V_{bi} - V)}} \qquad (3.2.56)$$

where N_{eff} is the effective doping density given by

$$N_{eff} = \left(\frac{1}{\sqrt{N_d\left(1 + \dfrac{N_d}{N_a}\right)}} + \frac{1}{\sqrt{N_a\left(1 + \dfrac{N_a}{N_d}\right)}}\right) = \frac{N_a N_d}{N_a + N_d} \qquad (3.2.57)$$

From Eq. (3.2.56), we notice that C_d is inversely proportional to $(V_{bi} - V)^{1/2}$ for an abrupt p–n junction.

For a linearly graded profile $(N_d - N_a = ax)$, we find

$$\cdots \qquad \left(\frac{a\epsilon_s^2 \cdots}{12(V_{bi} - V)}\right) \qquad (3.2.58)$$

where the built-in voltage is given by (see Sze, 1985)

$$V_{bi} = 2V_{th}\ln\left(\frac{ax_d}{2n_i}\right) \qquad (3.2.59)$$

In general, the $C_d(V)$ dependence can be used for extracting the doping profile of the junction. Assuming a one-sided p^+–n junction and differentiating Eq. (3.2.56) with respect to V, we obtain

$$N_d(x_d) = \frac{C_d^3}{q\epsilon_s S^2 |dC_d/dV|} = \frac{2}{q\epsilon_s S^2}\left|\frac{1}{d(1/C_d^2)/dV}\right| \qquad (3.2.60)$$

This expression may be used to determine the doping profile in the n-type region from measured capacitance–voltage characteristics at reverse and small forward bias.

Under a relatively large forward bias, a capacitance component related to the excess charge of the minority carriers in the neutral regions of a p–n junction—the so-called *diffusion capacitanc*—becomes very important or even dominant. To calculate the diffusion capacitance, we have to consider the total excess minority charges Q_p and Q_n in the neutral n- and p-regions, respectively [see Fig. 3.2.8 and Eq. (3.2.28)]:

$$|Q_p| = qS\int_{x_n}^{\infty}(p_n - p_{no})\,dx \approx qSL_p p_{no}\exp\left(\frac{V}{V_{th}}\right) \qquad (3.2.61)$$

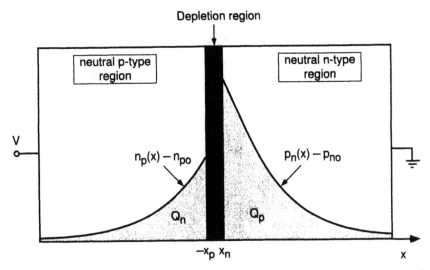

FIG. 3.2.8. Excess minority carrier charges in the neutral semiconductor adjacent to the depletion region.

$$|Q_n| = qS \int_{-\infty}^{-x_p} (n_p - n_{po})\, dx \approx qSL_n n_{po} \exp\!\left(\frac{V}{V_{th}}\right) \tag{3.2.62}$$

(Note that the excess minority carrier charge is always neutralized by an equally large excess majority carrier charge in each of these regions.) Hence, the small-signal capacitance associated with these charges becomes

$$C_{dif} = K_d \frac{d(|Q_p| + |Q_n|)}{dV} = \frac{K_d qS(L_p p_{no} + L_n n_{po})}{V_{th}} \exp\!\left(\frac{V}{V_{th}}\right) \tag{3.2.63}$$

where K_d is a constant that accounts for the contribution of the excess majority charges. A detailed analysis of the small signal admittance of the junction shows that $K_d = \frac{1}{2}$ (see, e.g., Lee et al., 1993).

For a p^+–n diode, $p_{no} \gg n_{po}$ and Eq. (3.2.63) becomes

$$C_{dif} \approx \frac{SJ\tau_p}{2V_{th}} = \frac{G_d \tau_p}{2} \tag{3.2.64}$$

where $G_d \equiv 1/R_d = SJ/V_{th}$ is the junction conductance. This result shows that the characteristic time constant $C_{dif} R_d$ of a forward-biased *p–n* junction is of the order of the recombination time, as could be expected.

A small-signal analysis of a forward-biased short p^+–n diode (where the length of the *n*-section X_n is much smaller than the hole diffusion length) shows that τ_p in Eq. (3.2.64) should be replaced with the hole transit time $t_{tp} = (X_n - x_n)^2/(2D_p)$.

Since for a short diode $L_n \gg X_n - x_n$, $t_{tp} \ll \tau_p$, the time response of such a diode is considerably faster than for a long diode.

The small-signal equivalent circuit of a p–n junction is shown in Fig. 3.2.9. Here we have taken into account that the above expression for the diffusion capacitance is no longer valid at large forward bias, since the densities of minority carriers are no longer proportional to $\exp(V/V_{th})$. This can be accounted for by introducing an effective diffusion capacitance C_{dife}:

$$C_{dife} = \frac{1}{\dfrac{1}{C_{dif}} + \dfrac{1}{C_{difmax}}}$$

(3.2.65)

where the maximum diffusion capacitance C_{difmax} can be estimated as

$$C_{difmax} \approx \frac{\epsilon_s S}{L_{Dn} + L_{Dp}}$$

(3.2.66)

Here $L_{Dn} = (\epsilon_s V_{th}/qN_d)^{1/2}$ and $L_{Dp} = (\epsilon_s V_{th}/qN_a)^{1/2}$ are the so-called Debye lengths in

or transition regions between the depletion region and the neutral regions of the p–n junction.)

The equivalent circuit includes the depletion and diffusion capacitances, the intrinsic diode conductance G_d, the series resistance R_s (comprised of the contact resistances and the resistance in the neutral regions of the semiconductor), as well as a parasitic inductance L_s and the geometric capacitance of the diode, $C_{geom} = \epsilon_s S/L$, where L is the device length.

At high frequencies and high forward biases, when the impedances related to C_d, C_{dif}, and R_d all become small, the parasitic inductance impedance ωL_s may become

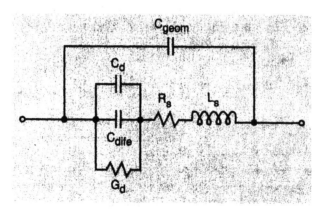

FIG. 3.2.9. Small-signal equivalent circuit of a p–n junction. (After Lee et al., 1993.)

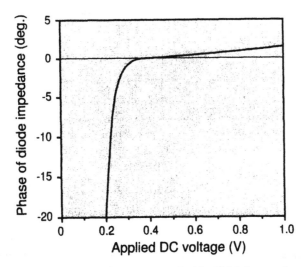

FIG. 3.2.10. Phase of diode impedance simulated with AIM-Spice at a frequency of 2 MHz. A suitable inductance L_s is connected in series with the diode.

very important and the impedance (which is typically measured at about 2 MHz) changes from capacitive to inductive, to give

$$Z_{meas} \approx R_s + i\omega L_s \qquad (3.2.67)$$

The transition from capacitive to inductive behavior is illustrated in terms of the phase of the diode impedance in Fig. 3.2.10. (A negative phase corresponds to the capacitive impedance and a positive phase to an inductive impedance.) This figure clearly shows that the measured capacitances of a *p–n* junction at large forward bias have to be interpreted with a great deal of caution, since the results can be dominated by parasitics.

3.2.5. SPICE Implementation of *p–n* Diode Model

So far, we have discussed most of the important physical mechanisms of the *p–n* diode, and we are ready to present a SPICE model for this device. This SPICE model consists of an intrinsic diode in series with an ohmic resistance R_s, as indicated in Fig. 3.2.11. The intrinsic diode is further divided into a current source I in parallel with a capacitance C.

The diode *I–V* characteristic is modeled in terms of a diffusion current I_D, a generation/recombination current I_{GR}, and a breakdown current I_B:

$$I = K_{HI}I_D + I_{GR} - I_B \qquad (3.2.68)$$

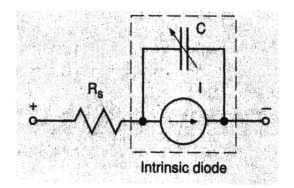

FIG. 3.2.11. SPICE model of p–n diode.

The diffusion current is that of the ideal p–n junction [see Eq. (3.2.34)] with a saturation current **IS** and adjusted by an ideality factor **N** in the exponent (note that all SPICE parameters are given in boldface):

$$I_\mathrm{D} = \mathbf{IS}\left[\exp\left(\frac{V_{di}}{\mathbf{N}V_\mathrm{th}} \right) - 1 \right] \qquad (3.2.69)$$

Here V_{di} is the voltage across the intrinsic diode. The purpose of the factor K_HI is to account for high-injection effects. The following form gives a smooth transition from the low-injection regime to the high-injection regime:

$$K_\mathrm{HI} = \begin{cases} \sqrt{\mathbf{IKF}/(\mathbf{IKF} + I_D)} & \text{for } \mathbf{IKF} > 0 \\ 1, & \text{otherwise} \end{cases} \qquad (3.2.70)$$

where the SPICE parameter **IKF** represents the onset current for significant high-injection effects.

For the generation/recombination current I_GR, we use expressions similar to those discussed in Section 3.2.3:

$$I_\mathrm{GR} = \mathbf{ISR}\left[\left(1 - \frac{V_{di}}{\mathbf{VJ}} \right)^2 + 0.001 \right]^{\mathbf{M}/2} \left[\exp\left(\frac{V_{di}}{\mathbf{NR}V_\mathrm{th}} \right) - 1 \right] \qquad (3.2.71)$$

where **ISR** is a saturation current parameter and **NR** is the emission factor (ideality factor) for the generation/recombination current. Note that the first bracket on the right-hand side of Eq. (3.2.71) gives the scaling of the depletion layer width with the potential V_{di} for an arbitrary grading of the junction doping profile, specified by the grading parameter **M** ($\frac{1}{2}$ for an abrupt junction and $\frac{1}{3}$ for a linearly graded junction). Here, **VJ** is the contact (built-in) potential of the junction. A small numerical constant is included to avoid convergence problems in SPICE.

The breakdown current I_B is modeled as follows:

$$I_B = \text{IBV} \exp\left(-\frac{V_{di} + \text{BV}}{N V_{th}}\right) \tag{3.2.72}$$

where **IBV** is the breakdown "knee" current and **BV** is the breakdown voltage [see Eq. (3.2.47)].

The p–n diode capacitance C consists of a depletion layer capacitance C_d and a diffusion capacitance C_{dif}, as discussed in Section 3.2.4. Equation (3.2.73) is a generalized version of the depletion capacitance expression of Eq. (3.2.56), used by SPICE to account for the grading of the junction doping profile (through the grading parameter **M**) and for high-level forward-bias effects (when $V_{di} > \text{FC·VJ}$, where **FC** is called the forward-bias depletion capacitance coefficient). The strong forward-bias expression is designed to avoid a divergence in C_d as V_{di} approaches **VJ**:

$$C_d = \begin{cases} \text{CJO}\left(1 - \dfrac{V_{di}}{\text{VJ}}\right)^{-\text{M}} & V_{di} \le \text{FC·VJ} \\[3mm] \text{CJO}(1 - \text{FC})^{-(1+\text{M})}\left[1 - \text{FC}(1 + \text{M}) + \text{M}\dfrac{V_{di}}{\text{VJ}}\right] & V_{di} > \text{FC·VJ} \end{cases} \tag{3.2.73}$$

The SPICE parameter **CJO** is the zero-bias depletion capacitance.

The diffusion capacitance is given by Eq. (3.2.63). This expression can be rewritten as

$$C_{dif} = \text{TT}G_d = \text{TT}\frac{dI}{dV_{di}} \tag{3.2.74}$$

where the parameter **TT** is called the transit time (C_{dif} is also called the transit time capacitance), G_d is the junction conductance, and I is the diode dc current given by Eqs. (3.2.68)–(3.2.72).

In addition to the model expressions presented here, a noise model for the p–n diode is also included. A comprehensive model may also include a description of the temperature dependence of several of the model parameters. The present model has been implemented in AIM-Spice and is very similar to models included in other versions of SPICE (e.g., PSpice). However, AIM-Spice has two diode models, level 1 (the present model) and level 2 (a heterostructure diode model that is unique for AIM-Spice; see Section 3.3). The level number should be specified on the model line of the circuit description as LEVEL=X, where X is the level number. However, level 1 is the default and the level specification can be omitted when this model is selected.

3.2.6. SPICE Example: Simple Voltage Regulator

Connecting one or more diodes in series with a resistor and a power supply provides a simple way of generating a relatively constant voltage, somewhat independent of

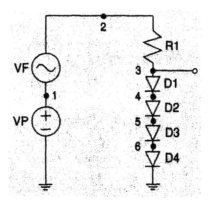

FIG. 3.2.12. Simple voltage regulator circuit.

fluctuations in the power supply level. One example of such a circuit is shown in Fig. 3.2.12, where four diodes are connected in series with a 1-kΩ resistor and a 10-V dc power supply VP. An ac voltage source VF is added to simulate fluctuations in the power supply. Since the forward voltage drop of each diode remains almost constant at approximately 0.7 V for a wide range of diode currents, the output voltage (node 3) of the circuit is about 2.8 V.

We now use AIM-Spice to investigate the effect of power supply fluctuations on the output voltage. We assume that the fluctuations are sinusoidal with a frequency of 50 Hz and an amplitude of 1 V. The circuit description for this case is shown in Fig. 3.2.13. To study circuit waveforms, we run a transient analysis from 0 to 100 ms using a maximum step size of 0.5 ms to generate smooth graphs. On completion of the AIM-Spice simulation, we plot the power supply and the output waveform (node 3) from the voltage regulator circuit as shown in Fig. 3.2.14. The average out-

```
Simple Voltage Regulator
Vp 1 0 dc 10
Vf 1 2 dc 0 sin(0 1 50)
r1 2 3 1k
d1 3 4 diode
d2 4 5 diode
d3 5 6 diode
d4 6 0 diode
.model diode d is=880.5E-18 rs=.25 ikf=0 n=1
+ xti=3 eg=1.11 cjo=175p m=0.5516 vj=.75 fc=.5
+ isr=1.859n nr=2 bv=4.7 ibv=20.245m
```

FIG. 3.2.13. Circuit description for the simple voltage regulator circuit shown in Fig. 3.2.12. The diode model parameters used are for Motorola 1N750.

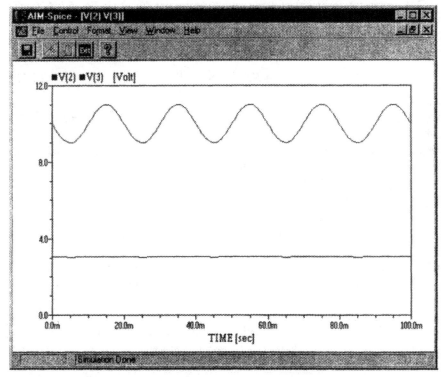

FIG. 3.2.14. Voltage waveforms from the voltage regulator circuit. The upper curve shows the power supply voltage signal with the superimposed ripple. The lower curve shows the voltage at output node 3.

put voltage is about 3 V, which is somewhat higher than estimated above. However, the ripple in the output waveform at node 3 is very much reduced compared to that of the power supply voltage. Using the cursor feature in AIM-Postprocessor, we find the output peak-to-peak ripple to be 32 mV, compared to 2 V in the power supply.

3.2.7. SPICE Example: Temperature Dependence of Diode Characteristics

The SPICE diode model incorporates explicit dependencies on temperature of the ideal diode current [Eq. (3.2.69)], the generation/recombination current [Eq. (3.2.71)], and the breakdown current [Eq. (3.2.72)]. Additional temperature dependencies of the SPICE model enter through the saturation current I_s and the built-in (contact) voltage V_{bi}. In the present example, we demonstrate how the I–V characteristic of a commercial *p–n* diode from Motorola (1N750) is affected by temperature.

The expression for the ideal diode saturation current density in Eq. (3.2.42) indicates a proportionality to n_i^2. The temperature dependence of the intrinsic carrier concentration has the following exponential form (see, e.g., Lee et al., 1993):

$$n_i = KT^{3/2} \exp\left(-\frac{E_{go}}{2k_BT}\right) \tag{3.2.75}$$

where K is a constant and E_{go} is the energy gap at 0 K. The energy gap is itself a function of temperature, and so are the diffusion coefficients and the diffusion lengths entering into Eq. (3.2.42). Recognizing the dominant influence of the intrinsic carrier concentration, the temperature dependence of the saturation current is modeled as follows in SPICE:

$$I_s = \mathbf{IS} \cdot \left(\frac{T}{T_o}\right)^{\mathbf{XTI}/\mathbf{N}} \exp\left[\frac{\mathbf{EG} \cdot (T - T_o)}{\mathbf{N} \cdot qV_{th}T_o}\right] \tag{3.2.76}$$

where the SPICE parameters **IS**, **EG**, **XTI**, and **N** are the room temperature $(T = T_o)$ values of the saturation current, the energy gap, the saturation current temperature exponent, and the ideality factor, respectively.

The built-in voltage V_{bi} enters (as SPICE parameter VJ) in the SPICE expressions for the generation/recombination current and for the capacitances [Eqs. (3.2.71), (3.2.73) and (3.2.74)]. The temperature dependence of V_{bi} for an ideal, abrupt p–n junction is contained in Eq. (3.2.5) (through V_{th} and n_i). In SPICE, the following model expression is used to describe V_{bi} as a function of T:

$$V_{bi}(T) = \frac{T}{T_o}\mathbf{VJ} - 2V_{th}\ln\left(\frac{T}{T_o}\right)^{1.5} - \frac{1}{q}\left[\frac{T}{T_o}E_{go} - E_g(T)\right] \tag{3.2.77}$$

where the SPICE parameter **VJ** is the built-in voltage at room temperature. The temperature dependence of the energy gap is modeled as follows:

$$E_g(T) = E_{go} - \frac{\alpha T^2}{\beta + T} \tag{3.2.78}$$

where, for silicon, we have the parameter values $\alpha = 7.02 \times 10^{-4}$ eV/K, $\beta = 1108$ K, and $E_{go} = 1.16$ eV.

In the present simulation example, we illustrate how the diode characteristic depends on temperature by calculating the diode current at six different temperatures. The SPICE description of the circuit is given in Fig. 3.2.15 for 27°C. The voltage source vd is swept from 0 to 1 V. The simulation is repeated for 50, 75, 100, 150, and 200°C by changing the parameter **TEMP** on the device line in the circuit description. Note that the purpose of the voltage source vid is to monitor the diode current in AIM-Spice. This voltage source is kept at zero voltage, and before starting the simulation, we have to select to plot the current through this

```
Temperature dependence of diode current
vd 1 0 dc 0
vid 1 2 dc 0
d1 2 0 dmod temp=27
* Motorola 1N750
.model dmod d Is=880.5E-18 Rs=.25 Ikf=0 N=1 Xti=3
+ Eg=1.11 Cjo=175p M=.5516 Vj=.75 Fc=.5 Isr=1.859n
+ Nr=2 Bv=4.7 Ibv=20.245m
```

FIG. 3.2.15. Circuit description for a simple diode circuit.

source. In general, whenever we are interested in measuring a branch current, we insert a voltage source with a zero value to use it as an ampere meter. In PSpice this is not necessary, since ampere meters are automatically added to all branches of a circuit

The results of the simulation shown in Fig. 3.2.16 clearly indicate a strong increase in the diode current with increasing operating temperature.

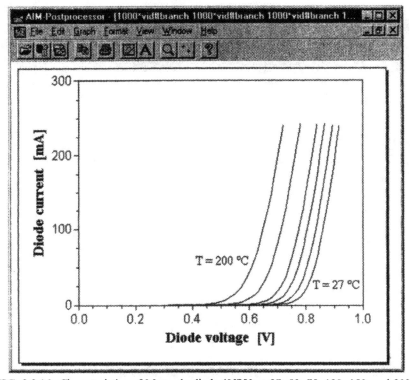

FIG. 3.2.16. Characteristics of Motorola diode 1N750 at 27, 50, 75, 100, 150, and 200°C modeled by AIM-Spice.

3.2.8. SPICE Example: Temperature Dependence of Voltage Regulator

In the previous example, we investigated how the diode characteristic is affected by temperature. The most important effect is a decrease in the turn-on voltage with increasing temperature, as shown in Fig. 3.2.16. To see how this temperature dependence affects the output of a diode circuit, we simulate the voltage regulator circuit from example 3.2.6 at different temperatures. The circuit temperature is changed by choosing the General Simulation Options command from the Options menu. In the dialog box, which is displayed in response to the command, the circuit temperature is set by specifying a value for the option Operating Temperature (TEMP).

In Fig. 3.2.17, the voltage waveform at node 3 is plotted for three different circuit temperatures. We observe that the dc level of the output voltage decreases rapidly with increasing temperature, as expected from the temperature dependence of the diode characteristic.

3.2.9. SPICE Example: Capacitance–Voltage Characteristics of Diode

Unfortunately, SPICE does not have the capability to directly calculate and plot capacitances versus voltage. However, as part of a small-signal analysis, SPICE calcu-

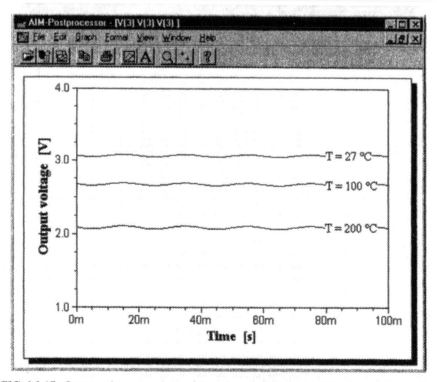

FIG. 3.2.17. Output voltage waveform of the voltage regulator circuit for three different circuit temperatures (Motorola diode 1N750).

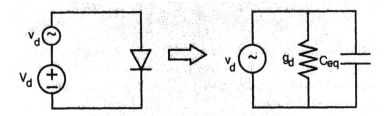

FIG. 3.2.18. Circuit containing an ideal diode and the corresponding small-signal equivalent circuit.

lates the small-signal parameters, including the diode capacitance. By repeating the small-signal analysis at different operating points, a capacitance–voltage plot can be generated. In the present example, we use this approach to calculate and plot the capacitance of a *p–n* diode versus the diode dc voltage V_d. The circuit in Fig. 3.2.18 was simulated over a narrow frequency range for V_d varying from –2.0 to 0.5 V.

The value of $C_{eq} = C_d + C_{dife}$ can be easily obtained from the output file of an AC Analysis in AIM-Spice using the simulated voltages and currents. A simple analysis of the small-signal equivalent in Fig. 3.2.17 gives

$$C_{eq} = \frac{\text{Im}(Y)}{2\pi f} = \frac{\text{Im}(i_d/v_d)}{2\pi f} \qquad (3.2.79)$$

where $Y = g_d + j2\pi f C_{eq}$ is the admittance of the parallel combination of C_{eq} and g_d. From this, we obtain the capacitance–voltage characteristics of the *p–n* diode shown in Fig. 3.2.19 using the default diode parameter values CJO=1pF and TT=12ns.

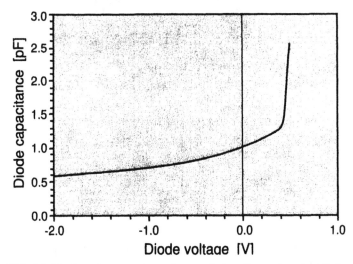

FIG. 3.2.19. Capacitance–voltage characteristics of a *p–n* junction diode.

From the earlier discussion, we recall that the total intrinsic capacitance of a diode consists of two separate contributions: a depletion layer capacitance C_d and a diffusion capacitance C_{dife} (see Fig. 3.2.9). The sudden increase in the $C-V$ characteristic at a forward-bias voltage of about 0.4 V is caused by the diffusion capacitance, which increases exponentially with forward diode bias. At voltages below 0.4 V, the depletion capacitance dominates.

3.3. HETEROJUNCTIONS

3.3.1. Basic Concepts

A key feature of $p-n$ junctions is the potential barrier between the p-region and the n-region, which can be controlled by the applied bias. However, potential barriers are also created at the interfaces between different materials. Electrons and holes in a piece of semiconductor may be considered as sitting in a potential box. This potential box is different for different semiconductors, as indicated in Fig. 3.3.1, and combining semiconductor layers of different compositions and thicknesses, we can design potential barriers, quantum wells, and superlattices. Using a term coined by Federico Capasso of AT&T Bell Labs, we can call this approach "band-gap engineering.

Not all combinations of semiconductors can be used to form such layered structures, also called heterostructures, with the quality needed for electronic devices. In fact, the semiconductors must have a very close matching of their lattice constants and, as a rule, the same crystal symmetry. Otherwise, broken bonds at the interface

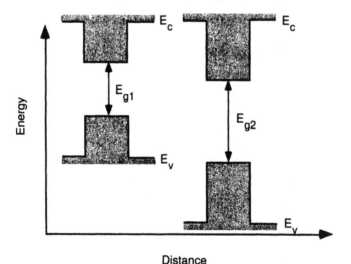

FIG. 3.3.1. Examples of band diagrams for layered semiconductor heterostructures with different energy gaps.

will create a huge density of interface states and imperfections, which will render the heterostructure unusable for devices. For example, AlAs and GaAs have almost the same lattice constants, but their energy gaps are quite different (approximately 2.2 eV for AlAs and 1.42 eV for GaAs).

Moreover, the solid-state solution $Al_xGa_{1-x}As$ also has nearly the same lattice constant as GaAs. Changing the composition x with distance, one can create semiconductor layers with abrupt junctions, such as shown in Fig 3.3.1, or with graded composition, as shown in Fig. 3.3.2. In a graded semiconductor, the built-in field may actually push electrons and holes in the same direction, as opposed to an applied electric field in a uniformly doped semiconductor, which pushes electrons and holes in opposite directions.

Heterostructure systems are widely used in novel semiconductor devices, including transistors, solar cells, light-emitting diodes, lasers, and other optoelectronic devices. Here we consider the heterojunction diode, but in subsequent chapters we will also discuss the heterojunction bipolar transistor (HBT) and the heterostructure field-effect transistor (HFET).

The first model for the ideal heterojunction was developed by Anderson (1962), who assumed that the vacuum energy level (i.e., the potential energy for free electrons that have escaped the semiconductor altogether) is continuous and that the conduction band discontinuity is determined by the difference of the electron affinities, X_1 and X_2, in the two regions:

$$\Delta E_c = X_1 - X_2 \tag{3.3.1}$$

The electron affinity is defined as the energy difference between the vacuum level and the conduction band edge, a material property of each semiconductor. Consequently, the valence band discontinuity is given by

$$\Delta E_v = \Delta E_g - \Delta E_c = \Delta E_g - X_1 + X_2 \tag{3.3.2}$$

where ΔE_g is the energy gap discontinuity.

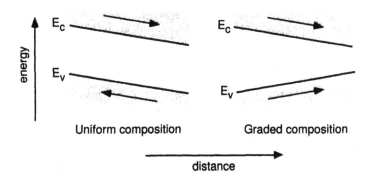

FIG. 3.3.2. Qualitative energy band diagrams for a uniform semiconductor and for a semiconductor with graded composition. (After Lee et al., 1993.)

Although the ideal Anderson model turns out to be in disagreement with experimental data, the model is still useful with a suitable adjustment between ΔE_c and ΔE_v. In principle, these discontinuities may be determined from experiments.

Figure 3.3.3 shows the band diagram of a p–n heterojunction at thermal equilibrium. It is drawn using exactly the same rules as for a conventional p–n junction (including the requirement that the Fermi level must be constant throughout the system at equilibrium). The only difference is the inclusion of the conduction and valence band discontinuities at the heterointerface. From Fig. 3.3.3, we find that the built-in voltage V_{bi} is given by

$$V_{bi} = \frac{E_{g1} - \Delta E_{Fn} - \Delta E_{Fp} + \Delta E_c}{q} \qquad (3.3.3)$$

Here E_{g1} is the energy gap of the narrow-gap p-type material (region 1), ΔE_{Fp} is the difference between the Fermi level and the top of the valence band far from the heterointerface in this material, and ΔE_{Fn} is the difference between the bottom of the conduction band and the Fermi level far from the heterointerface in the wide-gap n-type material (region 2). Note that ΔE_c must be included in Eq. (3.3.3) since the electrostatic potential must be continuous in a vacuum.

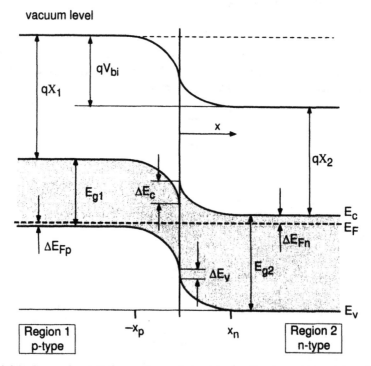

FIG. 3.3.3. Energy band diagram of a p–n heterojunction (lower part) shown together with the variation in the vacuum level (upper part). (After Lee et al., 1993.)

3.3.2. Current–Voltage Characteristic

With a reverse or small forward bias applied to the p-type region (region 1) relative to the n-type region (region 2), the junction voltage changes from V_{bi} to $V_{bi} - V$ (forward bias corresponds to a positive voltage V). Using the depletion approximation, as was done in Section 3.2 for the conventional p–n junction, we find that the potential drop in the heterojunction is divided between the regions as follows (region indicated in the subscript):

$$V_{d1} = \frac{\epsilon_2 N_d}{\epsilon_2 N_d + \epsilon_1 N_a} (V_{bi} - V) \tag{3.3.4}$$

$$V_{d2} = \frac{\epsilon_1 N_a}{\epsilon_2 N_d + \epsilon_1 N_a} (V_{bi} - V) \tag{3.3.5}$$

For an abrupt heterojunction with a spike ΔE_c in the conduction band, as indicated in Fig. 3.3.3, the transport of electrons through the depletion region will be impeded by the process of thermionic emission across the spike. Typically, this effect will be important only for $\Delta E_c \geq k_B T$. For $\Delta E_c < k_B T$, the transport in the depletion region itself is fully accounted for in terms of the voltage across the heterojunction, as in a conventional p–n junction (here we neglect generation and recombination processes).

Lee et al. (1993) showed that for a heterojunction of the shape indicated in Fig. 3.3.3, the current–voltage characteristic has the form

$$J = \left(\frac{q D_n n_{po}}{L_n \zeta_n} + \frac{q D_p p_{no}}{L_p \zeta_p} \right) \left[\exp\left(\frac{V}{V_{th}} \right) - 1 \right] \tag{3.3.6}$$

where, for long n- and p-type regions,

$$\zeta_n = 1 + \frac{D_n}{v_n L_n} \exp\left(\frac{\Delta E}{q V_{th}} \right) \tag{3.3.7}$$

$$\zeta_p = 1 \tag{3.3.8}$$

$$\Delta E_n = \Delta E_n - q V_{d1} \tag{3.3.9}$$

We note that in a typical diode the p-type region is very highly doped. Hence, V_{d1} is small so that $\Delta E_n > 0$. In this case, ΔE_n represents an additional barrier for the electrons entering region 1 from region 2. When $\Delta E_n < 0$, the conduction band discontinuity does not add to the barrier height since the conduction band spike at the heterointerface is below the conduction band edge in the neutral part of the p-type region.

In the above expressions, the effect of the conduction band spike is contained in the factor ζ_n. Usually, as for p-GaAs/n-Al$_x$Ga$_{1-x}$As diodes, the second term in ζ_n of Eq. (3.3.7) will be dominant. A similar derivation can be made for the hole current for material combinations that give a spike in the valence band. However, for the p-GaAs/n-Al$_x$Ga$_{1-x}$As system, there is no such spike and the hole current can be calculated as for a conventional p–n junction. In this case, the effect of the band gap discontinuity will enter only through the equilibrium minority carrier concentration p_{no} in the expression for the hole component of the current density.

In practice, for short diodes, where the length X_p of the p-type region and the length X_n of the n-type region are comparable to or smaller than the respective diffusion lengths, Eq. (3.3.6) is still valid if the factors ζ_n and ζ_p are changed to

$$\zeta_n = \tanh\left(\frac{X_p - x_1}{L_n}\right) + \frac{D_n}{v_n L_n}\exp\left(\frac{\Delta E_n}{qV_{th}}\right) \tag{3.3.10}$$

$$\zeta_p = \tanh\left(\frac{X_n - x_2}{L_p}\right) \tag{3.3.11}$$

where x_1 and x_2 are the depletion widths of the two regions.

For very short diodes, the hyperbolic tangents can be expanded to first order in $(X_p - x_1)/L_n$ and $(X_n - x_2)/L_p$. Furthermore, a practical heterostructure diode construction would be of the type p^+–n or n^+–p. In the case of a p^+–n diode, we will normally have $V_{d1} \ll \Delta E_c$ and $\Delta E_n \gg V_{th}$. Hence, the current density of such a diode can be written as

$$J \approx \left[qn_{po}v_n\exp\left(-\frac{\Delta E_n}{qV_{th}}\right) + \frac{qD_p p_{no}}{X_n - x_2}\right]\left[\exp\left(\frac{V}{V_{th}}\right) - 1\right] \tag{3.3.12}$$

In practical heterostructure diodes, the current level can be greatly enhanced by tunneling, by unintentional compositional grading at the heterointerface, and by other factors, such as self-heating at high current densities. Nevertheless, the above equations represent a useful framework for modeling heterostructure diodes, especially when parameters such as ΔE_c are allowed to be adjusted (to be extracted from experimental data).

3.3.3. Heterojunction Capacitance

From the voltage division between the two regions, given by Eqs. (3.3.4) and (3.3.5), we find the depletion widths

$$x_{d1} = \sqrt{\frac{2\epsilon_1 V_{d1}}{qN_a}} \tag{3.3.13}$$

$$x_{d2} = \sqrt{\frac{2\epsilon_2 V_{d2}}{qN_d}} \tag{3.3.14}$$

Hence, the depletion layer capacitance becomes

$$C_d = S\sqrt{\frac{q\epsilon_1\epsilon_2 N_a N_d}{2(V_{bi} - V)(\epsilon_1 N_a + \epsilon_2 N_d)}} \qquad (3.3.15)$$

where S is the junction cross section.

As for the conventional p–n junction, the diffusion capacitance in a heterostructure diode may become important or even dominant under forward-bias conditions. Lee et al. (1993) determined this capacitance from the small-signal device admittance using the continuity equations for the minority carriers in the two regions. For a short heterojunction p^+–n diode at forward bias, the expression for the diffusion capacitance reduces to that of the homojunction case:

$$C_{dif} = \frac{qS(X_n - x_1)p_{no}}{2V_{th}} \exp\left(\frac{V}{V_{th}}\right) \qquad (3.3.16)$$

We note that this expression is not valid at high bias voltages where, as discussed in Section 3.2, an effective diffusion capacitance should be used.

3.3.4. SPICE Implementation of Heterostructure Diode Model

The heterostructure diode model described above has been implemented in AIM-Spice as level 2 (choose LEVEL=2 in the model line, as in the AIM-Spice example below). We note that the capacitance expressions are structurally the same as those of the homojunction diode (see Section 3.2). Accordingly, similar models are used for both level 1 and level 2, the major difference being in the choice of SPICE parameter values.

Also, the current–voltage (I–V) model in level 2 is similar to that of level 1. But instead of using the saturation current **IS** as a parameter, level 2 requires that we specify this current in terms of additional SPICE parameters, such as the diffusion coefficients for electrons and holes (**DN** and **DP**), the corresponding diffusion lengths (**LN** and **LP**), the donor and acceptor doping densities (**ND** and **NA**), the widths of the n- and p-type regions (**XN** and **XP**), their dielectric constants (**EPSN** and **EPSP**), and the conduction band discontinuity (**DELTAEC**).

3.3.5. SPICE Example: Heterostructure Diode Characteristics

We would like to investigate the effect of the spike in the conduction band edge (see Fig. 3.3.3) on the I–V characteristic of the heterostructure diode. Figure 3.3.4 shows the circuit description for a DC Transfer Curve Analysis used for calculating the diode current versus the applied sweep source voltage vin. The simulation is repeated for several values of the series resistance rs.

The I–V characteristics of a typical heterostructure diode modeled as discussed above and calculated using AIM-Spice are shown in Fig 3.3.5. The plateau indicated in the magnified characteristic (in the small window) is caused by the presence of the spike in the conduction band edge.

```
HDIA I-V curve
vin 1 0 dc 0
vid 1 2 dc 0
d1 2 0 diode
.MODEL diode d level=2 dn=0.02 dp=0.000942 ln=7.21e-5
+ lp=8.681e-7 nd=7e24 na=3e22 deltaec=9.613150e-20
+ xp=1e-6 xn=1e-6 epsp=1.0593e-10 epsn=1.1594e-10
+ rs=20
```

FIG. 3.3.4. Circuit description for calculating heterostructure diode characteristics .

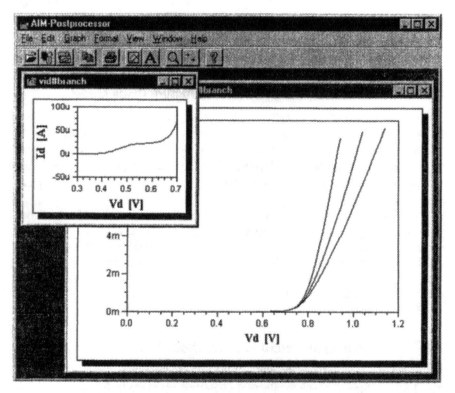

FIG. 3.3.5. AIM-Spice current–voltage characteristics of a heterostructure p^+–n diode with different series resistances. The small window shows, in magnification, details related to the conduction band spike. (After Lee et al., 1993.)

3.4. METAL–SEMICONDUCTOR JUNCTIONS

3.4.1. Schottky Barriers

As mentioned in Section 3.3, electrons in a semiconductor may be considered as being trapped in a potential box, and so is the case for electrons in a piece of a metal. However, for many metals, this potential box is deeper than that for semiconductors. Hence, when a semiconductor and a metal are brought into intimate contact, forming a so-called Schottky barrier, electrons from the semiconductor will transfer into the metal. Due to this transfer process, the metal acquires a net negative charge, while the semiconductor becomes positively charged. The net charge in the metal resides in a thin surface layer, while the opposite charge in the semiconductor is distributed between interface states (absent in the case of an ideal Schottky contact) and a space charge region, much like that found in the depletion layer of a p–n junction. The resulting electric field across the boundary counterbalances the charge transfer process. Equilibrium is reached when the Fermi level becomes constant throughout the metal–semiconductor system.

Figure 3.4.1 shows the resulting (simplified) energy band diagram of a metal–semiconductor (n-type) contact in thermal equilibrium. Here, Φ_m and Φ_s are called the metal and the semiconductor work functions. The work function is defined as the energy separation between the vacuum level and the Fermi level. (Note that the vacuum level is subject to the same effect of the interface polarization as the

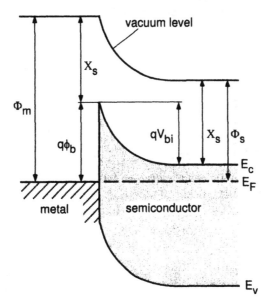

FIG. 3.4.1. Simplified energy band diagram of a metal–semiconductor barrier; $q\phi_b$ is the barrier height, X_s is the electron affinity in the semiconductor, Φ_s and Φ_m are the semiconductor and the metal work functions, and V_{bi} is the built-in voltage. (After Shur, 1990.)

energy bands of the materials constituting the junction.) The electron affinity X_s of the semiconductor corresponds to the energy separation between the vacuum level and the conduction band edge. This quantity is nearly independent of the doping (or the position of the Fermi level) in the semiconductor, at least at low doping densities.

The band diagram shown in Fig. 3.4.1 predicts that the following energy barrier forms between the metal and the semiconductor

$$q\phi_b = \Phi_m - X_s \tag{3.4.1}$$

The positive net space charge in the semiconductor gives rise to a total band bending (V_{bi} is the built-in voltage)

$$qV_{bi} = \Phi_m - \Phi_s = q\phi_b - (E_c - E_F) \tag{3.4.2}$$

Note that qV_{bi} is also identical to the difference between the Fermi levels in the metal and the semiconductor when separated by a large distance (no exchange of charge).

In practice, the Schottky barrier height is only weakly dependent on Φ_m, as shown for n-Si and n-GaAs in Fig. 3.4.2. Bardeen (1947) explained this difference by the effect of the surface states at the boundary between the semiconductor and a thin oxide layer (transparent for electron tunneling) at the interface. Bardeen assumed that these surface states are distributed continuously and with a constant density of states N_s within the energy gap of the semiconductor. Furthermore, he

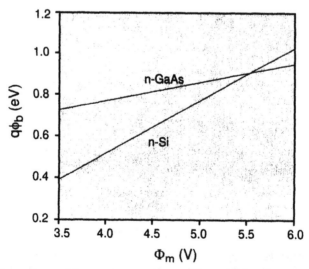

FIG. 3.4.2. Dependence of the metal–semiconductor barrier height; $q\phi_b$ on the metal work function Φ_m for n-type Si and GaAs. (See Cowley and Sze, 1965.)

suggested that the states closer to the conduction band edge are neutral when empty and those closer to the valence band edge are neutral when filled by electrons. The energy at the border between these two types of surface states is called the "neutral" level $q\phi_o$ (see Fig. 3.4.3). The following is an expression for the barrier height in the presence of interface states and an interfacial oxide layer (see Rhoderick, 1977):

$$q\phi_b = \gamma_s(\Phi_m - X_s) + (1 - \gamma_s)(E_c - q\phi_o) - \gamma_s\left(\frac{\epsilon_s}{\epsilon_i}\right)qF_m\delta_{ox} \qquad (3.4.3)$$

Here, E_g is the energy gap, ϵ_s and ϵ_i are the dielectric permittivities of the semiconductor and the interfacial layer, respectively, δ_{ox} is the thickness of the interfacial layer, N_s is the density of the surface states, and

$$\gamma_s = \frac{\epsilon_i}{\epsilon_i + q^2 N_s \delta_{ox}} \qquad (3.4.4)$$

$$F_m = \sqrt{\frac{2qN_dV_{bi}}{\epsilon_s}} \qquad (3.4.5)$$

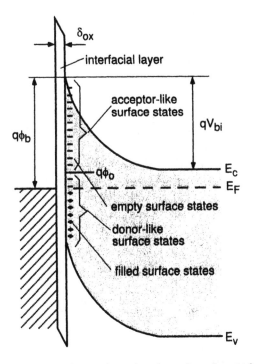

FIG. 3.4.3. Surface states at the metal–semiconductor boundary. (After Shur, 1990.)

where F_m is the maximum electric field inside the space charge region of the semiconductor and N_d is the donor concentration (n-type semiconductor). Equation (3.4.3) reduces to Eq. (3.4.1) when the density of the surface states is zero. In the opposite limit, when $N_s \rightarrow \infty$,

$$q\phi_b = E_c - q\phi_o \qquad (3.4.6)$$

In this case, the Fermi level in the semiconductor is said to be "pinned" to the neutral level since any deviation of E_F from this position will result in an infinitely large charge at the interface.

Bardeen's model is in better agreement with experimental data than the simplistic model discussed initially. Still, it cannot explain many properties of the Schottky barrier diodes. Even more modern and sophisticated theories cannot predict the magnitude of the Schottky barrier height. Therefore, this value must be extracted from experimental data.

The band bending in the semiconductor in Fig. 3.4.1 is very similar to that of a p^+–n junction. Therefore, once the Schottky barrier height is known, the variation of the space charge density ρ, the electric field F, and the potential ϕ in the semiconductor space charge region can be found using the depletion approximation, which yields

$$\rho = qN_d \qquad (3.4.7)$$

$$F = -\frac{qN_d (x_n - x)}{\epsilon_s} \qquad (3.4.8)$$

$$\phi = -\frac{qN_d (x_n - x)^2}{2\epsilon_s} = -V_{bi}\left(1 - \frac{x}{x_n}\right)^2 \qquad (3.4.9)$$

where $x = 0$ corresponds to the metal–semiconductor interface, N_d is a donor concentration, and x_n is a depletion layer width given by (at zero bias)

$$x_n = \sqrt{\frac{2\epsilon_s V_{bi}}{qN_d}} \qquad (3.4.10)$$

Under reverse and small forward bias, we have

$$x_n = \sqrt{\frac{2\epsilon_s(V_{bi} - V)}{qN_d}} \qquad (3.4.11)$$

(Note that a forward bias corresponds to a positive voltage applied to the metal relative to the semiconductor.)

3.4.2. Schottky Barrier Current–Voltage Characteristic

In a p–n junction under bias, the quasi–Fermi levels of majority carriers remain practically constant throughout the depletion region (see Fig. 3.2.4). A similar assumption—that the electron quasi–Fermi level remains constant throughout the depletion region in a Schottky diode—forms the foundation of the so-called thermionic emission model of electron transport in Schottky barriers. This model is valid when the interfacial barrier is the dominant impediment to current flow. However, in relatively lightly doped semiconductors with low mobility, the current through the Schottky barrier may be limited more by diffusion and drift processes in the space charge region than by the barrier itself. Under such circumstances, the electron quasi–Fermi level will vary inside the depletion region, and the thermionic emission model does not apply.

When the thermionic emission model is valid, the current–voltage characteristic of the Schottky diode is found by calculating the net electron flux through the metal–semiconductor interface. The electron flux $J_{s \to m}$ out of the semiconductor is proportional to the density n_s of electrons in the semiconductor bulk with energies higher than the peak of the barrier. With an applied voltage V, this flux becomes

$$J_{s \to m} \propto n_s = N_d \exp\left(-\frac{V_{bi} - V}{V_{th}}\right) = N_c \exp\left(-\frac{\phi_b - V}{V_{th}}\right) \qquad (3.4.12)$$

where N_c is the effective density of states in the conduction band. This flux is affected by the applied bias, since this bias changes the barrier for electrons in the semiconductor. The excess charge in the metal is strongly concentrated at the metal–semiconductor interface, and no band bending takes place in the metal (see Fig. 3.4.1). Therefore, the electron flux $J_{m \to s}$ from the metal into the semiconductor is taken to be independent of the applied voltage (assuming that the barrier height is independent of the applied voltage, which is a reasonable first approximation, but not strictly true). The total flux must be equal to zero when $V = 0$, corresponding to the condition $J_{m \to s} = J_{s \to m}(V = 0)$. Hence, we obtain the following expression for the electric current density:

$$J = J_{ss}\left[\exp\left(\frac{V}{V_{th}}\right) - 1\right] \qquad (3.4.13)$$

The saturation current is given by

$$J_{ss} = A^* T^2 \exp\left(-\frac{\phi_b}{V_{th}}\right) \qquad (3.4.14)$$

where A^* is known as Richardson's constant. Here, A^* depends on the effective mass in the direction of current flow and is therefore sensitive to the crystallographic orientation of the interface (see, e.g., Lee et al., 1993). For {111} surfaces of Si and GaAs, A^* is equal to 96 and 4.4 A/(cm^2 K^2), respectively.

In practical devices, additional mechanisms contribute to the current, such as the dependence of the barrier height on bias voltage and generation/recombination processes in the semiconductor depletion region. Hence, the current–voltage characteristic may be more accurately described by the following empirical expression:

$$J = J_{sd}\left[\exp\left(\frac{V}{\eta V_{th}}\right) - 1\right] \qquad (3.4.15)$$

where J_{sd} is the modified Schottky diode saturation current and the parameter η is the ideality factor. Additionally, as for the p–n junction, the Schottky barrier suffers breakdown at sufficiently high reverse bias and will enter a high-injection regime at large forward bias. All in all, the description of the Schottky barrier diode I–V characteristic is basically the same as for the p–n diode, and the same SPICE model applies to both. However, we should keep in mind that both the magnitudes and the temperature dependencies of the saturation currents for p–n junctions and Schottky diodes differ dramatically.

As seen from Fig. 3.4.4, the potential barrier becomes very thin near the top, and electrons may tunnel through. The tunneling becomes more and more important as the doping level increases and the depletion layer becomes thinner. This process is called *thermionic-field emission.*

In degenerate semiconductors, especially in semiconductors with a small electron effective mass such as GaAs, electrons can tunnel through the barrier near the Fermi level and the tunneling current becomes dominant. This mechanism is called *field emission.* The effective resistance of the Schottky barrier in the field emission regime is quite low and the metal–semiconductor contact loses its rectifying property. The resulting current–voltage characteristic at small voltage biases becomes quite linear, with a small effective impedance. Hence, a Schottky contact using a highly doped semiconductor behaves as an *ohmic contact* (see below).

3.4.3. Schottky Barrier Capacitance

The small-signal equivalent circuit of a Schottky barrier is basically the same as that of the p–n diode shown in Fig. 3.2.9, except for one important difference: the absence of a diffusion capacitance owing to the lack of minority carrier injection. We therefore avoid the recombination processes that tend to slow down the response of p–n junctions. Hence, the Schottky diodes are much faster than p–n diodes and are widely used as microwave mixers, detectors, and so on. But the capacitance associated with the semiconductor space charge region remains and can be estimated using the depletion approximation (just as for the p^+–n junction; see Section 3.2):

$$C_d = S\sqrt{\frac{qN_d\epsilon_s}{2(V_{bi} - V)}} \qquad (3.4.16)$$

In most cases, the characteristic time constant limiting the frequency response of a Schottky diode is given by

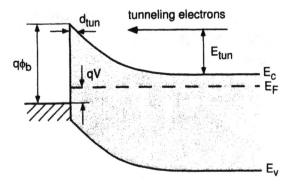

a) low doping $\Rightarrow$ thermionic-field emission

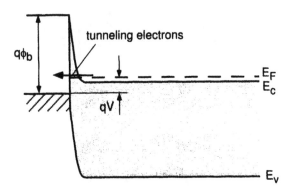

b) degenerate doping $\Rightarrow$ field emission

FIG. 3.4.4. Thermionic-field and field emission under forward bias: (*a*) low doping; (*b*) very high (degenerate) doping (d_{tun} indicates the characteristic tunneling length). (After Shur, 1990.)

$$\tau_{Sch} = R_s C_d \qquad (3.4.17)$$

where R_s is the series resistance associated with the contacts and the neutral semi-conductor region.

3.4.4. Ohmic Contacts

Current–voltage characteristics of a Schottky barrier diode and an ohmic contact are compared in Fig. 3.4.5. The ohmic contact has a linear current–voltage charac-teristic and a very small resistance. A good ohmic device contact is characterized by a resistance that is negligible compared to that of the active region of the device.

Ideally, the ohmic contact of an *n*-type semiconductor should be made using a metal with a lower work function than that of the semiconductor. Unfortunately,

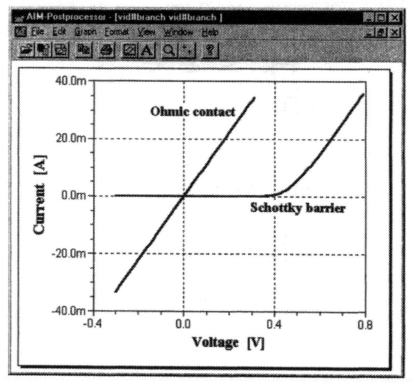

FIG. 3.4.5. Simulated current–voltage characteristics of metal–semiconductor contacts using AIM-Spice.

very few practical material systems satisfy this condition. Instead, metals usually form Schottky barriers at semiconductor interfaces. Therefore, as mentioned above, a practical way to obtain a low resistance ohmic contact is to dope the semiconductor degenerately near the metal–semiconductor interface in order to create a very thin barrier that allows current transport by tunneling in the field emission regime (see Fig. 3.4.4).

Band diagrams of a metal–n^+–n ohmic contact and a Schottky contact are compared in Fig. 3.4.6. This clearly illustrates the role played by the n^+ layer. A conventional approach to reducing the contact resistance is to form a very highly doped region near the surface by using alloyed ohmic contacts. Ion implantation or diffusion has also been used to create a highly doped region near the surface to facilitate the formation of ohmic contacts.

3.4.5. SPICE Example: Diode-Switching Transients

The metal–semiconductor junction is a majority carrier device, with no buildup of minority carrier charge in the forward direction such as in p–n junctions. Conse-

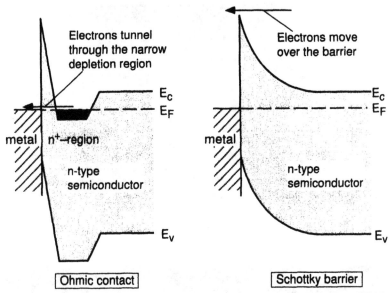

FIG. 3.4.6. Band diagrams of metal–semiconductor contacts. (After Shur, 1990.)

quently, Schottky diodes should have a shorter switching time than conventional p–n diodes. We illustrate this by simulating switching transients in both types of diodes.

Figure 3.4.7 shows the circuit description for a simple diode circuit with a p–n diode. Initially, the diode is biased in the forward direction with a voltage of 1.0 V. At the time $t = 5$ ns, the voltage is set to zero and the diode starts to turn off. The corresponding AIM-Spice Transient Analysis parameters are shown in Fig. 3.4.8.

The resulting transient diode current in Fig. 3.4.9 indicates that a finite but negative current is retained for some time after the input voltage has been set to zero. This negative diode current serves to create a voltage drop in the series resistance that is equal and opposite that of the forward voltage of the junction, which is maintained until the stored minority charge is removed. The time required to remove the stored charge—the *storage delay time*—is an important figure of merit in switching applications. The delay is related to the SPICE parameter **TT**, the diode transit time

```
Switching diode
vd 1 0 dc 0 pwl (0,1 5n,1 5.01n,0 50n,0)
vid 1 2 dc 0
d1 2 0 dmod
.model dmod d IS=0.1p RS=16 CJO=2p TT=12n
```

FIG. 3.4.7. Circuit description for simulation of diode switching.

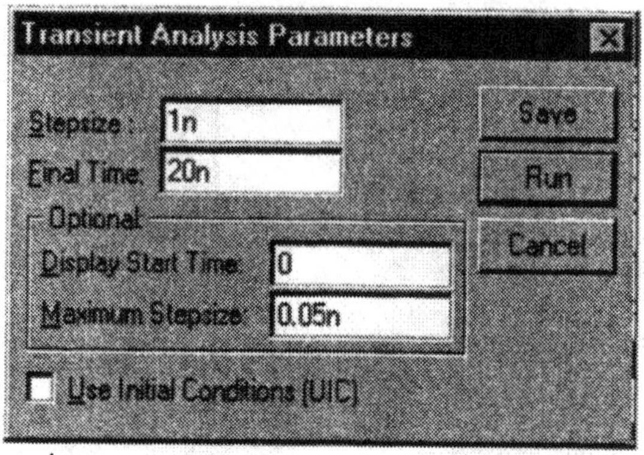

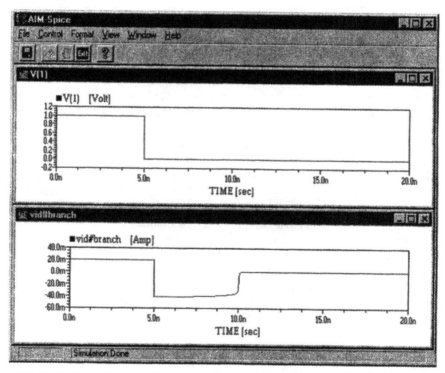

FIG. 3.4.8. Transient Analysis parameters for simulation of diode switching.

FIG. 3.4.9. Turn-off transient for a conventional *p–n* diode. The plots show the input voltage waveform (top) and the diode current (bottom).

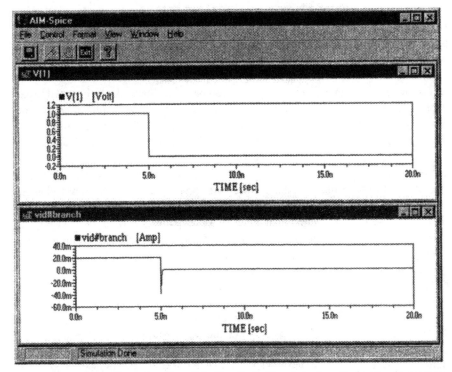

FIG. 3.4.10. Turn-off transient for a Schottky diode. The plots show the input voltage waveform (top) and the diode current (bottom).

indicated in the circuit description (TT=12n in our example, resulting in a storage delay of about 5 ns).

To simulate the switching of a Schottky diode, we set the diode transit time to zero in the circuit description. This excludes the minority carrier storage effect. As expected, the simulation result presented in Fig. 3.4.10 shows that the Schottky diode has a much faster turn-off than the corresponding *p–n* diode.

3.5. METAL–INSULATOR–SEMICONDUCTOR CAPACITOR

3.5.1. Interface Charge

The metal–insulator–semiconductor (MIS) capacitor is a two-terminal semiconductor device of practical interest in its own right and is widely used in integrated circuits. However, of far greater significance is its use as the current control element—the gate unit—in the most popular semiconductor device of them all: the MOSFET. Hence, an understanding of how the MIS capacitor operates is a prerequisite to understanding MOSFETs.

As indicated in Fig. 3.5.1, the MIS capacitor consists of a metal contact separated from the semiconductor by a dielectric insulator. An additional ohmic contact is provided at the semiconductor substrate. Almost universally, the MIS structure utilizes doped silicon as the substrate and its native oxide—silicon dioxide—as the insulator. In the silicon–silicon dioxide system, the density of surface states at the oxide–semiconductor interface is very low compared to the typical channel carrier density in a MOSFET. Also, the insulating quality of the oxide is quite good. On the other hand, in MIS structures based on GaAs, numerous surface states at the insulator–semiconductor interface usually prevent us from making useful MIS devices.

We assume that the insulator layer has infinite resistance, preventing any charge carrier transport across the dielectric layer when a bias voltage is applied between the metal and the semiconductor. Instead, the applied voltage will induce charges and counter charges in the metal and in the surface layer of the semiconductor, similar to what we expect in the metal plates of a conventional parallel-plate capacitor. However, what sets the semiconductor apart from a metal is that we may use the applied voltage to control the type of interface charge we induce: majority carriers, minority carriers, or depletion charge. Indeed, the ability to induce and modulate a conducting sheet of minority carriers at the semiconductor–metal interface is the basis for the operation of the MOSFET.

Our analysis starts by recognizing that the induced interface charge in the MIS capacitor is closely linked to the shape of the energy bands of the semiconductor near the interface. At zero applied voltage, the bending of the energy bands is determined by the difference in the work functions of the metal and the semiconductor, as was the case for the metal–semiconductor Schottky junction (see Section 3.4). This band bending changes with the applied bias, and the bands become flat when we apply the so-called flat-band voltage given by (see Fig. 3.5.2)

$$V_{\text{FB}} = \frac{\Phi_m - \Phi_s}{q} = \frac{\Phi_m - X_s - E_c + E_F}{q} \tag{3.5.1}$$

where Φ_m and Φ_s are the work functions of the metal and the semiconductor, respectively, X_s is the electron affinity for the semiconductor, E_c is the energy of the

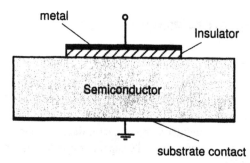

FIG. 3.5.1. Schematic diagram of the cross section of the MIS capacitor.

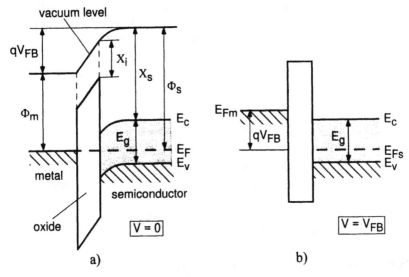

FIG. 3.5.2. Band diagrams: (*a*) of an MIS capacitor at zero bias; (*b*) with an applied voltage equal to the flat-band voltage. The flat-band voltage is negative in this example.

conduction band edge, and E_F is the Fermi level at zero applied voltage. The various energies involved are indicated in Fig. 3.5.2, where we show typical band diagrams of an MIS capacitor at zero bias and with the voltage $V = V_{FB}$ applied to the metal contact relative to the semiconductor. (Note that in real devices the flat-band voltage may be affected by surface states at the interface and/or by fixed charges in the insulator layer.)

At stationary condition, no net current flows in the direction perpendicular to the interface owing to the infinite resistance of the insulator layer. Hence, according to Eqs. (2.2.27) and (2.2.28), the Fermi level will remain constant inside the semiconductor, independent of the biasing conditions. However, between the semiconductor and the metal contact, the Fermi level is shifted by $E_{Fm} - E_{Fs} = qV$ (see Fig. 3.5.2b). Hence, we have a quasi-equilibrium situation where the semiconductor can be treated as if in thermal equilibrium.

In the case of a *p*-type semiconductor, when the voltage applied between the metal and the semiconductor is more positive than the flat-band voltage ($V_{FB} < 0$ in Fig. 3.5.2), the region close to the insulator–semiconductor interface becomes depleted of holes and we enter the so-called *depletion regime* of operation.

At even larger applied voltages, the band bending becomes so large that the energy difference between the Fermi level and the bottom of the conduction band at the insulator–semiconductor interface becomes smaller than that between the Fermi level and the top of the valence band. This is the situation indicated in Fig. 3.5.2a, for $V = 0$ V. Carrier statistics [see Eqs. (3.2.1) and (3.2.2)] tells us that the electron concentration will exceed the hole concentration near the interface in this so-called *inversion* regime.

Let us use the symbol ψ to signify the potential at an arbitrary position x in the semiconductor measured relative to the intrinsic Fermi level deep inside the semiconductor. Note that ψ becomes positive when the bands bend down, as in the example of a p-type semiconductor shown in Fig. 3.5.3. From equilibrium electron statistics, we find that the intrinsic Fermi level E_i in the bulk corresponds to an energy $q\varphi_b$ relative to E_F, where

$$\varphi_b = V_{\text{th}} \ln\left(\frac{N_a}{n_i}\right) \tag{3.5.2}$$

and N_a is the shallow acceptor density in the p-type semiconductor. According to the usual definition, *strong inversion* is reached when the total band bending equals $2q\varphi_b$, corresponding to the surface potential $\psi_s = 2\varphi_b$. Values of the surface potential such that $0 < \psi_s < 2\varphi_b$ correspond to the *depletion* and *weak-inversion* regimes, $\psi_s = 0$ is the flat-band condition, and $\psi_s < 0$ corresponds to the *accumulation* mode.

In terms of the surface potential, the surface concentrations of holes and electrons are given by equilibrium statistics as (p-type semiconductor)

$$p_s = N_a \exp\left(-\frac{\psi_s}{V_{\text{th}}}\right) \tag{3.5.3}$$

$$n_s = \frac{n_i^2}{p_s} = n_{po} \exp\left(\frac{\psi_s}{V_{\text{th}}}\right) \tag{3.5.4}$$

where $n_{po} = n_i^2/N_a$ is the equilibrium concentration of the minority carriers (electrons) in the bulk.

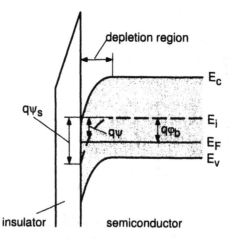

FIG. 3.5.3. Band diagram for MIS in weak inversion ($\varphi_b < \psi_s < 2\varphi_b$).

As was done for the p–n junction, the potential distribution $\psi(x)$ in the semiconductor can be determined from a solution of the one-dimensional Poisson equation:

$$\frac{d^2\psi(x)}{dx^2} = -\frac{\rho(x)}{\epsilon_s}$$

(3.5.5)

where the space charge density $\rho(x)$ is given by

$$\rho(x) = q(p - n - N_a)$$

(3.5.6)

The position-dependent hole and electron concentrations may be expressed as

$$p = N_a \exp\left(-\frac{\psi}{V_{th}}\right)$$

(3.5.7)

$$n = n_{po} \exp\left(\frac{\psi}{V_{th}}\right)$$

(3.5.8)

Note that deep inside the semiconductor, we have $\psi(\infty) = 0$.

The above equations do not, in general, have an analytical solution for $\psi(x)$. However, the following expression can be derived for the electric field F_s at the insulator–semiconductor interface in terms of the surface potential (see, e.g., Lee et al., 1993):

$$F_s = \sqrt{2}\,\frac{V_{th}}{L_{Dp}} f\left(\frac{\psi_s}{V_{th}}\right)$$

(3.5.9)

where the function f is defined by

$$f(u) = \pm \sqrt{[\exp(-u) + u - 1] + \frac{n_{po}}{N_a}\,[\exp(u) - u - 1]}$$

(3.5.10)

and

$$L_{Dp} = \sqrt{\frac{\epsilon_s V_{th}}{qN_a}}$$

(3.5.11)

is the Debye length. In Eq. (3.5.10), a positive sign should be chosen for a positive ψ_s and a negative sign corresponds to a negative ψ_s.

Using Gauss's law, we can relate the total charge Q_s per unit area (carrier charge and depletion charge) in the semiconductor to the surface electric field by

$$Q_s = -\epsilon_s F_s$$

(3.5.12)

At the flat-band condition ($V = V_{FB}$), the surface charge is equal to zero. In accumulation ($V < V_{FB}$) the surface charge is positive, and in depletion and inversion ($V > V_{FB}$) the surface charge is negative. From Eqs. (3.5.9) and (3.5.10), we find that the surface charge is dominated by mobile charge and is proportional to $\exp(|\psi_s|/2V_{th})$ in accumulation (when $|\psi_s|$ exceeds a few times V_{th}) and in strong inversion. In depletion and weak inversion, the depletion charge is dominant and varies as $\psi_s^{1/2}$. Figure 3.5.4 shows $|Q_s|$ versus ψ_s for p-type silicon with a doping density of 10^{16} cm^{-3}.

To complete this analysis, we have to relate the surface potential to the applied voltage. We note that this voltage is divided between the insulator and the semiconductor. Using the condition of continuity of the electric flux density at the semiconductor–insulator interface, we find

$$\epsilon_s F_s = \epsilon_i F_i \qquad (3.5.13)$$

where ϵ_i is the dielectric permittivity of the insulator layer and F_i is the constant electric field in the insulator (assuming no space charge).

Hence, with an insulator thickness d_i, the voltage drop across the insulator be-

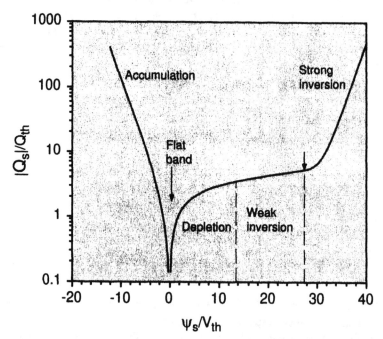

FIG. 3.5.4. Normalized total semiconductor charge per unit area versus normalized surface potential for p-type Si with $N_a = 10^{16}$ cm^{-3}: $Q_{th} = (2\epsilon_s q N_a V_{th})^{1/2} \approx 9.3 \times 10^{-9}$ C/cm^2 and $V_{th} \approx 0.026$ V at $T = 300$ K. The arrows indicate flat-band condition and onset of strong inversion.

comes $F_i d_i$. Accounting for the flat-band voltage, the applied voltage can be written as

$$V = V_{FB} + \psi_s + \frac{\epsilon_s F_s}{c_i} \tag{3.5.14}$$

where $c_i = \epsilon_i/d_i$ is the insulator capacitance per unit area.

3.5.2. Threshold Voltage

Equation (3.5.14) allows us to calculate the threshold voltage V_T corresponding to the onset of the strong-inversion regime. Here, V_T is one of the most important parameters characterizing metal–insulator–semiconductor devices. As discussed above, the onset of strong inversion occurs when the surface potential ψ_s equals $2\varphi_b$. For this surface potential, the charge of the free carriers induced at the insulator–semiconductor interface is still small compared to the charge in the depletion layer, which is given by

$$Q_{dT} = -qN_a d_{dT} = -\sqrt{4\epsilon_s q N_a \varphi_b} \tag{3.5.15}$$

where $d_{dT} = \sqrt{4\epsilon_s \varphi_b/qN_a}$ is the width of the depletion layer at threshold. Hence, at the onset of strong inversion, the electric field at the semiconductor–insulator interface becomes

$$F_{sT} = -\frac{Q_{dT}}{\epsilon_s} = \sqrt{\frac{4qN_a\varphi_b}{\epsilon_s}} \tag{3.5.16}$$

and from Eq. (3.5.14), we find the following expression for the threshold voltage:

$$V_T = V_{FB} + 2\varphi_b + \frac{\sqrt{4\epsilon_s q N_a \varphi_b}}{c_i} \tag{3.5.17}$$

Typical calculated dependencies of V_T on doping level and dielectric thickness are shown in Fig. 3.5.5.

For the MIS structure shown in Fig. 3.5.1, the application of a bulk bias, V_B, is simply equivalent to changing the applied voltage from V to $V - V_B$. Hence, for this situation, the threshold for strong inversion is simply shifted by V_B. However, the situation will be different if the conducting layer of mobile electrons in the inversion layer is maintained at some constant potential. This situation occurs in a MOSFET. If we assume that the inversion layer is grounded, V_B biases the effective junction between the inversion n-channel layer and the p-type substrate, changing the negative charge in the depletion layer. Under such conditions, the threshold voltage becomes

$$V_T = V_{FB} + 2\varphi_b + \frac{\sqrt{2\epsilon_s q N_a(2\varphi_b - V_B)}}{c_i} \tag{3.5.18}$$

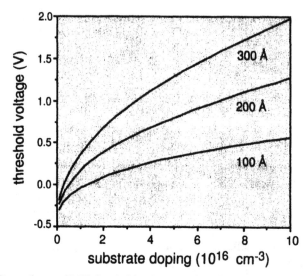

FIG. 3.5.5. Dependence of MIS threshold voltage on the substrate doping level for different thicknesses of the dielectric layer. Parameters used in calculation: energy gap, 1.12 eV; effective density of states in the conduction band, 3.22×10^{25} m^{-3}; effective density of states in the valence band, 1.83×10^{25} m^{-3}; semiconductor dielectric permittivity, 1.05×10^{-10} F/m; insulator dielectric permittivity, 3.45×10^{-11} F/m; flat-band voltage, -1 V; temperature, 300 K. (After Lee et al., 1993.)

The threshold voltage may also be affected by so-called fast surface states at the insulator-semiconductor interface and by fixed charges in the insulator layer (see, e.g., Shur, 1990).

The threshold voltage separates the subthreshold regime, where the mobile carrier charge increases exponentially with increasing applied voltage, from the above-threshold regime, where the mobile carrier charge is linearly dependent on the applied voltage. However, there is no clear distinction between the two regimes, and different definitions and experimental techniques have been used to determine V_T. Sometimes Eqs. (3.5.17) and (3.5.18) are taken to indicate the onset of so-called moderate inversion, while the onset of strong inversion is defined to be a few thermal voltages higher (see Lee et al., 1993).

3.5.3. MIS Capacitance

In an MIS capacitor, the metal contact and the neutral region in the doped semiconductor substrate are separated by the insulator layer and by the depletion region. Hence, the capacitance C_{mis} of the MIS structure can be represented as a series connection of the insulator capacitance $C_i = S\epsilon_i/d_i$, where S is the area of the MIS capacitor, and the capacitance of the semiconductor layer C_s, that is,

$$C_{mis} = \frac{C_i C_s}{C_i + C_s} \tag{3.5.19}$$

The semiconductor capacitance can be calculated as

$$C_s = S \left| \frac{dQ_s}{d\psi_s} \right| \tag{3.5.20}$$

where Q_s is the total charge density per unit area in the semiconductor and ψ_s is the surface potential. Using Eqs. (3.5.9)–(3.5.12) for Q_s and performing the differentiation, we obtain

$$C_s = \frac{C_{so}}{\sqrt{2} f(\psi_s/V_{th})} \left\{ 1 - \exp\left(-\frac{\psi_s}{V_{th}}\right) + \frac{n_{po}}{N_a} \left[\exp\left(\frac{\psi_s}{V_{th}}\right) - 1 \right] \right\} \tag{3.5.21}$$

Here, $C_{so} = S\epsilon_s/L_{Dp}$ is the semiconductor capacitance under the flat-band condition (i.e., for $\psi_s = 0$) and L_{Dp} is the Debye length given by Eq. (3.5.11). The relationship between the surface potential and the applied bias is given by Eq. (3.5.14).

This capacitance can be represented as the sum of two capacitances: a depletion layer capacitance C_d and a capacitance C_{fc} that is related to a delay in the generation/recombination of free minority carriers at the interface. The depletion layer capacitance is given by

$$C_d = S\epsilon_s/d_{dep} \tag{3.5.22}$$

where

$$d_{dep} = \sqrt{\frac{2\epsilon_s \psi_s}{qN_a}} \tag{3.5.23}$$

is the depletion layer width for applied voltages corresponding to the subthreshold regime. In strong inversion, any shift in the applied voltage will primarily affect the minority carrier population at the interface, owing to the strong dependence of this charge on the surface potential in this regime. This means that the depletion width reaches a maximum value at threshold, with no significant further increase in the depletion charge. This maximum depletion width can be determined by using the interface potential corresponding to threshold ($\psi_s = 2\varphi_b$) in Eq. (3.5.23).

The free carrier contribution to the semiconductor capacitance can be formally expressed as

$$C_{fc} = C_s - C_d \tag{3.5.24}$$

In an MIS structure, a variation in the minority carrier charge at the interface comes from the processes of generation and recombination of electron–hole pairs, primarily owing to recombination–generation centers in the semiconductor but also as a result of the thermal generation and recombination of electron–hole pairs. Once an electron–hole pair is generated, the majority carrier (a hole in p-type material and an electron in n-type material) is swept from the space charge region into the substrate by the electric field of this region. The minority carrier is swept in the opposite direction toward the semiconductor–insulator interface. The variation in minority carrier charge at the semiconductor–insulator interface proceeds at a rate limited by the processes of electron–hole pair generation and recombination. This finite rate corresponds to a delay that may be represented electrically in terms of an RC product consisting of the capacitance C_{fc} and a resistance R_{GR}, as reflected in the equivalent circuit of the MIS structure shown in Fig. 3.5.6. The capacitance C_{fc} becomes important in the inversion regime, especially in strong inversion where the mobile charge is dominant. The resistance R_s in the equivalent circuit is the series resistance of the semiconductor layer.

As seen from Fig. 3.5.6, the equivalent MIS capacitance is frequency dependent. In the low-frequency limit ($\omega \to 0$), we can neglect the effect of R_{GR} and find (taking into account that $C_s = C_d + C_{fc}$)

$$C_{mis}^o = \frac{C_s C_i}{C_s + C_i} \tag{3.5.25}$$

In strong inversion, we usually have $C_s \gg C_i$, which results in

$$C_{mis}^o \approx C_i \tag{3.5.26}$$

in the low-frequency limit.

In the high-frequency limit ($\omega \to \infty$), the time constant of the generation/recombination mechanism will be much longer than the signal period ($R_{GR}C_{fc} \gg 2\pi/\omega$) and C_d effectively shunts the lower branch of the parallel subcircuit in Fig. 3.5.6. Hence, the high-frequency series capacitance of the equivalent circuit becomes

$$C_{mis}^\infty = \frac{C_d C_i}{C_d + C_i} \tag{3.5.27}$$

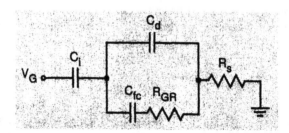

FIG. 3.5.6. Equivalent circuit of the MIS capacitor. (After Shur, 1990.)

The calculated dependence of C_{mis} on the applied voltage for different frequencies is shown in Fig. 3.5.7. At negative applied voltages, C_{mis} is equal to C_i because the device is in the accumulation mode of operation. The calculated curves clearly show how the MIS capacitance in the strong-inversion regime changes from $C_d C_i/(C_d + C_i)$ at high frequencies to C_i at low frequencies. However, it is important to realize that in a MOSFET, where the highly doped source and drain regions act as reservoirs of minority carriers for the inversion layer, the time constant $R_{GR}C_{fc}$ must be substituted by a much smaller time constant corresponding to the time needed for transporting carriers from these reservoirs to the central portions of the MOS-FET gate area. Consequently, MOSFET gate–channel characteristics in the depletion and inversion regimes will resemble the zero-frequency MIS characteristics even at very high frequencies.

Since the MIS capacitance in the strong-inversion regime is constant, the inversion charge induced into the MOSFET channel can be approximated by

$$qn_s \approx c_i(V - V_T) \tag{3.5.28}$$

This equation serves as the basis of a simple charge control model that allows us to calculate MOSFET current–voltage characteristics in strong inversion.

From measured MIS C–V characteristics, we can easily determine important parameters of the MIS structure, including the gate insulator thickness, the semiconductor substrate doping density, and the flat-band voltage. The maximum measured capacitance C_{max} (capacitance C_i in Fig. 3.5.7) yields the insulator thickness

$$d_i \approx \frac{S\epsilon_i}{C_{max}} \tag{3.5.29}$$

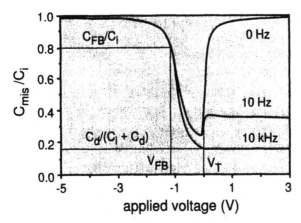

FIG. 3.5.7. Calculated dependence of C_{mis} on the applied voltage for different frequencies: insulator thickness, 2×10^{-8} m; semiconductor doping density, 10^{15} cm^{-3}; generation time, 10^{-8} s. (After Shur, 1990.)

The minimum measured capacitance C_{min} (at high frequency) allows us to find the doping concentration in the semiconductor substrate. First, we determine the depletion capacitance in the strong-inversion regime from Eq. (3.5.27):

$$\frac{1}{C_{min}} = \frac{1}{C_d} + \frac{1}{C_{max}} \tag{3.5.30}$$

On the other hand, the thickness of the depletion region is given by

$$d_{dep} = \frac{S\epsilon_s}{C_d} \tag{3.5.31}$$

Then, we calculate the doping density N_a using Eq. (3.5.23) for d_{dep} in strong inversion (i.e., with $\psi_s = 2\varphi_b$) and Eq. (3.5.2) for φ_b. This results in the following transcendental equation for N_a:

$$N_a = \frac{4\epsilon_s V_{th}}{q d_{dep}^2} \ln\left(\frac{N_a}{n_i}\right) \tag{3.5.32}$$

This equation can easily be solved by iteration or by an approximate analytical technique described by Fjeldly et al. (1991).

Once d_i and N_a have been obtained, the device capacitance C_{FB} under flat-band conditions can be determined using $C_s = C_{so}$ [Eq. (3.5.21) at flat-band condition] in combination with Eq. (3.5.19):

$$C_{FB} = \frac{C_i C_{so}}{C_i + C_{so}} = \frac{\epsilon_i \epsilon_s}{\epsilon_i L_{Dp} + \epsilon_s d_i} \tag{3.5.33}$$

The flat-band voltage V_{FB} is simply equal to the applied voltage corresponding to this value of the device capacitance.

Strictly speaking, this characterization technique applies to an ideal MIS structure. Different nonideal effects that may change the frequency response of the MIS capacitor are discussed, for example, by Lee et al. (1993).

3.5.4. Unified Charge Control Model for MIS Capacitors

Well above threshold, the charge of mobile carriers in the channel can be calculated using the standard charge control model of Eq. (3.5.28). This model gives an adequate description for the strong-inversion regime of the MIS capacitor but fails for applied voltages near and below threshold (i.e., in the weak-inversion and depletion regimes). Byun et al. (1990) proposed a new unified charge control model (UCCM). According to this model, the inversion charge qn_s per unit area in the MIS capacitor is related to the applied voltage as follows:

$$V - V_T = \frac{q(n_s - n_o)}{c_a} + \eta V_{th} \ln\left(\frac{n_s}{n_o}\right) \tag{3.5.34}$$

where $c_a \approx c_i$ is approximately the insulator capacitance per unit area [with a small correction; see Lee et al. (1993)], $n_o = n_s(V = V_T)$ is the density of minority carriers per unit area at threshold, and η is the so-called subthreshold ideality factor. The ideality factor accounts for the subthreshold division of the applied voltage between the gate insulator and the depletion layer and $1/\eta$ represents the fraction of this voltage that contributes to the interface potential. A simplified analysis gives (see Lee et al., 1993)

$$\eta = 1 + \frac{C_d}{C_i} \tag{3.5.35}$$

and

$$n_o = \frac{\eta V_{th} c_a}{2q} \tag{3.5.36}$$

In the subthreshold regime, Eq. (3.5.34) rapidly approaches the limit

$$n_s = n_o \exp\left(\frac{V - V_T}{\eta V_{th}}\right) \tag{3.5.37}$$

We note that Eq. (3.5.34) does not have an exact analytical solution for the inversion charge in terms of the applied voltage. However, the UCCM expression can be

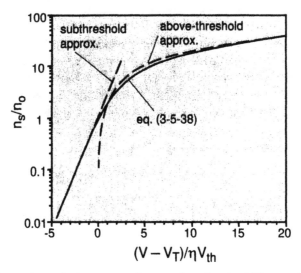

FIG. 3.5.8. Comparison of charge control expressions for the MIS capacitor. Equation (3.5.38) is a close approximation to UCCM while Eqs. (3.5.28) and (3.5.37) are approximation to the above- and below-threshold regimes, respectively.

converted to the form of the nonideal diode equation for which a very precise, approximate solution has been derived (see Fjeldly et al., 1991). For many purposes, the following less precise approximate solution is suitable:

$$n_s = 2n_o \ln\left[1 + \frac{1}{2}\exp\left(\frac{V - V_T}{\eta V_{th}}\right)\right] \tag{3.5.38}$$

This expression reproduces the correct limiting behavior both in strong inversion and in the subthreshold regime, although it deviates slightly from Eq. (3.5.34) near threshold [see Lee et al. (1993) for further details.] Various charge control expressions of the MIS capacitor are compared in Fig. 3.5.8.

REFERENCES

R. A. Anderson, *Solid State Electronics*, vol. 5, pp. 341–351 (1962).

J. Bardeen, *Physical Review*, vol. 71, p. 717 (1947).

Y. Byun, K. Lee, and M. Shur, "Unified Charge Control Model and Subthreshold Current in Heterostructure Field Effect Transistors," *IEEE Electron Device Letters*, vol. EDL-11, no. 1, pp. 50–53 (1990). Erratum *IEEE Electron Device Letters*, vol. EDL-11, no. 6, p. 273 (1990).

M. Cowley and S. M. Sze, "Surface States and Barrier Height of Metal-Semiconductor Systems," *Journal of Applied Physics*, vol. 36, p. 3212 (1965).

T. A. Fjeldly, B. Moon, and M. Shur, "Analytical Solution of Generalized Diode Equation," *IEEE Transactions on Electron Devices*, vol. ED-38, no. 8, pp. 1976–1977 (1991).

K. Lee, M. Shur, T. A. Fjeldly and T. Ytterdal, *Semiconductor Device Modeling for VLSI*, Prentice-Hall, Englewood Cliffs, NJ (1993).

E. H. Rhoderick, *Metal-Semiconductor Contacts*, Clarendon, Oxford (1977).

M. Shur, *Physics of Semiconductor Devices*, Prentice-Hall, Englewood Cliffs, NJ (1990).

S. M. Sze, *Physics of Semiconductor Devices*, Wiley, New York (1981).

S. M. Sze, *Semiconductor Devices. Physics and Technology*, Wiley, New York (1985).

A. van der Ziel, *Solid State Physical Electronics*, Prentice-Hall, Englewood Cliffs, NJ (1976).

PROBLEMS

3.2.1. (a) Simulate forward *p–n* diode characteristics and plot the results in a logarithmic scale (similar to the plot shown in Fig. 3.2.6). Use the following model parameters:

```
.model diode d is=192.1p n=1 xti=3 eg=1.11 cjo=893.8f
+ m=98.29m vj=.75 fc=.5 bv=5 ibv=10u nr=2
+ isr=16.91n ikf=1e-2 rs=0.1
```

(b) Repeat the simulation using series resistances of 1 and 10 Ω. Compare and discuss the results. Also, compare with the results for an ideal diode.

3.2.2. Calculate and plot p–n diode current–voltage characteristics using default SPICE parameters for temperatures of –50, 0, 25, and 100°C.

3.2.3. Calculate and plot the output voltage versus the input voltage V_{in} for –3 V ≤ V_{in} ≤ 3 V of the circuit shown in Fig. P3.2.3. Use default SPICE parameters.

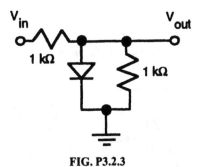

FIG. P3.2.3

3.2.4. Calculate and plot the output voltages V_1, V_2, and V_3 versus the input voltage V_{in} for –3 V ≤ V_{in} ≤ 3 V of the circuit shown in Fig. P3.2.4. Use default SPICE parameters.

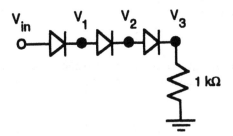

FIG. P3.2.4

3.2.5. Calculate and plot the output voltage V_3: (a) versus the input voltage 0 ≤ V_1 ≤ 3 V for V_2 = 0; (b) versus the input voltage V_1 = V_2 = V for 0 ≤ V ≤ 3 V of the circuit shown in Fig. P3.2.5. Use default SPICE parameters

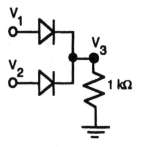

FIG. 3.2.5

3.2.6. Calculate the current–voltage characteristic of a *p–n* junction diode using default SPICE parameters. Suggest how to extract the value of the built-in voltage from this characteristic.

3.2.7. Simulate current–voltage characteristics of two *p–n* junction diodes connected in series and "back to back" using default parameters. Comment on the results.

3.2.8. From the *I–V* characteristics of the *p–n* junction shown in Fig. P3.2.8, estimate the diode saturation current, the ideality factor, and series resistance.

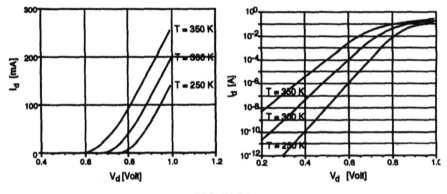

FIG. P3.2.8

Estimate the energy gap of the diode semiconductor material using the following empirical equation for the temperature dependence of the saturation current:

$$I_s(T) = I_s(T_o)\left(\frac{T}{T_o}\right)^{\kappa/\eta} \exp\left[\frac{E_g}{k_B T_o}\left(1 - \frac{T_o}{T}\right)\right]$$

3.2.9. Simulate the operation of the so-called "basic buck" voltage conversion circuit shown in Fig. P3.2.9 for duty cycle $D = 5{:}1$.

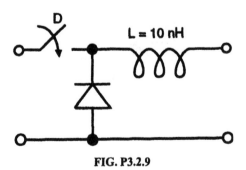

FIG. P3.2.9

3.2.10. Use Eqs. (3.2.1) and (3.2.2) to calculate and plot the Fermi level position in neutral n-type Si and p-type Si for doping concentrations in the range 10^{13}–10^{18} cm^{-3} at room temperature ($T = 300$ K). Assume $n_i = 10^{10}$ cm^{-3}.

3.2.11. A p–n junction with a series resistance R is biased by a current source. At time $t = 0$, the current is doubled. What happens to the voltage across the junction at $t = 0$? Check your answer using SPICE simulations.

3.2.12. Assume that Si p–n junction diodes experience breakdown when the maximum electric field in the junction reaches 300 kV/cm. Calculate the breakdown voltage of a Si p^+–n junction for electron concentration in the range 10^{13}–10^{18} cm^{-3}.

3.4.1. (a) Review the example in Section 3.4.5.
(b) Choose a suitable operating point and calculate and compare the small-signal frequency response of both the p–n junction diode and the Schottky diode. Use the same SPICE parameters as in the example.

3.4.2. (a) Review the example in Section 3.2.9.
(b) Do a similar calculation and plot the capacitance–voltage characteristics of a Schottky diode.

3.4.3. Use the SPICE diode model to simulate the current–voltage characteristics of a GaAs Schottky diode at 300 and 400 K by choosing the parameters such that they correspond to a diode with Richardson's constant A* = 4.4 A/cm^2, a barrier height of 0.8 V, and an ideality factor of 1.25. The diode cross section is 25 μm^2.

3.4.4. Sketch a small-signal equivalent circuit of a Schottky diode. Compare this circuit with the one for a p–n junction diode. What is the difference? Estimate parameters of the equivalent circuit for a GaAs Schottky diode with Richardson's constant A* = 4.4 A/cm^2, a barrier height of 0.8 V, a carrier concentration $n = 10^{16}$ cm^{-3}, and an ideality factor of 1.25. The diode cross section is 25 μm^2 and the length is 3 μm.

3.5.1. (a) Review the example in Section 3.2.9.
(b) Calculate and plot the capacitance–voltage characteristics of an MIS capacitor for different small-signal frequencies. Compare the results with the theory presented in Section 3.5.

Hint: Use the SPICE circuit description shown below to calculate the MIS capacitance. Since SPICE does not include the MIS capacitor, we model it using a MOSFET with source and drain contacts floating (the resistance `rf` accomplishes this). If we neglect the effect of generation/recombination in Fig. 3.5.6, we can write $C_{mis} = [2\pi f \operatorname{Im}(Z)]^{-1}$.

```
MIS capacitance
x1 2 0 dut
vin 1 0 dc=1 ac 1
vi 1 2 dc 0
```

```
.subckt dut 1 2
m1 10 1 10 2 mn l=100u w=100u
rf 10 0 10g
.ends

.model mn nmos level=2 vto=1
```

CHAPTER 4

BIPOLAR JUNCTION TRANSISTORS

4.1. INTRODUCTION

The purpose of this chapter is to provide basic information on bipolar junction transistors (BJTs) and to present bipolar transistor models implemented in SPICE. The BJT was the first transistor type that came into practical use, challenging and eventually replacing the vacuum tube electronics that prevailed in the first decades after World War II. Since the first published paper on the BJT [Bardeen and Brattain (1948)], this device developed into one of the most essential components of modern electronics. Even though FETs are dominant in most electronics applications today, bipolar transistors are still very important, particularly in high-speed systems. In fact, new bipolar technologies utilizing material systems such as AlGaAs/GaAs and AlInAs/InAs—the so-called heterostructure bipolar transistors (HBTs)—are projected to be among the front runners in the high-speed gamut, opening up new opportunities for bipolar devices.

In Section 4.2, we consider the principle of operation of BJTs. The BJT consists, in reality, of two back-to-back p–n junctions sharing a common doped region. Hence, an understanding of this device can be derived from our knowledge of how p–n junctions operate. Using simple and intuitive arguments, we readily establish fundamental relationships governing the behavior of the BJT. As a basis for bipolar device models implemented in SPICE, we present in Section 4.3 a more comprehensive description of the BJT in terms of the so-called Ebers–Moll and Gummel–Poon models. Finally, in Section 4.4, we discuss a model for HBTs.

4.2. BJT BASICS

As mentioned, the BJT is comprised of two back-to-back p–n junctions, as shown schematically for an n–p–n transistor in Fig. 4.2.1a. An emitter region and a base region form the first junction. Sharing the base with the emitter–base junction, the second p–n junction is formed between the base and a collector region. Figure 4.2.1b shows the band structure through the transistor for typical operating conditions, where the emitter–base junction is forward biased and the collector–base junction is reverse biased

The base is made to be much shorter than the diffusion length of the minority carriers. Therefore, excess minority carriers injected into the base at the emitter–base junction can traverse the base region and reach the collector without much loss due to recombination. The minority carrier injection into the base is caused by a lowering of the energy barrier at the emitter–base junction using a forward bias. Hence, the collector current is effectively controlled by the applied volt-

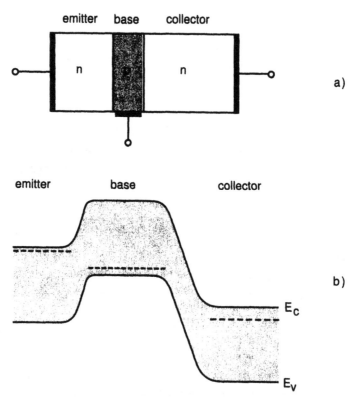

FIG. 4.2.1. Schematics of (a) an n–p–n bipolar junction transistor and (b) the band diagram in the forward active mode. The positions of the quasi–Fermi levels in the three regions (dashed lines) also reflect the relative doping levels.

age across this junction (see Section 3.2 on p–n junctions for details). But, as will be seen later, the controlling voltage can be translated into a base current, which is a more convenient parameter for expressing the control function in BJTs.

Except for the depletion zones associated with the two junctions, the base region remains electrically neutral, and the minority carriers are transported through the base by diffusion (except for cases where a built-in electric field is present). The interaction between the emitter–base and the collector–base junctions, achieved by using a sufficiently short base region, is crucial to the operation of the BJT. This interaction, combined with the control of the injection barrier exercised via the base contact, sets the BJT clearly apart from a pair of ordinary p–n diodes wired back to back externally. This difference is illustrated in the equivalent circuit of an ideal BJT shown in Fig. 4.2.2. Here, the controlled current sources account for the interaction between the emitter–base and the collector–base junctions, absent in the case of a wide-base device and in externally wired p–n diodes.

A schematic view of a modern n–p–n BJT compatible with integrated circuit technology is shown in Fig. 4.2.3. However, for our simplified analysis, we use the one-dimensional representation in Fig. 4.2.1.

4.2.1. Modes of Operation

Under typical operating conditions, the emitter–base junction is forward biased while the collector–base junction is reverse biased. This is called the *forward active mode* of operation and is used for amplification and other small-signal applications. We note from Fig. 4.2.2 that a degree of symmetry exists between emitter and collector, indicating that transistor action also can be achieved using the collector as emitter and vice versa. However, in order to optimize performance in real devices, the emitter is normally more highly doped than the collector, and the base has an intermediate doping level. This results in an asymmetry that makes the so-called *inverse active mode* of operation unsuitable in most cases. Nonetheless, forward biasing of both junctions, that is, injecting minority carriers from emitter and collector simultaneously, is a very useful mode of operation called the *saturation mode*. The *cutoff mode* is a fourth

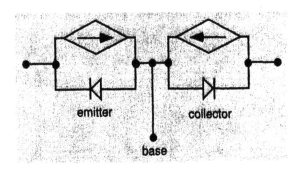

FIG. 4.2.2. Equivalent circuit of bipolar transistor.

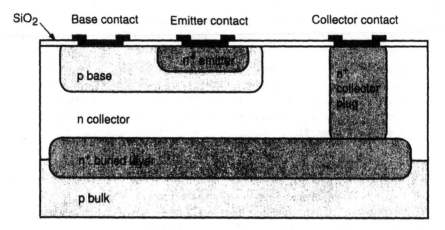

FIG. 4.2.3. Schematic view of a modern n–p–n BJT compatible with integrated circuit technology.

mode of operation where both junctions are reverse biased. The saturation mode and the cutoff mode may serve as logical states in digital circuits.

In the forward active mode, the minority carrier concentration n_{be} in the base at the depletion zone edge of the emitter–base junction (electrons in an n–p–n transistor) is obtained from the *law of the junction* (see Section 3.2)

$$n_{be} = n_{bo} \exp\left(\frac{V_{be}}{V_{th}}\right) \gg n_{bo} \tag{4.2.1}$$

where V_{be} ($\gg V_{th}$) is the positive base–emitter voltage, $V_{th} = k_B T/q$ is the thermal voltage, $n_{bo} = n_{ib}^2/N_{ab}$ is the equilibrium electron concentration, n_{ib} is the intrinsic carrier concentration, and N_{ab} is the acceptor doping density in the base region. Correspondingly, for the reverse-biased base–collector junction, the electron concentration n_{bc} in the base at the depletion zone edge of the collector–base junction becomes

$$n_{bc} = n_{bo} \exp\left(\frac{V_{bc}}{V_{th}}\right) \ll n_{bo} \tag{4.2.2}$$

where V_{bc} ($|V_{bc}| \gg V_{th}$) is the negative collector–base voltage.

Equation (4.2.2) shows that n_{bc} becomes very small in the forward active mode and can often be approximated by $n_{bc} \approx 0$. If the width W of the base is small compared to the electron diffusion length L_{nb} in the base, the carrier recombination in the base will also be small. To a lowest order approximation, the electron concentration n_b in the base then varies linearly with distance. Assuming that the widths of the depletion regions at the emitter–base and collector–base interfaces are sufficiently small and can be neglected compared to the base width W, we have, for the forward active mode,

$$n_b \approx n_{be}\left(1 - \frac{x}{W}\right) \qquad (4.2.3)$$

where $x = 0$ corresponds to the position of the emitter–base junction. This result is based on the solution of the continuity equation for electrons in the base and is obtained in the same way as for a short p–n diode, that is, by expanding $\exp(x/L_{nb})$ and $\exp(-x/L_{nb})$ into Taylor series and retaining only up to the linear term in x.

Figure 4.2.4 shows qualitatively, in a semilog plot, the minority carrier densities of the various regions of the BJT for the four modes of operation. This figure also shows schematically how the widths of the depletion zones depend on the doping and the bias conditions of the junctions. Note that the minority carrier concentration n_b in the base has a near-linear distribution (in a linear plot) in both the forward and the inverse active modes, and $n_b \gg n_{bo}$ in most of the base region. In saturation, where both junctions are forward biased, we still have a near-linear distribution of

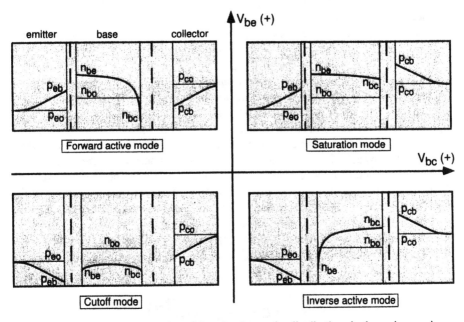

FIG. 4.2.4. Qualitative illustration of the minority carrier distributions in the various regions of an n–p–n BJT for the four modes of operation (note that the magnitudes of concentrations are drawn in a logarithmic scale). The minority carrier concentrations at the depletion zone edges and at equilibrium are indicated. The depletion widths are drawn to reflect the doping and bias conditions of the junctions.

minority carriers in the base, with $n_b \gg n_{bo}$ everywhere. Finally, in cutoff, we have $n_b \ll n_{bo}$ everywhere. Hence, the total excess minority carrier charge in the base can be expressed approximately as follows for the four modes of operation:

$$\Delta Q_b \approx qS \int_0^W [n_b(x) - n_{bo}]\, dx$$

$$\approx \frac{qSWn_{bo}}{2} \begin{cases} \exp(V_{be}/V_{th}) & \text{(forward active)} \\ \exp(V_{bc}/V_{th}) & \text{(inverse active)} \\ \exp(V_{be}/V_{th}) + \exp(V_{bc}/V_{th}) & \text{(saturation)} \\ 0 & \text{(cutoff)} \end{cases} \qquad (4.2.4)$$

We now consider the electron diffusion current in the base. The magnitude of this current can be written as

$$I_n = SqD_n \frac{\partial n_b}{\partial x} \qquad (4.2.5)$$

where D_n is the electron diffusion coefficient and S is the device cross section. The diffusion current remains practically constant throughout the base and is nearly identical to the emitter and the collector currents.

In the forward active mode, practically all electrons injected into the base region from the emitter diffuse to the collector. In the collector, they are carried away toward the collector contact by the strong electric field of the reverse-biased collector–base junction. Of course, some electrons do recombine with holes in the base region, setting up a small base current that serves to maintain neutrality in the base. However, the number of such recombinations is small since the base region is much shorter than the electron diffusion length.

In saturation, where electrons are injected into the base from both emitter and collector, the gradient in the electron concentration will decrease compared to that of the forward active mode, causing a reduction in the diffusion current. In fact, the forward biases of the two junctions can be balanced in such a way that no net current will flow between emitter and collector. In this case, all injected electrons will recombine in the base and thus contribute to the base current.

Substituting Eq. (4.2.1) into Eq. (4.2.3) and the resulting equation into Eq. (4.2.5), for the emitter current I_e and the collector current I_c in the forward active mode, we obtain

$$I_e \approx I_c \approx I_n \approx \frac{SqD_n n_{bo}}{W} \exp\left(\frac{V_{be}}{V_{th}}\right) \qquad (4.2.6)$$

This expression shows that the operation of the BJT is based on the exponential variation of the injected carrier density with the height of the potential barrier between emitter and base, controlled by the emitter–base voltage. As a consequence, the transconductance, which is given by

$$g_m = \frac{\partial I_c}{\partial V_{be}} \approx \frac{I_c}{V_{th}} \qquad (4.2.7)$$

can become very large compared with the corresponding parameter for an FET. Large transconductances and large current swings are factors that make the BJT a device of choice for many high-speed and high-power applications both in discrete and in integrated circuits.

In modern BJTs, the base current I_b primarily consists of the hole injection current I_{pe} from the base into the emitter (n–p–n transistor in the forward active mode). This hole current in the n-type emitter is a diffusion current, just like the electron current in the p-type base. Hence, we have, in full analogy with Eq. (4.2.6),

$$I_b \approx I_{pe} \approx SqD_p \frac{\partial p_e}{\partial x}\bigg|_{x=0} \approx \frac{SqD_p p_{eo}}{X_e} \exp\left(\frac{V_{be}}{V_{th}}\right) \qquad (4.2.8)$$

Here, X_e is the width of the emitter region and $p_{eo} = n_{ie}^2/N_{de}$ is the equilibrium minority carrier concentration, where N_{de} is the donor density and n_{ie} is the intrinsic carrier concentration in the emitter. In Eq. (4.2.8), we assumed that $X_e \ll L_{pe}$, where L_{pe} is the diffusion length of holes in the emitter region. This is always true for practical BJTs.

Let us now compare Eqs. (4.2.6) and (4.2.8). Usually, X_e is of the same order of magnitude as W. However, the emitter region is much more heavily doped than the base region, and hence, $p_{eo} \ll n_{bo}$. This means that the base current is much smaller than the emitter current, and the ratio between I_c and I_b, called the *common-emitter current gain* (see Section 4.2.2), is much greater than unity and can be written as

$$\beta \approx \frac{D_n N_{de} X_e}{D_p N_{ab} W} \qquad (4.2.9)$$

We also note that since, generally, $D_n/D_p > 1$ (approximately 2.5 in silicon), a higher β can be achieved for n–p–n BJTs than for p–n–p BJTs.

The base current also includes other contributions, such as the recombination current in the depletion region of the emitter–base junction, the generation current in the depletion region of the collector–base junction, and the recombination current in the neutral base region. The latter contribution will be proportional to the total excess minority charge in the base given by Eq. (4.2.4). A comparison of the expression for ΔQ_b in the forward active mode with that of the hole injection current from base to emitter in Eq. (4.2.8) shows that these two contributions to I_b have the same dependence on V_{be}. Hence, the effect of the neutral base recombinations simply means a reduction in the value of β. However, for a crude estimate of the transistor current gain, all contributions to the base current except I_{pe} may be neglected.

4.2.2. Current–Voltage Characteristics of BJTs

From the above discussion of the various currents in the BJT, we may conclude that the base current is a very suitable parameter for expressing the control function in

BJTs. However, in order to further analyze the electrical characteristics of the BJT in terms of the terminal currents, let us consider the current–voltage characteristics of some basic circuit configurations of BJTs.

Figure 4.2.5 shows an *n–p–n* BJT connected in the so-called *common-base configuration* along with typical current–voltage characteristics. In this configuration, a load resistance R_c is connected in series with a dc voltage supply V_{cc} between the collector and the base terminals, and a resistance R_e is connected in series with a voltage supply V_{ee} between the emitter and base terminals.

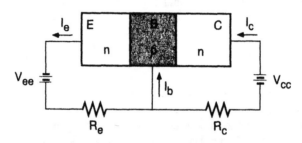

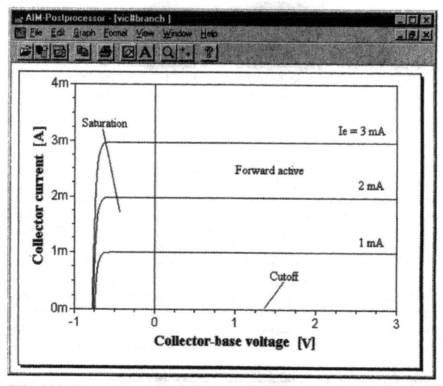

FIG. 4.2.5. Common-base (CB) configuration of an *n–p–n* BJT with typical CB current–voltage characteristics simulated with AIM-Spice using default parameters. Various modes of operation are indicated.

With the emitter–base junction forward biased and the collector–base junction reverse biased, the transistor will be in the forward active mode of operation with $I_c \approx I_e$, as indicated in the figure. More precisely, we can write

$$I_c = \alpha I_e + I_{cbo} \qquad (4.2.10)$$

where I_{cbo} is the small saturation current of the reverse-biased collector–base junction at $I_e = 0$ and the coefficient α, called the *common-base current gain*, is close to unity. A typical value for α is 0.99.

As discussed earlier, the transistor enters the saturation mode when both junctions are forward biased, corresponding to $V_{cb} < 0$ in the *I–V* characteristics of Fig. 4.2.5. We notice a sharp drop in the collector current as we approach the bias configuration where the minority carrier injection from both emitter and collector to the base become equal in magnitude.

The cutoff mode, corresponding to both junctions being reverse biased, is indicated as a state where very little current flows in the transistor.

Kirchhoff's current law applied to the transistor in Fig. 4.2.5 gives $I_e = I_c + I_b$, noting the convention indicated for the current directions. This expression combined with Eq. (4.2.10) allows us to express the collector current in the forward active mode in terms of the base current as follows:

$$I_c = \beta I_b + I_{ceo} \qquad (4.2.11)$$

where $\beta = \alpha/(1 - \alpha)$ is the common-emitter current gain discussed earlier and $I_{ceo} = I_{cbo}/(1 - \alpha)$ is the collector-emitter leakage current at $I_b = 0$. Eq. (4.2.11) corresponds to the type of *I–V* characteristics found in the *common-emitter configuration* of the BJT shown in Fig. 4.2.6, where the collector current is plotted versus the voltage applied between the collector and the emitter terminals, using the base current as a parameter. We note that in the forward active mode, the emitter–base junction will be forward biased by $V_{be} \approx 0.7$ V (typical turn-on voltage for a silicon *p–n* junction) and $V_{cb} \approx V_{ce} - V_{be}$ is the reverse bias of the collector–base junction.

Another noteworthy feature of Fig. 4.2.6 is the finite slope observed in the characteristics in the forward active mode. This so-called base-width modulation effect (also called Early effect) can be explained as follows: As V_{ce} increases, the reverse bias of the collector–base junction also increases, resulting in a widening of the depletion zone of this junction. Hence, the effective neutral base width will be reduced and, according to Eq. (4.2.9), causes β to increase with increasing V_{ce}, yielding the observed effect. When extrapolating these portions of the characteristics to $I_c = 0$, they are found to converge reasonably well at some negative value of V_{ce}. The magnitude of this voltage is called the Early voltage and is used as a parameter for characterizing BJTs, as will be discussed later.

The operating point of the circuit in Fig. 4.2.6 is determined by the intersection of a given characteristic determined by the applied base current and the load line indicated in the diagram. This load line is simply the characteristic of the load resistor R_c, indicating how the current in the collector loop depends on the voltage $V_{cc} - V_{ce}$

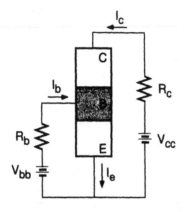

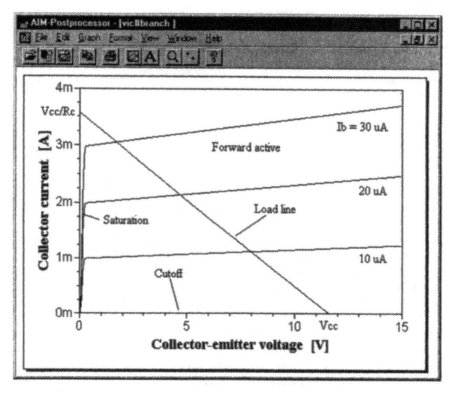

FIG. 4.2.6. Common-emitter (CE) configuration of an *n–p–n* BJT with typical CE current–voltage characteristics simulated with AIM-Spice using default parameters. Various modes of operation are indicated. The figure also shows a loadline corresponding to the load resistance in the collector loop.

across the resistor. Note that this current is zero when $V_{ce} = V_{cc}$ and reaches the value V_{cc}/R_c when $V_{ce} = 0$. In saturation, the operating point obviously becomes nearly independent of the base current at a collector current level of $I_{csat} \approx V_{cc}/R_c$, while in cutoff, the operating point is defined by $I_c \approx 0$, $V_{ce} = V_{cc}$. In digital applications, the circuit is switched between these two extremes by means of the base current. For small-signal applications, the operating point is chosen at an intermediate location within the forward active region.

Finally, we briefly describe how to operate a BJT as an amplifier. As an example, we consider the common-base configuration shown in Fig. 4.2.5 (see also the schematic representation of this configuration in Fig. 4.2.7a). The input signal may be represented by adding an ac voltage source in the emitter loop of the common-base circuit (i.e., the input). A small variation of the base–emitter voltage ΔV_{be}, caused by the input signal, leads to nearly identical variations of the emitter and collector currents, that is, $\Delta I_c \approx \Delta I_e$. However, the collector current flows in the output loop containing the power supply ($V_{cc} \gg V_{be}$) and the variation of the voltage drop $\Delta V_c = \Delta I_c R_c$ across R_c will be much greater than ΔV_{be}. (With an appropriate value of load resistance, the maximum value of ΔV_c is limited by the collector voltage supply.) Hence, we can have a voltage gain of $\Delta V_c / \Delta V_{be}$.

The ac power supplied by the input signal is of the order of $\Delta I_e \Delta V_{be}$, and the ac power generated in the output loop is of the order of $\Delta I_c \Delta V_c \gg \Delta I_e \Delta V_{be}$. Hence, this circuit provides a power gain that is approximately the same as the voltage gain. Physically, the gain in the common-base configuration occurs because electrons are

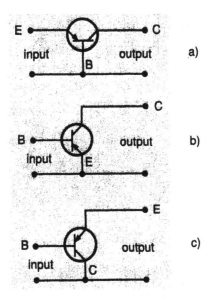

FIG. 4.2.7. (a) Common-base, (b) common-emitter, and (c) common-collector circuit configurations for a p–n–p BJT. (After Lee et al., 1993.)

injected from the emitter region, through the base, and into the reverse-biased collector region where the voltage drop is large compared to the small forward emitter–base voltage.

A similar analysis can also be performed for the other two BJT configurations shown in Fig. 4.2.7.

4.3. BJT MODELING

The simplest large-signal equivalent circuit of an ideal ("intrinsic") BJT includes two diodes and two current-controlled current sources describing the interaction between the emitter–base and collector–base junctions, as indicated in Fig. 4.2.2. This coupled-diode model is identical to the basic Ebers–Moll model. However, in order to represent a real BJT, the model should be extended by additional circuit elements to describe, for example, effects of recombination/generation processes, shown in Fig. 4.3.1, and series resistances.

Even the improved Ebers–Moll model does not account for important mechanisms such as high-injection effects, base-width modulation (Early effect), and effects related to current crowding and band gap narrowing. To describe such phenomena, we have to turn to the more sophisticated analysis afforded by the Gummel–Poon model. In fact, the BJT model implemented in SPICE is the Gummel–Poon model, of which the Ebers–Moll model can be considered a special case.

4.3.1. Ebers–Moll Model

The simple model for the "intrinsic" transistor, called the Ebers–Moll model after Ebers and Moll (1954), is based on the theory for p–n junctions developed in Sec-

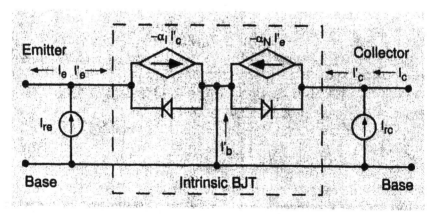

FIG. 4.3.1. Equivalent circuit of an n–p–n bipolar transistor. (After Shur, 1990.)

tion 3.2. For the narrow-base region, the continuity equation is solved using the excess minority carrier concentrations at the edges of the neutral base as boundary conditions. By straightforward calculations, the following "intrinsic" emitter current I_e' and collector current I_c' are obtained:

$$I_e' = a_{11}\left[\exp\left(\frac{V_{be}}{V_{th}}\right) - 1\right] + a_{12}\left[\exp\left(\frac{V_{bc}}{V_{th}}\right) - 1\right] \qquad (4.3.1)$$

$$I_c' = a_{21}\left[\exp\left(\frac{V_{be}}{V_{th}}\right) - 1\right] + a_{22}\left[\exp\left(\frac{V_{bc}}{V_{th}}\right) - 1\right] \qquad (4.3.2)$$

The coefficients a_{11}, a_{12}, a_{21}, and a_{22} are related to the material parameters, the transistor dimensions, and the doping levels. For n–p–n transistors, using the sign convention for the currents I_e' and I_c' in Fig. 4.3.1, we have

$$a_{11} = -qS\left[\frac{D_n n_{bo}}{L_{nb}}\coth\left(\frac{W_{eff}}{L_{nb}}\right) + \frac{D_p p_{eo}}{X_e}\right]$$

$$\approx -qSn_i^2\left[\frac{D_n}{N_{ab}W_{eff}} + \frac{D_p}{N_{de}X_e}\right] \qquad (4.3.3)$$

$$a_{12} = a_{21} = \frac{qSD_n n_{bo}}{L_{nb}\sinh(W_{eff}/L_{nb})} \approx \frac{qSD_n n_i^2}{N_{ab}W_{eff}} \qquad (4.3.4)$$

$$a_{22} = -qS\left[\frac{D_n n_{bo}}{L_{nb}}\coth\left(\frac{W_{eff}}{L_{nb}}\right) + \frac{D_p p_{co}}{X_c}\right]$$

$$\approx -qSn_i^2\left[\frac{D_n}{N_{ab}W_{eff}} + \frac{D_p}{N_{dc}X_c}\right] \qquad (4.3.5)$$

Here W_{eff} is the effective base width, reduced in comparison to the metallurgical width W by the finite extents of the junction depletion regions into the base. The terms X_e and X_c are the widths of the emitter and collector regions, assumed to be much less than the corresponding minority carrier diffusion lengths; L_{nb} is the minority carrier diffusion length in the base; N_{de}, N_{ab}, and N_{dc} are the doping densities of the emitter, the base, and the collector, respectively; and D_n and D_p are the diffusion constants for minority electrons and minority holes. The other symbols are defined in Section 4.2. The approximations indicated in Eqs. (4.3.3)–(4.3.5) are obtained by assuming that $W_{eff} \ll L_{nb}$.

Equations (4.3.1) and (4.3.2) can easily be rearranged in the following manner:

$$I_c' = -\alpha_N I_e' - I_{co}\left[\exp\left(\frac{V_{bc}}{V_{th}}\right) - 1\right] \tag{4.3.6}$$

$$I_e' = -\alpha_I I_c' - I_{eo}\left[\exp\left(\frac{V_{be}}{V_{th}}\right) - 1\right] \tag{4.3.7}$$

where $\alpha_N = -a_{12}/a_{11}$ and $\alpha_I = -a_{21}/a_{22}$ are the forward (normal) and inverse common-base current gains, respectively, and $I_{co} = a_{12}a_{21}/a_{11} - a_{22}$ and $I_{eo} = a_{12}a_{21}/a_{22} - a_{11}$ are the common-base reverse saturation currents of the collector and emitter junctions, respectively. Note that $\alpha_N I_e'$ and $\alpha_I I_c'$ represent the current sources on the collector and emitter sides of the "intrinsic" BJT model of Fig. 4.3.1. In the forward active mode, V_{bc} is negative with a magnitude much larger than V_{th}, and Eq. (4.3.6) reduces to Eq. (4.2.10);. In this approximation, we neglect the generation/recombination currents I_{re} and I_{rc} so that $I_e \approx -I_e'$, $\alpha_N \approx \alpha$, and $I_{co} \approx I_{cbo}$. However, Eq. (4.2.10) may still account for the recombination current by letting α depend on the collector current. Hence, the current gains α_N and α and the saturation currents I_{co} and I_{cbo} do not have to be exactly equal.

The Ebers–Moll model for the intrinsic transistor has the four parameters a_{11}, a_{12}, a_{21}, and a_{22}, or equivalently, α_N, α_I, I_{eo}, and I_{co}. However, the reciprocity relationship for an ideal two-port device requires that $a_{12} = a_{21}$. Hence, only three parameters are required for this basic transistor model (e.g., α_N, I_{eo}, and I_{co}). In practice, however, the reciprocity is not precisely satisfied for real transistors.

The basic Ebers–Moll model may be improved somewhat by adding "extrinsic" effects, such as current sources accounting for the emitter–base and collector–base generation/recombination currents, as indicated in Fig. 4.3.1. These current sources are equivalent to those discussed earlier for the p–n diode. Also, series resistances associated with the three terminals and capacitances associated with the junctions can be incorporated (see Sze, 1981). A feedback loop with an additional current source is sometimes added between the collector and the emitter to account for the base-width modulation effect, specified in terms of the Early voltage (see Section 4.2).

Even this extended Ebers–Moll model does not account for the dependence of the current gain on the injection level and several other effects related to the intrinsic transistor that are very important for practical regimes of transistor operation. At high injection levels, some basic assumptions in the above formalism break down, including the assumption that the minority carrier concentration is always much less than that of the majority carriers. Accordingly, the model should only be used for moderate emitter and collector currents. However, the more accurate and realistic model proposed by Gummel and Poon (1970) addresses this and other important issues not included in the Ebers–Moll model.

To prepare for our discussion of the Gummel–Poon model, we rewrite the

Ebers–Moll equations in a more symmetrical form. Using the relationships for the Ebers–Moll model discussed earlier, we can express the coefficients a_{11}, $a_{12} = a_{21}$, and a_{22} through the emitter reverse saturation current I_{eo}, the normal common-base current gain α_N, and the inverse common-base current gain α_I as follows:

$$a_{11} = -\frac{I_{eo}}{1 - \alpha_N \alpha_I} \tag{4.3.8}$$

$$a_{12} = \frac{\alpha_N I_{eo}}{1 - \alpha_N \alpha_I} = -I_i \tag{4.3.9}$$

$$a_{22} = -\frac{\alpha_N I_{eo}}{\alpha_I (1 - \alpha_N \alpha_I)} \tag{4.3.10}$$

In Eq. (4.3.9), we introduced the so-called intercept current $I_i = -a_{12}$. This parameter derives its name from the fact that $\ln(|I_i|)$ can be determined as the intercept with $V_{bc} = 0$ of the extrapolation of $\ln(I_e)$ versus V_{bc} in the region $V_{bc} \gg V_{th}$, taking $V_{be} = 0$.

Using this notation and the relationships between the common-emitter current gains and the common-base current gains, that is, $\beta_N = \alpha_N/(1 - \alpha_N)$ and $\beta_I = \alpha_I/(1 - \alpha_I)$, we can rewrite the Ebers–Moll equations in the following form:

$$I_e' = I_{cc} + I_{be} \tag{4.3.11}$$

$$I_c' = -I_{cc} + I_{bc} \tag{4.3.12}$$

where

$$I_{cc} = I_i \left[\exp\left(\frac{V_{be}}{V_{th}}\right) - \exp\left(\frac{V_{bc}}{V_{th}}\right) \right] \tag{4.3.13}$$

$$I_{be} = \frac{I_i}{\beta_N} \left[\exp\left(\frac{V_{be}}{V_{th}}\right) - 1 \right] \tag{4.3.14}$$

$$I_{bc} = \frac{I_i}{\beta_I} \left[\exp\left(\frac{V_{bc}}{V_{th}}\right) - 1 \right] \tag{4.3.15}$$

Here I_{cc} is the principal component of the emitter and collector currents. Note that with the sign convention used here I_i is a negative current for n–p–n transistors.

Figure 4.3.2 shows an alternative form of the intrinsic BJT model that better reflects the essence of Eqs. (4.3.11)–(4.3.15).

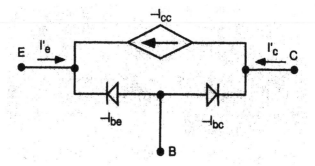

FIG. 4.3.2. Intrinsic BJT model corresponding to Eqs. (4.3.11) and (4.3.12).

4.3.2. SPICE Example: Emitter-Coupled Logic

Emitter-coupled logic is still one of the most popular logic families in mainframe computers and supercomputers, where speed is the most important issue. Emitter-coupled logic is a nonsaturating family of bipolar digital circuits and, hence, provides higher speed than logic circuits that incorporate transistors biased in saturation during parts of the logic cycle. In the present example, we investigate an ECL OR/NOR gate both in static and dynamic operation using AIM-Spice with the Ebers–Moll BJT model. Figures 4.3.3 and 4.3.4 show the basic circuit configuration and the circuit description, respectively.

AIM-Spice always uses a complete Gummel–Poon BJT model. However, with the limited set of parameters specified in the `.model` statement of Fig. 4.3.4, the

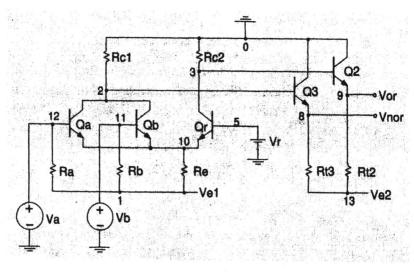

FIG. 4.3.3. Dual-input ECL OR gate with a complementary NOR output.

```
ECL OR gate with NOR output
Ve1 1 0 dc -5.2
Ve2 13 0 dc -2.0
Va 12 0 dc 0 pwl(0,-1.77V 2ns,-1.77V 3ns,
+ -0.884V 30ns,-0.884V)
Vb 11 0 dc -1.77
Qa 2 12 10 npnmod
Qb 2 11 10 npnmod
Qr 3 5 10 npnmod
Q2 0 3 9 npnmod
Q3 0 2 8 npnmod
Ra 12 1 50k
Rb 11 1 50k
Re 10 1 779
Rc1 0 2 220
Rc2 0 3 245
Rt2 9 13 50
Rt3 8 13 50
Vr 5 0 -1.32V
.model npnmod npn IS=0.30fA BF=120 BR=1
+ TF=0.15ns CJE=1.5pF CJC=1.5pF
```

FIG. 4.3.4. Circuit description for a Transient Analysis of the ECL gate shown in Fig. 4.3.3.

Gummel–Poon model automatically simplifies to the Ebers–Moll model. The SPICE parameters specified can be identified as follows (for some of the parameters, we refer to the discussion of the *p–n* diode SPICE model in Section 3.2): **IS** = $|I_i|$ (intercept current), **BF** = β_N (normal common-emitter current gain), **BR** = β_I (inverse common-emitter current gain), **TF** (forward transit time, corresponding to minority carrier delay in the base), **CJE** (emitter–base depletion capacitance at zero bias), and **CJC** (collector–base depletion capacitance at zero bias).

In Fig. 4.3.5, we show the transfer characteristics obtained from a DC Transfer Curve Analysis of the ECL gate. The results of a Transient Analysis are shown in Fig. 4.3.6, where the input voltage Va is switched from −1.77 to −0.884 V at time *t* = 2 ns. Using the cursor feature, we find that the 10–90% rise time of the OR output for this switching operation is 2.894 ns.

4.3.3. Gummel–Poon Model

The improved BJT model developed by Gummel and Poon (1970) adds several important features, including a description of the high-injection regime. The basic idea of the model is to accurately describe the majority carrier population in the base by accounting for the depletion regions of the base–emitter and base–collector junctions and by including the added density of majority carriers in the rest of the base needed

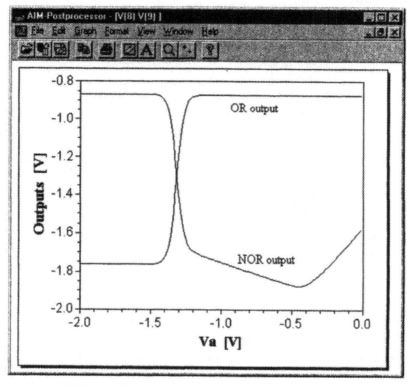

FIG. 4.3.5. Transfer characteristics of the ECL gate shown in Fig. 4.3.3.

for neutralizing the injected minority carriers. The mathematics of the problem is simplified by accounting for high-injection effects only through the most important current component I_{cc}. In this model, Eq. (4.3.13) is replaced by (n–p–n transistor)

$$I_{cc} = \frac{Q_{bo}I_i}{Q_b}\left[\exp\left(\frac{V_{be}}{V_{\text{th}}}\right) - \exp\left(\frac{V_{bc}}{V_{\text{th}}}\right)\right] \qquad (4.3.16)$$

where Q_{bo} and Q_b are the total majority carrier (hole) charges in the base per unit area at zero bias ($V_{be} = V_{bc} = 0$) and with applied bias, respectively. This equation can be derived by assuming a constant majority carrier quasi–Fermi level in the base and neglecting the recombinations in the base (see, e.g., Sze, 1981). If the injection level is low, we have

$$Q_b \approx Q_{bo} \qquad (4.3.17)$$

and the Gummel–Poon model reduces to the Ebers–Moll model.

However, in the general case, the base hole charge Q_b is affected by the changes in the emitter–base and collector–base space charge regions with bias and by the

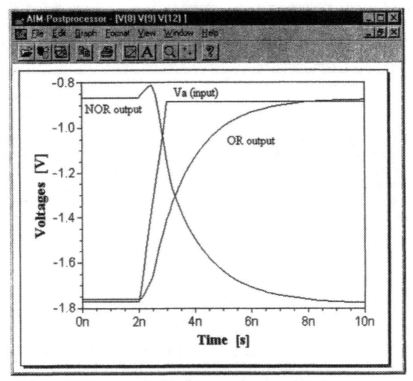

FIG. 4.3.6. Transient response of the ECL gate shown in Fig. 4.3.3.

charge of the majority carriers drawn to the base to neutralize the injected minority carriers:

$$Q_b = Q_{bo} + Q_{be} + Q_{bc} + Q_{\text{dife}} + Q_{\text{difc}} \qquad (4.3.18)$$

Here Q_{dife} and Q_{difc} are the hole charges per unit area in the base associated with the electrons injected across the emitter–base and collector–base junctions, respectively; Q_{be} and Q_{bc} are the contributions to the hole charge of the base associated with changes in the depletion regions of the emitter–base and collector–base junctions, respectively. The latter can be determined using the depletion approximation (see Section 3.2). For practical transistors with nonuniform doping in the base, the following empirical and linearized expression for $Q_{bc} + Q_{be}$ may be used:

$$Q_{be} + Q_{bc} = Q_{bo}\left(\frac{V_{bc}}{V_A} + \frac{V_{be}}{V_B}\right) \qquad (4.3.19)$$

where V_A and V_B are the forward and reverse Early voltages, respectively. These Early voltages correspond approximately to the collector–emitter voltages found by extrapolating the forward and inverse active mode common-emitter characteristics of the BJT to zero collector current.

The neutralizing hole charges associated with minority carrier injection may be expressed as (see, e.g., Sze, 1981)

$$Q_{\text{dife}} + Q_{\text{difc}} = -I_i \frac{Q_{bo}}{SQ_b}\left\{\tau_F\left[\exp\left(\frac{V_{be}}{V_{\text{th}}}\right) - 1\right] + \tau_R\left[\exp\left(\frac{V_{bc}}{V_{\text{th}}}\right) - 1\right]\right\} \quad (4.3.20)$$

where τ_F and τ_R are effective minority carrier lifetimes associated with the forward and reverse currents, respectively. At high injection, these charges cause a strong dependence of transistor parameters, such as the common-emitter current gain β, on the injection level.

Using Eqs. (4.3.19) and (4.3.20) in combination with Eq. (4.3.18), we obtain the following expression for Q_b:

$$\frac{Q_b}{Q_{bo}} = \frac{1}{2}\left(1 + \frac{V_{bc}}{V_A} + \frac{V_{be}}{V_B}\right) + \left\{\frac{1}{4}\left(1 + \frac{V_{bc}}{V_A} + \frac{V_{be}}{V_B}\right)^2 + \frac{|I_i|}{I_{KF}}\left[\exp\left(\frac{V_{be}}{V_{\text{th}}}\right) - 1\right]\right.$$

$$\left. + \frac{|I_i|}{I_{KR}}\left[\exp\left(\frac{V_{bc}}{V_{\text{th}}}\right) - 1\right]\right\}^{1/2} \quad (4.3.21)$$

Here $I_{KF} = SQ_{bo}/\tau_F$ and $I_{KR} = SQ_{bo}/\tau_R$ are called the knee currents for high-level injection. We notice that in the forward active mode of operation, Eq. (4.3.21) in combination with Eqs. (4.3.16) and (4.3.12) describes a collector current of the form I_c' $\sim \exp(V_{be}/2V_{\text{th}})$ at high forward emitter bias, that is, with a reduced rate of increase, similar to what takes place in the p–n diode current at high injection. On the other hand, the high injection will result in a corresponding increase in the majority carrier density in the base because of the neutrality condition. This will promote additional injection from base to emitter, to maintain a steady rise in the base current. Hence, the Gummel–Poon model predicts that the common-emitter current gain β' $\approx I_c'/I_b'$ will drop in the high-injection regime. This is illustrated in the so-called Gummel plot in Fig. 4.3.7 and in the SPICE simulation example in Section 4.3.5.

In practice, we also observe a reduction in β at small injection levels. This is caused by current contributions associated with generation/recombination processes in the junction depletion layers, at the junction interfaces and at the outer surface. Fortunately, these contributions have a similar form, and we can write a combined generation/recombination term for each of the two junctions of the form

$$I_{re} = I_{res}\left[\exp\left(\frac{V_{be}}{m_{re}V_{\text{th}}}\right) - 1\right] \quad (4.3.22)$$

$$I_{rc} = I_{rcs}\left[\exp\left(\frac{V_{bc}}{m_{rc}V_{\text{th}}}\right) - 1\right] \quad (4.3.23)$$

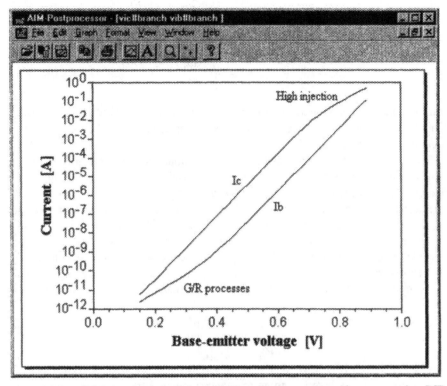

FIG. 4.3.7. AIM-Spice simulation of variation in collector current and base current in a BJT (Gummel plot) using the Gummel–Poon model. The effects of high-injection and generation/recombination mechanisms are indicated.

where I_{res} and I_{rcs} are appropriate saturation currents and m_{re} and m_{rc} are ideality factors (values close to 2) for the generation/recombination currents associated with the emitter and collector junctions, respectively. The generation/recombination currents may be a significant fraction of the base current at low-level injection, as indicated in Fig. 4.3.7.

Note that the recombination in the base itself also gives rise to a contribution to the base current. But since this contribution has the same form as the injection terms discussed previously, it can be absorbed in the intrinsic Ebers–Moll model by a trivial adjustment of the basic Ebers–Moll parameters.

Under dynamic conditions, the various terminal currents of the BJT can be written as

$$I_b = -I_{be} - I_{bc} + I_{re} + I_{rc} + S\frac{dQ_b}{dt} \qquad (4.3.24)$$

$$I_e = -I_{cc} - I_{be} + I_{re} + S \frac{dQ_{\text{dife}}}{dt} + C_{de} \frac{dV_{be}}{dt} \tag{4.3.25}$$

$$I_c = -I_{cc} + I_{bc} - I_{rc} - S \frac{dQ_{\text{difc}}}{dt} - C_{dc} \frac{dV_{bc}}{dt} \tag{4.3.26}$$

Here C_{de} and C_{dc} are depletion capacitances of the emitter–base and collector–base junctions, respectively. The modeling of these capacitances is similar to that of the p–n junction (see Section 3.2).

The above results constitute the basis of the Gummel–Poon model. The important issue in this model is the inclusion of high-injection effects, although the Early effect and generation/recombination effects are also directly included in the model. Other effects, such as Auger recombination and band gap narrowing, may be indirectly accounted for by an appropriate choice of model parameters. In addition, parasitic emitter and collector series resistances and the base spreading resistance have to be included in a comprehensive modeling of BJTs. In order to account for the emitter and collector series resistances R_e and R_c, we can replace the applied voltages V_{eb} and V_{cb} in the model expressions by their "intrinsic" counterparts V'_{eb} and V'_{cb} as follows:

$$V'_{be} = V_{be} - |I_e|R_e \tag{4.3.27}$$

$$V'_{cb} = V_{cb} - |I_c|R_c \tag{4.3.28}$$

The effects of the base spreading resistance is more difficult to include since it leads to a nonuniform distribution of the emitter and collector current densities and to emitter current crowding (see, e.g., Sze, 1985). However, within the framework of the Gummel–Poon model, the degradation of the common-emitter current gain β caused by current crowding at high emitter currents may be reproduced to some extent by choosing an appropriate value of the forward knee current I_{KF}.

4.3.4. Implementation of Gummel–Poon Model in SPICE

The Gummel–Poon model implemented in SPICE can be represented by the equivalent circuit shown in Fig. 4.3.8, where the various circuit elements are defined by the equations discussed above. We note that C_{dife} and C_{difc} are the diffusion capacitances associated with the neutralizing base charges Q_{dife} and Q_{difc}, respectively [see Eqs. (4.3.25) and (4.3.26)], and correspond to the diffusion capacitances of the p–n junction (see Section 3.2). The parameters C_{dife} and C_{difc} are defined in terms of the effective minority carrier lifetimes $\mathbf{TF} = \tau_F$ and $\mathbf{TR} = \tau_R$ (SPICE parameters are shown in boldface). The depletion capacitances C_{de} and C_{dc} are implemented in the same way as for the p–n diode in terms of the SPICE parameters **CJE** and **CJC** (zero-bias capacitance values for the two junctions), **MJE** and **MJC** (grading parameters), **VJE** and **VJC** (built-in voltages), and the forward-bias depletion capacitance coefficient **FC**.

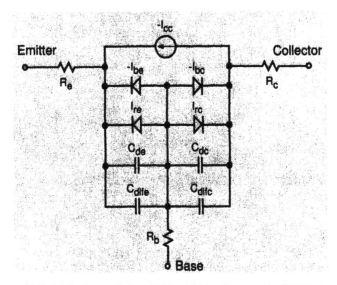

FIG. 4.3.8. Gummel–Poon BJT model implemented in SPICE.

The central idea of the Gummel–Poon model is to correct for the majority carrier charge Q_b in the base at high injection levels. In terms of SPICE parameters, this charge can be written as

$$\frac{Q_b}{Q_{bo}} = \frac{1}{2}\left(1 + \frac{V_{bc}}{\mathbf{VAF}} + \frac{V_{be}}{\mathbf{VAR}}\right) + \left\{\frac{1}{4}\left(1 + \frac{V_{bc}}{\mathbf{VAF}} + \frac{V_{be}}{\mathbf{VAR}}\right)^2 + \frac{\mathbf{IS}}{\mathbf{IKF}}\left[\exp\left(\frac{V_{be}}{\mathbf{NF}V_{th}}\right) - 1\right]\right.$$

$$\left. + \frac{\mathbf{IS}}{\mathbf{IKR}}\left[\exp\left(\frac{V_{bc}}{\mathbf{NR}V_{th}}\right) - 1\right]\right\}^{1/2} \qquad (4.3.29)$$

where **VAF** and **VAR** are the forward and reverse Early voltages, $\mathbf{IS} = |I_i|$ is the intercept current, **NF** and **NR** are the ideality (emission) factors of the forward and reverse diffusion currents, and **IKF** and **IKR** are the forward and reverse knee currents for onset of high-level injection. The charge Q_b is then used to correct the principal term I_{cc} in the collector current while keeping the Ebers–Moll versions of the secondary diffusion terms:

$$-I_{cc} = \frac{Q_{bo}}{Q_b}\mathbf{IS}\left[\exp\left(\frac{V_{be}}{\mathbf{NF}V_{th}}\right) - \exp\left(\frac{V_{bc}}{\mathbf{NR}V_{th}}\right)\right] \qquad (4.3.30)$$

$$-I_{be} = \frac{\mathbf{IS}}{\mathbf{BF}}\left[\exp\left(\frac{V_{be}}{\mathbf{NF}V_{th}}\right) - 1\right] \qquad (4.3.31)$$

$$-I_{bc} = \frac{\text{IS}}{\text{BR}}\left[\exp\left(\frac{V_{bc}}{\text{NR}V_{\text{th}}}\right) - 1\right] \qquad (4.3.32)$$

Here **BF** and **BR** are the forward (normal) and reverse (inverse) common-mode current gains.

Finally, we have the generation/recombination currents associated with the emitter and the collector junctions:

$$I_{re} = \text{ISE}\left[\exp\left(\frac{V_{be}}{\text{NE}V_{\text{th}}}\right) - 1\right] \qquad (4.3.33)$$

$$I_{rc} = \text{ISC}\left[\exp\left(\frac{V_{bc}}{\text{NC}V_{\text{th}}}\right) - 1\right] \qquad (4.3.34)$$

where **ISE** and **ISC** are the corresponding saturation currents and **NE** and **NC** are the ideality (emission) factors.

We note that in addition to the model expressions presented here, the full Gummel–Poon model implemented in SPICE contains several more refinements, such as a noise model and models for the temperature variation of key parameters, for effects related to the distributed nature of the base spreading resistance, and for the Kirk effect (base push-out) associated with velocity saturation in the collector depletion region (see Sze, 1981).

4.3.5. SPICE Example: Common-Emitter Current Gain

It is of interest to compare the dependence of the common-emitter current gain β on the collector current in the Ebers–Moll and the Gummel–Poon models. Figure 4.3.9 shows the AIM-Spice circuit description of the DC Transfer Curve Analysis used for calculating the collector current with the base current ib as the sweep source. AIM-Postprocessor is used to calculate the common-emitter current gain versus collector current. With the model parameters shown, the Ebers–Moll model is se-

```
Circuit for calculating the common-emitter current gain
vb 1 0 dc 0
vib 1 2
vic 4 3
vce 4 0 dc 4
q1 3 2 0 qn
.model qn npn Is=14.34f Xti=3 Eg=1.11 Bf=100 Ne=2
+ Ise=143.4f Xtb=1.5 Br=9.715 Nc=2 Isc=0 Rc=1
+ Itf=.6 Vtf=1.7 Xtf=3
```

FIG. 4.3.9. Circuit description for calculating the BJT common-emitter current gain.

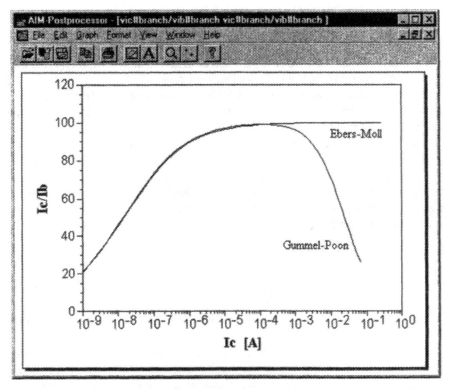

FIG. 4.3.10. AIM-Spice dependencies of the common-emitter current gain on the collector current calculated using the Ebers–Moll and Gummel–Poon models.

lected. To select the Gummel–Poon model, the following model parameters are added: Vaf=74.03 and Ikf=0.2385e-1.

The simulated common-emitter current gain using the Ebers–Moll model and the Gummel–Poon model are shown in Fig. 4.3.10. We observe that the two models give similar results for β at low collector currents. However, at large collector currents, the Gummel–Poon model correctly predicts the decrease in β resulting from high-injection effects.

4.3.6. BJT Small-Signal Modeling

One of the important modes of operation of bipolar transistors is the small-signal regime, where the input ac signal amplitude is relatively small so that the resulting ac currents and voltages vary only in the vicinity of the operating point. Consider, for example, the common-emitter characteristics of Fig. 4.2.6. Here we can identify the operating point as the intersection between the characteristics of an applied dc base current I_b and the load line defined by

$$I_c = \frac{V_{cc} - V_{ce}}{R_c} \qquad (4.3.35)$$

where I_c is the dc collector current, V_{ce} is the dc collector–emitter voltage, V_{cc} is the applied dc collector voltage, and R_c is a load resistance.

The ac transistor response may be described in terms of a linear two-port network. Such a description allows the transistor to be represented by different equivalent circuits, depending on the choice of dependent and independent current and voltage variables. For example, the so-called *h-parameter equivalent circuit* defined by the linear network in Fig. 4.3.11 is frequently used for transistor characterization at low frequencies (below about 100 MHz). However, the different choices of networks are equivalent and their parameters can be expressed through each other (see, e.g., Shur, 1990). Using this representation, the relationship between the dependent and independent variables are given by

$$v_1 = h_{11}i_1 + h_{12}v_2 \qquad (4.3.36)$$

$$i_2 = h_{21}i_1 + h_{22}v_2 \qquad (4.3.37)$$

where the small-signal currents and voltages are defined in Fig. 4.3.11. Similar equations can be written for all other equivalent circuits.

The small-signal parameters may be determined from a set of short-circuit and open-circuit measurements at the input and output ports. For example, the parameter $h_{11} = v_1/i_1|_{v_2=0}$ is called the short-circuit input impedance (h_i), $h_{12} = v_1/v_2|_{i_1=0}$ is called the open-circuit reverse-voltage ratio (h_r), $h_{21} = i_2/i_1|_{v_2=0}$ is called the short-circuit forward-current ratio (h_f), and $h_{22} = i_2/v_2|_{i_1=0}$ is called the open-circuit output admittance (h_o). In the alternate notation shown in parentheses, a second subscript is often used to denote the transistor configuration. Thus, the *h*-parameters for the common-emitter transistor circuit configuration are denoted h_{ie}, h_{re}, h_{fe}, and h_{oe}.

Out of a total of 12 *h*-parameters for 3 transistor configurations, the 4 common-emitter parameters are usually provided by transistor manufacturers in their data sheets. However, we note that the parameters of the small-signal equivalent circuits vary with temperature, measuring frequency, and operating point.

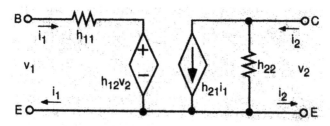

FIG. 4.3.11. Two-port linear network for BJTs in terms of an *h*-parameter equivalent circuit.

Some of the *h*-parameters can be directly related to the parameters of the Ebers–Moll or the Gummel–Poon model. For example, at low frequencies, $h_{fe} = \beta$ and $h_{fb} = -\alpha$. For other *h*-parameters, such relationships are less straightforward. On the other hand, the so-called *hybrid-π equivalent circuit* shown in Fig. 4.3.12 often provides a convenient representation of the common-emitter circuit configuration, since its circuit elements more easily can be linked to the model parameters used in SPICE. Hence, the following relationships are found:

$$g_m = \frac{\partial I_c}{\partial V_{b'e}} \approx \frac{I_c}{V_{th}} \tag{4.3.38}$$

$$r_{b'e} = \frac{\partial V_{b'e}}{\partial I_b} = \frac{\partial V_{b'e}}{\partial I_c/\beta_N} = \frac{\beta_N}{g_m} \tag{4.3.39}$$

$$r_{ce} = \frac{\partial V_{ce}}{\partial I_c} \approx \frac{|V_A|}{I_c} \approx \frac{|V_A|}{g_m V_{th}} \tag{4.3.40}$$

$$r_{b'c} = \frac{\partial V_{cb'}}{\partial I_b} \approx \frac{\partial V_{ce}}{\partial I_c/\beta_N} = \beta_N r_{ce} = \frac{\beta_N |V_A|}{g_m V_{th}} \tag{4.3.41}$$

$$C_{b'e} = C_{de} + C_{dife} = C_{de}(V_{b'e}) + g_{mF}\tau_F \tag{4.3.42}$$

$$C_{b'c} = C_{dc} + C_{difc} = C_{dc}(V_{b'c}) + g_{mR}\tau_R \tag{4.3.43}$$

Note that in Eqs. (4.3.42) and (4.3.43), $g_{mX}\tau_X = (\partial I_c/\partial V'_{bx})\tau_X = \partial Q_{difx}/\partial V'_{bx} = C_{difx}$, where *X* means *F* or *R* when *x* is *e* or *c*.

The simplified hybrid-π equivalent circuit with shorted output shown in Fig. 4.3.13 can be used for calculating the frequency dependence of the short-circuit

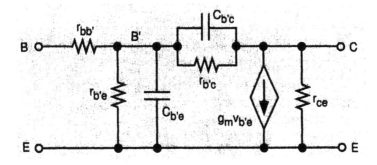

FIG. 4.3.12. Hybrid-π equivalent circuit.

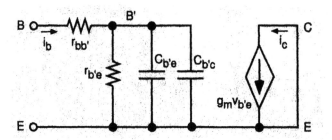

FIG. 4.3.13. Hybrid-π equivalent circuit with shorted output used in the calculation of β_ω.

current gain β_ω. From this circuit, we find that the ac input (base) current can be expressed in terms of the voltage drop $v_{b'e}$ as

$$i_b = v_{b'e}\left[\frac{1}{r_{b'e}} + j\omega(C_{b'e} + C_{b'c})\right] \tag{4.3.44}$$

The short circuit ac collector current $i_c = g_m v_{b'e}$ is determined by the current source. Hence, using Eq. (4.3.39), we find

$$\beta_\omega = \frac{i_c}{i_b} = \frac{g_m}{1/r_{b'e} + j\omega(C_{b'e} + C_{b'c})} = \frac{\beta_N}{1 + j\omega/\omega_\beta} \tag{4.3.45}$$

This defines the so-called *beta cutoff frequency;*

$$\omega_\beta = 2\pi f_\beta = \frac{1}{r_{b'e}(C_{b'e} + C_{b'c})} = \frac{g_m}{\beta_N(C_{b'e} + C_{b'c})} \tag{4.3.46}$$

At $f = f_\beta$, the common-emitter current gain drops to 0.707 of its zero-frequency value.

We also introduce the *cutoff frequency* f_T, defined as the frequency at which the magnitude of the short-circuit common-emitter current gain equals unity, that is, $|\beta_\omega| = 1$. From Eq. (4.3.45), we find

$$f_T = f_\beta\sqrt{\beta_N^2 - 1} \approx f_\beta\beta_N \approx \frac{g_m}{2\pi(C_{b'e} + C_{b'c})} \tag{4.3.47}$$

where we assumed that $\beta_N \gg 1$.

A more accurate expression for f_T may be obtained by including into Eq. (4.3.47) additional time delays associated with the collector depletion layer transit time, the collector charging time, and the parasitic capacitance C_p.

The collector transit time may be estimated as

$$\tau_{cT} \approx \frac{x_{dcb}}{v_s} \tag{4.3.48}$$

where x_{dcb} is the width of the collector–base depletion region and v_s is the electron saturation velocity (for n–p–n transistors). The collector charging time is

$$\tau_c = r_c C_{b'c} \qquad (4.3.49)$$

where r_c is the collector series resistance.

The cutoff frequency may now be rewritten as

$$f_T \approx \frac{1}{2\pi\tau_{\text{eff}}} \qquad (4.3.50)$$

where $\tau_{\text{eff}} = \tau_e + \tau_c + \tau_{cT}$ is the effective delay time and

$$\tau_e = \frac{C_{b'e} + C_{b'c} + C_p}{g_m} \approx \frac{(C_{b'e} + C_{b'c} + C_p)V_{\text{th}}}{I_c} \qquad (4.3.51)$$

In most practical transistors, τ_e is the dominant contribution to the effective delay time. As can be seen from Eq. (4.3.51), this time can be reduced by increasing the collector current. However, at very large collector currents, a displacement of the effective base-collector boundary into the collector region, caused by the so-called Kirk (base push-out) effect, leads to an increase in the effective base width and to an increase in the emitter diffusion capacitance C_{dife}. This may explain the decrease in the cutoff frequency f_T always observed at large collector currents.

An analysis of the above shows that in order to achieve a large cutoff frequency, narrow emitter stripes, large emitter currents, thin base regions, high base doping, and low parasitic capacitances are required. Very high cutoff frequencies (over 200 GHz) have been achieved in heterojunction bipolar transistors where high doping in the base region may be combined with high emitter injection efficiency (see Section 4.4).

Another important characteristic of the BJT is the maximum oscillation frequency f_{max}, defined as the frequency at which the power gain of the transistor is equal to unity under optimum matching conditions of the input and output impedances. Using a simplified π-equivalent circuit where we neglect $r_{b'c}$ and r_{ce}, we can obtain the following expression for f_{max} (see, e.g., Sze, 1981):

$$f_{\text{max}} = \sqrt{\frac{f_T}{8\pi r_{bb} C_{b'c}}} \qquad (4.3.52)$$

4.3.7. SPICE Example: Small-Signal Frequency Response

In this example, we consider a BJT in the common-emitter configuration acting as a current amplifier. We investigate the frequency dependence of the small-signal current gain at different levels of emitter current, assuming a shorted output (see Fig. 4.3.13). In the circuit description of Fig. 4.3.14, the base current is defined in terms

```
Simple common-emitter voltage amplifier (Ib = 1 uA)
vc 1 0 dc 2
vic 1 2 dc 0
ib 0 3 dc 0.001m ac 1
q1 2 3 0 npnmod
.model npnmod npn Is=14.34f Xti=3 Eg=1.11 Vaf=74.03
+ Bf=65.62 Ne=1.208 Ise=19.48f Ikf=.2385 Xtb=1.5
+ Br=9.715 Nc=2 Isc=0 Ikr=0 Rc=0 Cjc=9.393p Mjc=.3416
+ Vjc=.75 Fc=.5 Cje=22.01p Mje=.377 Vje=.75 Tr=58.98n
+ Tf=408.8p Itf=.6 Vtf=1.7 Xtf=3 Rb=10
```

FIG. 4.3.14. Circuit description of a simple BJT common-emitter amplifier with shorted output.

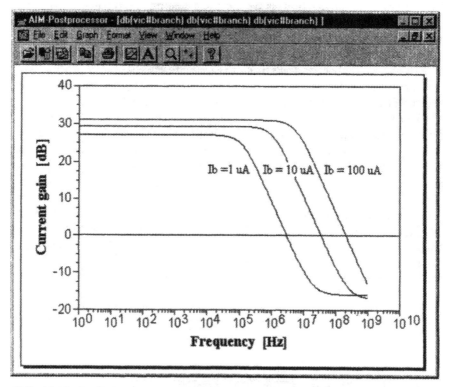

FIG. 4.3.15. Small-signal current gain versus frequency at different levels of dc emitter (base) current for a simple common-emitter amplifier with shorted output.

of a current source ib at the input, with a dc and a small-signal component. The dc emitter current is varied by adjusting the dc base current. In the collector loop, we have a constant dc voltage source vc. The statement in line 3 allows us to monitor the small-signal response at the output. The choice of parameters in the .model statement corresponds to the use of a full Gummel–Poon model.

The simulated small-signal gain versus frequency is shown in Fig. 4.3.15. In agreement with our earlier analysis, we observe that both the gain and the cutoff frequency (unity-gain frequency) increases with increasing dc base current.

4.4. HETEROJUNCTION BIPOLAR TRANSISTOR

4.4.1. Principle of Operation

The idea of the heterojunction bipolar transistor (HBT) was proposed by Shockley (1951) and was later developed by Kroemer (1982). This idea can be explained as follows: In a modern conventional BJT, the common-emitter current gain β is limited by the emitter injection efficiency. For the BJT, we have

$$\beta = \frac{I_c}{I_b} < \beta_{max} = \frac{v_{nb} N_{de}}{v_{pe} N_{ab}} \exp\left(-\frac{\Delta E_g}{k_B T}\right) \tag{4.4.1}$$

where I_c and I_b are the collector and base currents, respectively, β_{max} is the maximum value of the common-emitter current gain, N_{de} and N_{ab} are the donor concentration in the emitter region and acceptor concentration in the base region, respectively, and v_{nb} and v_{pe} are effective velocities of electrons in the base and holes in the emitter, respectively, which include contributions from both carrier diffusion and drift. The maximum gain is limited by the hole component of the emitter current, caused by injection of holes from the base into the emitter region. The exponential factor in Eq. (4.4.1) results from band gap narrowing in the highly doped emitter region and contributes to a reduction of the gain. One possible mechanism of the band gap narrowing is the formation of an impurity band when the wave functions of the donor electrons overlap.

In an HBT, the emitter region has a wider band gap than the base. Many HBTs are made using AlGaAs/GaAs heterostructures since the lattice constants of these two materials are very close. Consequently, the heterointerface quality is excellent with very few dislocations and a low density of interface states. The band diagram of an n–p–n HBT biased in the active mode is shown in Fig. 4.4.1. With an abrupt heterojunction, we typically obtain a "spike-and-notch" region in the conduction band at the interface. This may be used as a "launching pad" for injection of hot electrons into the base, allowing the electrons to traverse the base near ballistically at high speed. However, by grading the material composition of the heterointerface, the spike in the conduction band can be removed, resulting in a lowering of the barrier for electron injection into the base. With a total band discontinuity ΔE_g divided

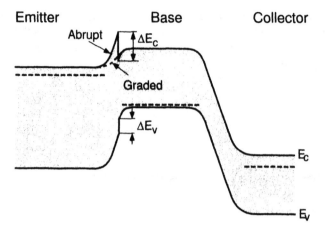

FIG. 4.4.1. Schematic band structure; of an n–p–n HBT in the active mode for an abrupt (solid line) and a graded (dashed line) heterojunction. The quasi–Fermi levels in the emitter, base, and collector are indicated as horizontal dashed lines

between ΔE_c in the conduction band and ΔE_v in the valence band, the expression for β_{max} becomes for the graded junction:

$$\beta_{max} \approx \frac{v_{nb}N_{de}}{v_{pe}N_{ab}} \exp\left(\frac{\Delta E_g}{k_BT}\right) \qquad (4.4.2)$$

Hence, a considerable value of β_{max} may be achieved even when N_{de} is much smaller than N_{ab}.

This discussion shows that the emitter injection efficiency of an HBT may be made quite large. In practice, however, the HBT gain is limited less by the emitter efficiency than by the recombination current. The recombinations primarily take place at the emitter–base interface, in the base region, and at the emitter surface periphery (surface recombination). The latter is especially important in devices with narrow emitter stripes (used to avoid the phenomenon of emitter crowding; see Sze, 1985). In this case, the common-emitter transistor gain can be written as

$$\beta = \frac{I_c}{I_b} \approx \frac{J_{ne}}{J_{re} + J_{rb} + j_{rs}/Z} \qquad (4.4.3)$$

Here J_{ne} is the electron component of the emitter current density, and J_{re}, J_{rb}, and j_{rs}/Z are the current densities associated with recombinations at the emitter–base interface, in the base region, and at the emitter edges, respectively. The parameter Z is the emitter stripe width (narrow dimension).

A large recombination current may severely limit the maximum current gain of an HBT. However, as Kroemer (1982) pointed out, an even greater advantage of the

HBT compared to the homojunction BJT is the smaller base spreading resistance. Indeed, we may not need a very high current gain for many applications, but the wide-band-gap emitter allows us to dope the base region very strongly, drastically reducing the base resistance and the base-width modulation.

In devices where the overall recombination current is small, the current gain could be as high as a few thousand or more. The electron component of the emitter current density can be estimated as follows:

$$J_{ne} \approx q n_p(0) v_{nb} \qquad (4.4.4)$$

and the recombination current density associated with the base region can be estimated as

$$J_{rb} \approx \frac{c q n_p(0) W}{\tau} \qquad (4.4.5)$$

where c is a numerical constant of the order of unity, $n_p(0)$ is the concentration of minority carriers (electrons) at the emitter side of the base, W is the base width, and τ is the minority carrier lifetime. Thus, if the other recombination mechanisms can be neglected (which may be possible at relatively high forward bias and with relatively wide emitter stripes), we obtain

$$\beta \approx \frac{J_{ne}}{J_{rb}} \approx \frac{\tau}{t_B} \qquad (4.4.6)$$

where $t_B = W/v_{nb}$ is the electron transit time across the base. For a sufficiently short base (say, $W \sim 0.1~\mu m$), $\beta > 1000$ can be obtained even if the lifetime is of the order of a nanosecond. Common-emitter gains in access of 1600 have been achieved for AlGaAs/GaAs HBTs. For comparison, the band gap narrowing of the strongly doped emitter in conventional homojunction silicon transistors leads to an exponential reduction of the maximum common-emitter current gain [see Eq. (4.4.1)], by up to a factor of 20 for $N_{de} = 10^{19}$ cm^{-3}. This energy gap shrinkage represents one of the dominant performance limitations for conventional Si BJTs (see, e.g., Sze, 1981).

In addition to high injection efficiencies and correspondingly high common-emitter current gains, HBTs have a number of other advantages over conventional bipolar transistors. By allowing a higher base doping, the base spreading resistance or, alternatively, the base width could be reduced. Because of a relatively low doping of the emitter region, the emitter–base capacitance could be made quite small. All these factors result in a higher speed of operation.

The simplest model for the descriptions of HBTs is the Ebers–Moll model, which is identical to one of the models used for conventional BJTs. There are, however, many important effects that are specific for HBTs, such as surface recombination near the emitter edges that greatly reduces the gain in devices with narrow emitter stripes. Hot electron, overshoot, and ballistic effects play an important role both in the base and in the collector region.

A more accurate analysis of the HBT operation shows that the propagation delay of an HBT logic gate is often limited by the time constant given by the product of the collector–base capacitance and the load resistance. For example, in the schematic structure of an AlGaAs/GaAs HBT shown in Fig. 4.4.2, a proton-implanted area is used to reduce the collector capacitance.

As mentioned, the lateral dimensions of the emitter stripes are also very important in practical transistors. Parasitic capacitances and resistances and power dissipation are all dependent on these dimensions. Using emitter stripe widths of only 1.2 μm in the HBT design of Fig. 4.4.2, Chang et al. (1987) demonstrated HBTs with a cutoff frequency f_T = 67 GHz and a maximum oscillation frequency f_{max} = 105 GHz. Another important feature of these devices was the very high base-region doping (up to 10^{20} cm^{-3}), which gave a small base spreading resistance.

The first HBTs were implemented using the AlGaAs/GaAs material system. However, the use of In$_x$Ga$_{1-x}$As as base material has several advantages over GaAs. The low-field mobility of InGaAs is higher than that of GaAs, and the energy separation between the central Γ minimum and the satellite L minima is higher than in GaAs. This allows electrons to achieve higher velocities without being transferred to heavy-mass satellite valleys, making ballistic and overshoot transport in the base much more efficient. Also, the composition of InGaAs in AlGaAs/InGaAs/GaAs HBTs can be graded, with the percentage of In increasing from the emitter toward the base. This creates an additional built-in field that speeds up the electron transport across the base.

4.4.2. Thermionic Emission–Diffusion Model for HBT

Considering a single heterojunction n–p–n HBT without interface grading and using the expressions derived in Section 3.3 for the transport across heterojunctions,

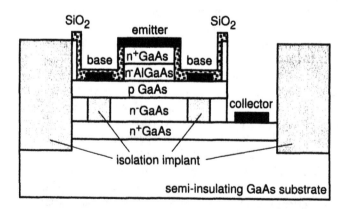

FIG. 4.4.2. Schematic structure of a high-speed AlGaAs/GaAs HBT. (After Chang et al., 1987. Copyright 1987 by IEEE.)

the following set of Ebers–Moll equations can be derived for the emitter and collector currents of the intrinsic transistor (see Lee et al., 1993):

$$I_e' = A_{11}\left[\exp\left(\frac{V_{be}}{V_{th}}\right) - 1\right] + A_{12}\left[\exp\left(\frac{V_{bc}}{V_{th}}\right) - 1\right] \tag{4.4.7}$$

$$I_c' = A_{21}\left[\exp\left(\frac{V_{be}}{V_{th}}\right) - 1\right] + A_{22}\left[\exp\left(\frac{V_{bc}}{V_{th}}\right) - 1\right] \tag{4.4.8}$$

For a typical device, the diffusion lengths of electrons in the base (L_{nb}) and of holes in the emitter (L_{pe}) and the collector (L_{pc}) can be taken to be much larger than the respective widths of the base (W_B), the emitter (X_E), and the collector (X_C) regions. Also, we assume that the additional barrier created by the spike, $\Delta E_n \equiv \Delta E_c - qV_{db}$, is much larger than qV_{th}, where ΔE_c is the band offset in the conduction band and qV_{db} is the band bending of the base at the base–emitter interface corresponding to the "notch" in Fig. 4.4.1. (When $\Delta E_n < 0$, the conduction band discontinuity does not create an additional barrier because the conduction band spike at the heterointerface is below the bottom of the conduction band in the neutral base.) These assumptions lead to the following simplified coefficients of the above equations [see Lee et al. (1993), where also the more general case is treated]:

$$A_{12} = A_{21} \approx Sqn_{bo}v_n \exp\left(-\frac{\Delta E_n}{qV_{th}}\right) \tag{4.4.9}$$

$$A_{11} \approx -Sqn_{bo}v_n \exp\left(-\frac{\Delta E_n}{qV_{th}}\right) - \frac{SqD_{pe}p_{eo}}{X_e} \tag{4.4.10}$$

$$A_{22} \approx -\frac{Sqn_{bo}D_{nb}W_B}{L_{nb}^2} - \frac{SqD_{pc}p_{co}}{X_C} \tag{4.4.11}$$

Here S is the emitter area, v_n is the mean thermal electron velocity in the direction from the emitter to the base, and n_{bo}, p_{eo}, and p_{co} are the equilibrium electron density in the base and the hole densities in the emitter and collector, respectively.

We note that the term $\exp(-\Delta E_n/qV_{th})$ in A_{11}, A_{12}, and A_{21} will result in a voltage offset in the I–V characteristics for both the common-base and the common-emitter configurations.

By rearranging Eqs. (4.4.10) and (4.4.11), the intrinsic emitter and collector currents can also be expressed as

$$I_e' = -\alpha_I I_c' - I_{eo}\left[\exp\left(\frac{V_{be}}{V_{th}}\right) - 1\right] \tag{4.4.12}$$

$$I_c' = -\alpha_N I_e' - I_{co}\left[\exp\left(\frac{V_{bc}}{V_{th}}\right) - 1\right] \tag{4.4.13}$$

where $I_{eo} = A_{12}A_{21}/A_{22} - A_{11}$ is the emitter reverse saturation current, $I_{co} = A_{12}A_{21}/A_{11} - A_{22}$ is the common-base collector reverse saturation current, $\alpha_N = -A_{21}/A_{11}$ is the normal (or forward) common-base current gain, and $\alpha_I = -A_{12}/A_{22}$ is the inverse common-base current gain (see Section 4.3). We may also define the normal and inverse common-emitter current gains as $\beta_N = \alpha_N/(1-\alpha_N)$ and $\beta_I = \alpha_I/(1-\alpha_I)$, respectively.

4.4.3. Nonideal Effects in the HBT Model

So far, we have considered a thermionic-emission model for an ideal HBT structure. However, in a realistic device model, several additional effects have to be accounted for, including the various recombination mechanisms discussed earlier (see Section 4.4.1), tunneling near the top of the barrier, grading of the heterojunction, emitter series resistance, and base push-out (Kirk effect).

As mentioned, the recombination current consists of contributions from recombination of minority carriers in the neutral base, surface recombination at the emitter periphery, and recombination at the emitter–base junction. The corresponding contributions to the emitter current, I_{rb}, I_{rs}, and I_{re}, are indicated in the equivalent circuit in Fig. 4.4.3. This equivalent circuit also includes a current source I_{rc} at the collector–base interface representing the generation current of a reverse-biased junction (forward active mode).

The various components of the recombination current can be approximated by the following diode-type expression (see Lee et al., 1993):

$$I_{rx} = I_{rx}^o \left[\exp\left(\frac{V_{be}}{m_{rx}V_{th}} \right) - 1 \right] \tag{4.4.14}$$

where m_{rx} is an ideality factor for mechanism x. Typically, $m_{rb} = 1$, $m_{rs} = 1, \ldots, 2$, and $m_{re} \approx 2$. Note that the "saturation current" I_{rb}^o is proportional to the volume of the neutral base, I_{re}^o is proportional to the area of the base–emitter interface, and I_{rs}^o

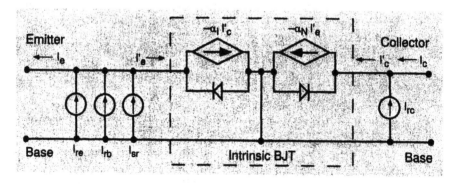

FIG. 4.4.3. Equivalent circuit of HBT accounting for recombination and generation currents and for emitter series resistance. (After Lee et al., 1993.)

is proportional to the emitter periphery. The generation/recombination current I_{rc} at the collector interface can be represented by an expression similar to that for I_{re}. Tunneling of electrons through the conduction band spike at the emitter–base heterojunction can be estimated by defining a barrier transparency (see Lee et al., 1993). However, the effect of the tunneling may be incorporated in the thermionic-emission model by using an effective barrier height or by adjusting the velocity parameter v_n in the equations for the emitter and collector currents (see Section 4.4.2).

A qualitative analysis of the effects of grading the AlGaAs composition at the heterointerface was given by Kroemer (1982) and by Grinberg et al. (1984). The effect of grading is to make the barrier spike in the conduction band smaller. Often, for practical values of grading length, the spike is abrupt enough to justify the use of the thermionic-field emission model. Hence, the only necessary change in the theory given above is to substitute the additional barrier height ΔE_n by its value for the graded junction.

Assuming that the composition x of $Al_xGa_{1-x}As$ in the emitter region varies linearly over the grading length W_g, and considering only the case of uniform doping profiles in the emitter and base regions, the following expression can be derived for the barrier height seen from the emitter side of the interface:

$$E_{bg} = \frac{q^2 N_{de}(x_e - W_g)^2}{2\epsilon_e} \qquad (4.4.15)$$

Here, x_e is the depletion width, ϵ_e is the dielectric permittivity, and N_{de} is the donor density on the emitter side. The additional barrier height with grading then becomes

$$\Delta E_{ng} = \Delta E_n - E_b + E_{bg} \qquad (4.4.16)$$

where E_b is the barrier height seen from the emitter side with a nongraded interface. These equations show how the grading causes a reduction of the additional barrier for the electrons entering the base from the emitter region. When the grading length approaches the depletion length x_e, E_{bg} vanishes and the conduction band of the graded junction can be approximated as indicated in Fig. 4.4.1.

Note that the quality of the heterointerface may depend on the grading. In fact, grading may lead to an increase in the recombination current and actually cause a sharp decrease in the gain.

At high current levels, we have to consider the effects of voltage drops across the emitter and the base resistances. These give rise to a reduction of the intrinsic base-emitter voltage V_{be} compared to the externally applied voltage V_{be}^{ext}, given by

$$V_{be} = V_{be}^{ext} - r_e I_e - r_b I_b \qquad (4.4.17)$$

Here r_e and r_b are the emitter and base resistances, including contact resistances, and I_e and I_b are emitter and base terminal currents, respectively. In HBTs, the base is highly doped and, hence, r_b is usually less than or comparable to r_e. Furthermore, $I_b \approx I_e/\beta \ll I_e$ so that the voltage drop $r_b I_b$ will normally be small compared to $r_e I_e$.

(This voltage drop may still be important because it may lead to emitter current crowding.)

Another important effect at high current levels is the so-called base push-out, or Kirk, effect. This occurs when the collector current density J_c exceeds a "critical" value J_{crit} that is slightly in excess of $J_{sat} = qN_{dc}v_s$, where J_{sat} corresponds to electron velocity saturation in the collector region, v_s is the electron saturation velocity, and N_{dc} is the donor density in the low-doped collector. An attempt to increase the collector current beyond J_{crit} is counteracted by the extension of an electron–hole plasma into the collector region, increasing the effective width of the base region. According to Schrenk (1978), this base-width enlargement can be written approximately as

$$\Delta W_B = X_c\left(1 - \frac{J_{crit} - qv_sN_{dc}}{J_c - qv_sN_{dc}}\right) \tag{4.4.18}$$

where X_c is the width of the low-doped collector region.

4.4.4. Implementation of HBT Model in AIM-Spice

In Section 4.4.2, we discussed an idealized semianalytical HBT model. However, practical devices often do not exhibit such an ideal behavior. Several possible nonideal mechanisms associated with the heterointerface were discussed in conjunction with the heterojunction diode in Section 3.3. Above, we also noted that the effect of grading, for example, may be exactly the opposite of what is expected from the ideal theory. Therefore, for the purpose of HBT circuit simulation, it is more practical to rely on a simplified device model with parameters extracted from experimental data.

We can summarize the results of the HBT modeling discussed above in terms of a system of equations similar to those of the Ebers–Moll model [see Eqs. (4.3.11)–(4.3.15)]. The recombination terms are also very similar to those of the BJT model [see Eqs. (4.3.22) and (4.3.23)]. Hence, by a judicious choice of parameter values, the BJT model discussed earlier can also be used for HBTs. Nonetheless, a separate HBT model has been established in AIM-Spice as a modification of the BJT model. In this model, we include effects that are important in HBTs, such as base push-out, and recombination current associated with the emitter periphery.

At very high collector currents, the common-emitter gain in HBTs should reach a plateau if the critical current I_{crit} for base push-out is large enough. However, in practical devices, where the collector is doped relatively weakly in order to obtain a higher breakdown voltage, I_{crit} is not very large, and β may start dropping at large collector currents owing to the increased effective base width. For the purpose of circuit modeling, we may account for the base push-out by introducing a saturation in the intrinsic collector current of the following form:

$$I_c' = \frac{I_{co}'}{[1 + (I_{co}'/\text{ICSAT})^M]^{1/M}} \tag{4.4.19}$$

where I'_{co} is the intrinsic collector current in the absence of base push-out, **ICSAT** = I_{crit}, and **M** is a shape parameter.

The additional recombination current associated with the emitter periphery is modeled as

$$I_{rs} = \mathbf{IRS1}\left[\exp\left(\frac{V_{be}}{V_{th}}\right) - 1\right] + \mathbf{IRS2}\left[\exp\left(\frac{V_{be}}{2V_{th}}\right) - 1\right] \qquad (4.4.20)$$

where the "saturation" currents **IRS1** and **IRS2** are proportional to the length of the emitter periphery.

Figure 4.4.4 shows the dependence of the common-emitter current gain β of an AlGaAs/GaAs HBT on the collector current calculated using this AIM-Spice model.

In summary, the main differences between HBT modeling and conventional BJT modeling are the following:

1. The important recombination mechanisms are different in the two cases. In particular, the surface recombination is important, or even dominant, in HBTs.

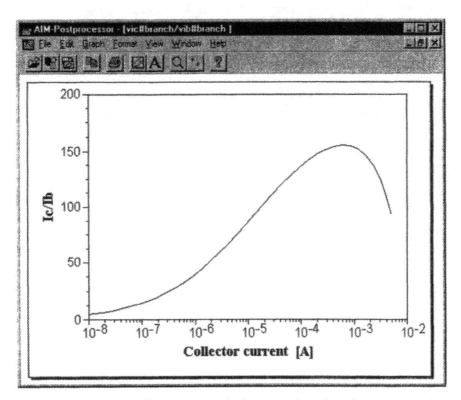

FIG. 4.4.4. Dependence of the common-emitter current gain on the collector current calculated for an AlGaAs/GaAs HBT using the AIM-Spice HBT model.

2. The preexponential current densities A_{11} and A_{22} of the emitter–base and the collector–base characteristics, respectively, are much more different for a single HBT than for a conventional BJT. When $\Delta E_n \gg V_{th}$, we have approximately $A_{11}/A_{22} \propto \exp(-\Delta E_n/qV_{th})$, leading to an offset voltage $\Delta E_n/q$ in the output current–voltage characteristics.

3. High-injection effects in HBTs are not as important as in conventional BJTs owing to the very high doping level in the HBT base region. This is the reason why an Ebers–Moll model is applicable for HBTs, while a Gummel–Poon model is needed for reasonable accuracy in BJTs.

4. The effects of base push-out may be very important in HBTs and may strongly degrade the overall common-emitter gain at high collector currents.

REFERENCES

J. Bardeen and W. H. Brattain, "The Transistor, a Semiconductor Triode," *Physical Review,* vol. 74, p. 230 (1948).

M. F. Chang, P. M. Asbeck, K. C. Wang, G. J. Sullivan, N. H. Sheng, J. A. Higgins, and D. L. Miller, "AlGaAs/GaAs Heterojunction Bipolar Transistors Fabricated Using a Self-Aligned Dual-Lift-Off Process," *IEEE Electron Device Letters,* vol. EDL-8, pp. 303–305 (1987).

J. J. Ebers and J. L. Moll, "Large-Signal Behavior of Junction Transistors," *Proceedings of the IRE,* vol. 42, p. 1761 (1954).

A. A. Grinberg, M. Shur, R. J. Fisher, and H. Morkoç, "Investigation of the Effect of Graded Layers and Tunneling on the Performance of AlGaAs/GaAs Heterojunction Bipolar Transistors," *IEEE Transactions on Electron Devices,* ED-31, no. 12, pp. 1758–1765 (1984).

H. K. Gummel and H. C. Poon, "An Integral Charge Control Model of Bipolar Transistors," *Bell System Technical Journal,* vol. 49, p. 827 (1970).

H. Kroemer, "Heterostructure Bipolar Transistors and Integrated Circuits," *Proceedings of the IEEE,* vol. 70, pp. 13–25 (1982).

K. Lee, M. Shur, T. A. Fjeldly, and T. Ytterdal, *Semiconductor Device Modeling for VLSI,* Series in Electronics and VLSI, Prentice-Hall, Englewood Cliffs NJ (1993).

H. Schrenk, *Bipolare Transistoren,* Springer-Verlag, Berlin (1978).

W. Shockley, U.S. Patent No. 2,569,347 (issued 1951).

M. Shur, *Physics of Semiconductor Devices,* Prentice-Hall, Englewood Cliffs NJ (1990).

S. M. Sze, *Physics of Semiconductor Devices,* 2nd ed., Wiley, New York (1981).

S. M. Szc, *Semiconductor Devices. Physics and Technology,* Wiley, New York (1985).

PROBLEMS

4.2.1. (a) Assume an energy scale of 1 eV/cm in Fig. 4.2.1 and determine the values of the emitter–base and collector–base biases. Simulate BJT *I–V* characteristics using the default parameters in SPICE, find the point corresponding to these bias values, comment on their magnitudes, and suggest more realistic bias values for the active forward mode of operation.

(b) What does the position of the Fermi levels in the three regions suggest regarding the relative magnitudes of the doping levels. Explain why the doping levels are chosen this way.

4.2.2. Simulate the temperature dependence of the collector current and the transconductance of an n–p–n silicon transistor in the common-emitter configuration for the temperature range 0–100°C. Use default SPICE parameters.

4.2.3. Simulate the common-emitter current–voltage characteristics of an n–p–n BJT transistor with the following SPICE model parameters: forward Early voltage vaf = 80 V, corner for forward and reverse β high current roll-off if = 0.24 A and ikr = 0 A. With this choice of parameters the Gummel–Poon model is automatically selected. Perform the simulation for base currents of 5, 10, 15, and 20 μA and the emitter–collector voltage varying from –0.5 to 5 V. Discuss the transistor characteristics.

4.3.1. A typical circuit for measuring transistor switching times is shown in Fig. P4.3.1 along with the input voltage waveform. The common-emitter gain β = 100. Simulate the collector current waveform for V_1 = 5 V and V_1 = 10 V with the following BJT parameter listing:

```
.model qn npn is=14.34f xti=3 eg=1.11 vaf=74.03 bf=100 ne=2
+ ise=143.4f ikf=.02385e xtb=1.5 br=9.715 nc=2 isc=0 ikr=0
+ rc=1 itf=.6 vtf=1.7 xtf=3 cjc=9.393p mjc=.3416 vjc=.75
+ fc=.5 cje=22.01p mje=.377 vje=.7 tr=10n tf=408.8p itf=.6
+ vtf=1.7 xtf=3 rb=10
```

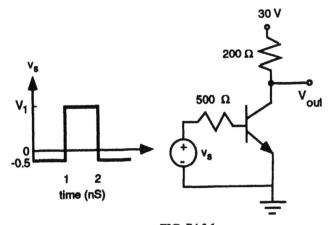

FIG. P4.3.1

4.3.2. (a) Using default SPICE parameters of a Si n–p–n BJT, calculate and plot the transfer characteristic of the inverter shown in Fig. P4.3.2.

(b) Repeat the calculation using default parameters of a Si p–n–p BJT.

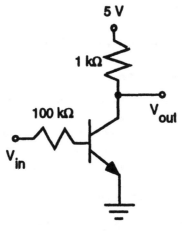

FIG. P4.3.2

4.3.3. (a) Review the example in Section 4.3.2.

(b) Connect two NOR gates in series and set input V_a to logic zero level at each gate. Calculate the propagation delay t_d from the input port to the output port of the first NOR gate by applying a suitable voltage waveform at the input V_b of the first gate. The propagation delay is defined as shown in the Fig. P4.3.3.

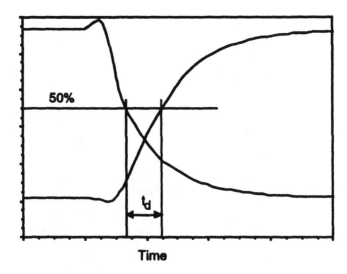

FIG. P4.3.3

(c) Apply a 50% duty cycle pulse train with a period much larger than the propagation delay found in (b). Then reduce the period of the pulse train until the output of the circuit is not able to follow the input. Discuss the result.

4.3.4. Use SPICE to simulate the frequency response of the BJT emitter current to an ac voltage source connected between emitter and base. Compare the results with the following expression for the ac emitter current i_e:

$$i_e = \frac{\beta_N v_{b'e}(1 + j\omega C_{b'e} r_{b'e}/\beta_N)}{r_{b'e}} = i_{eo}(1 + j\omega C_{b'e} r_{b'e}/\beta_N)$$

Here C_e and r_e are small-signal emitter resistance and capacitance.

4.3.5. A silicon BJT operates at room temperature (300 K) with a collector current of 20 mA. The short-circuit common-emitter current gain $\beta = 50$. Estimate the parameters of the simplified hybrid-π transistor equivalent circuit shown in Fig. P4.3.5 and the h-parameters h_{ie} and h_{fe}.

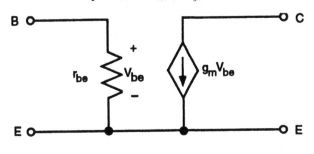

FIG. P4.3.5

4.3.6. Simulate the ac response of BJT amplifiers for the three basic biasing configurations (common emitter, common base, and common collector). Use SPICE default parameters. Then simulate two BJT stages connected in series with the first stage in the common-base configuration and the second in the common-collector configuration. From the simulation results, complete the table below (A_i is the current amplification, A_v is the voltage amplification, and A_p is the power amplification):

	Common Base	Common Collector	Common Emitter	Two-Stage Amplifier
A_i				
A_v				
A_p				

4.3.7. Which parameters in the SPICE circuit description of Example 3.4.5 should you change, and how, in order to obtain a 3-dB decrease in the current gain at the Ebers–Moll collector current of 100 mA? Check your answer by doing the SPICE simulation.

4.4.1. Use the AIM-Spice HBT mode (see Model Parameter Specifications N in Appendix A2) and calculate the output HBT I–V characteristics for the common-emitter configuration for values of the base current from 1 to 11 μA in steps of 2 μA. Use default parameter values.

4.4.2. Repeat the calculation of the small-signal frequency response of the common-emitter configuration in Example 4.3.7, where the BJT is replaced by an HBT. Use default parameters for the HBT. Compare the results with those presented in Fig. 4.3.15 and comment on the differences.

4.4.3. (a) Calculate the temperature dependence of the maximum common-emitter current gain β for a Si n–p–n BJT in the temperature interval from -75 to $125°$C. Use default SPICE parameters.
(b) Repeat the calculation for an HBT, also using default AIM-Spice parameters, and compare the results.

4.4.4. (a) Calculate and plot the transfer characteristic of the Si BJT current mode logic (CML) inverter shown in Fig. P4.4.4. Use default SPICE parameters.
(b) Repeat the calculation using default HBTs.

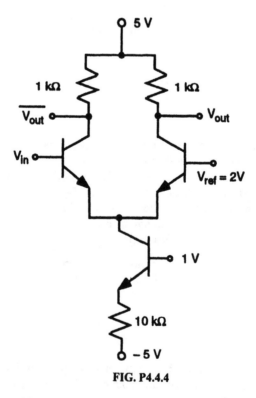

FIG. P4.4.4

4.4.5. Calculate and plot the response of the CML inverter described in Problem 4.4.4 to the voltage pulse shown in Fig. P4.4.5 for the following two cases:

(a) using conventional Si n–p–n BJTs and (b) using AlGaAs/GaAs HBTs. Compare the characteristic delay and switching times for these two cases.

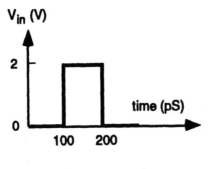

FIG. P4.4.5

CHAPTER 5

FIELD-EFFECT TRANSISTORS

5.1. INTRODUCTION

A field-effect transistor (FET) operates as a conducting semiconductor channel with two ohmic contacts—the *source* and the *drain*—where the number of charge carriers in the channel is controlled by a third contact—the *gate*. Ideally, the gate contact is isolated from the conducting channel and controls the channel charge by capacitive coupling (field effect). In practice, the gate–channel structure is usually one of the two-terminal devices discussed in Chapter 3, that is, either an MOS structure or a reverse-biased rectifying junction, such as a *p–n* junction, a Schottky barrier, or a heterojunction. The corresponding FETs are called MOSFET (metal–oxide–semiconductor FET), JFET (junction FET), MESFET (metal–semiconductor FET), and HFET (heterostructure FET), respectively. Note that in all cases the stationary gate–channel impedance is very large at normal operating conditions, a major advantage compared to BJTs. The basic FET structure is shown schematically in Fig. 5.1.1.

The most important FET is the MOSFET. In a silicon MOSFET, the gate contact is separated from the channel by a silicon dioxide (SiO_2) layer. The charge carriers of the conducting channel constitute an inversion charge, that is, electrons in the case of a *p*-type substrate (*n*-channel device) or holes in the case of an *n*-type substrate (*p*-channel device), induced in the semiconductor at the silicon–insulator interface by the voltage applied to the gate electrode (see Section 3.5 for details). The electrons enter the channel from n^+ source and drain contacts in the case of an *n*-channel MOSFET and from p^+ contacts in the case of a *p*-channel MOSFET.

Silicon MOSFETs are used both as discrete devices and as active elements in monolithic integrated circuits (ICs). In recent years, the device feature size in such

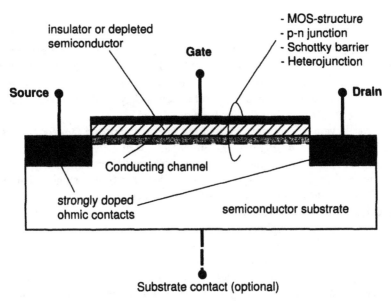

FIG. 5.1.1. Schematic illustration of a generic FET. This device can be viewed as a combination of two orthogonal two-terminal devices. The lateral device is a semiconductor resistive channel with ohmic contacts. The vertical device is either an MOS structure or a reverse-biased rectifying device that controls the mobile charge in the channel by capacitive coupling (field effect).

circuits has been scaled down into the deep submicrometer range. We can expect quarter-micrometer gate lengths to become typical for VLSIs only a few years from now, and by the turn of the century, 0.1 μm will be used, with a commensurate increase in speed and in integration scale (many millions of transistors on a single chip are already used in memory ICs today). Complementary MOSFET (CMOS) technology that combines both *n*-channel and *p*-channel MOSFETs provides very low power consumption along with high speed. New silicon-on-insulator (SOI) technology may help to achieve three-dimensional integration, that is, packing of devices into many layers, with a dramatic increase in integration density. New improved device structures and the combination of bipolar and field-effect technologies (BiCMOS) may lead to further advances yet unforeseen.

Silicon technology is the most developed and advanced semiconductor technology due to the high purity that can be achieved for this material (since it is an elemental semiconductor) and to the excellent properties of its "native" oxide— SiO_2—which can easily be grown thermally on the surface of silicon wafers. However, many other FETs have emerged and are challenging the supremacy of silicon MOSFETs, primarily at important technological fringes such as in ultra-high-speed circuits, in circuits operating at high temperatures and in harsh environments, in radiation hard circuits, and in very large scale thin film integrated circuits (used primarily in liquid crystal displays). These devices utilize compound semiconductor materials, semiconductor heterostructures, and amorphous or polycrystalline silicon thin films.

Compound semiconductor FETs occupy an important niche in the electronics industry. Schottky barrier MESFETs and HFETs based on GaAs are the most commonly used devices of this type. MESFET amplifiers, oscillators, mixers, switches, attenuators, modulators, and current limiters are widely used, and high-speed ICs based on MESFETs and HFETs have been developed.

There are several advantages of GaAs and related compound semiconductors for applications in submicrometer devices. First of all, the effective mass of electrons in GaAs is smaller than in Si. This leads to a much higher electron mobility in GaAs. Moreover, in high electric fields, the electron velocity in GaAs is higher than in Si, which is even more important in submicrometer devices where such high electric fields exist. When the electrons in GaAs move through the short active region of a high-speed device, they experience so-called "overshoot" effect, which is related to the finite time it takes for an electron to relax its energy. This may cause the electron velocity to be boosted considerably beyond its normal stationary saturation value. Another important advantage of materials such as GaAs and InP, for applications in high-speed submicrometer devices, is the availability of semi-insulating substrates that eliminate parasitic capacitances related to junction isolation (present in silicon circuits) and allow the fabrication of microstrip lines with small losses. The latter is especially important for applications in microwave monolithic integrated circuits (MMICs).

We should also mention that GaAs is a direct-gap material that is widely used in optoelectronic applications. The direct gap makes possible a monolithic integration of ultra-high-speed submicrometer transistors together with lasers or light-emitting diodes (LEDs) on the same chip for use in optical communication. The direct gap also results in high electron–hole recombination rates that may lead to a better radiation hardness. The availability of excellent heterostructure systems such as AlGaAs/GaAs, GaInAs/InP, and InGaAs/AlGaAs, and new crystal fabrication technologies such as MBE and MOCVD has opened up rich opportunities for development of devices such as the HFET [also called high electron mobility transistor (HEMT)], the HBT, and many other novel devices.

Conventional semiconductor devices are usually fabricated on crystalline wafers of either silicon or a compound semiconductor such as GaAs. These wafers are fragile, relatively expensive, and limited in size. However, in many applications, amorphous or polycrystalline materials that may be deposited in an inexpensive, continuous process on a variety of different large-area substrates can compete with crystalline silicon or even open up opportunities for completely new applications. Amorphous silicon films can be prepared by glow discharge decomposition of silane gas (SiH_4) and exhibit a relatively low density of defect states in the energy gap. The material obtained by this process is, in fact, an amorphous silicon–hydrogen alloy (a-Si:H) with a fairly large concentration of hydrogen. The hydrogen atoms tie up dangling bonds that are present in amorphous silicon in large numbers and decrease the density of localized states in the energy gap. However, these states still play a dominant role in determining the transport properties of the material. As a consequence, the behavior of a-Si:H devices is quite different from that of their crystalline counterparts.

Very large area (2 × 4 feet and larger) high-quality amorphous silicon (a-Si:H) films may be inexpensively produced in a continuous process, making this material very attractive for applications in electronics and photovoltaics. Amorphous silicon alloy thin-film transistors (TFTs) have the potential to become a viable and important technology for large-area, low-cost integrated circuits. These circuits are currently being used to drive large-area liquid crystal displays. Basic ICs and addressable image sensing arrays have also been implemented.

The main drawback of amorphous silicon and related amorphous materials is the very small mobility of electrons and holes. In polycrystalline silicon (poly-Si) TFTs, the electron and hole mobilities are much larger. Integrated circuits based on poly-Si TFTs have a smaller area than a-Si ICs, and this technology competes with a-Si for applications in flat-panel display devices, printers, scanners, and three-dimensional large-scale integrated (LSI) circuits.

The development of accurate device models is an integral part of modern semiconductor technology and is almost as essential as the fabrication equipment. Device models are important for circuit simulation and design but also as a tool for development of better fabrication processes. On the other hand, as the fabrication technology advances toward ever-increasing sophistication, new and tough challenges are constantly being presented to the model developer. Hence, the emphasis in MOSFET technology development is on further miniaturization of device features into the deep submicrometer range, which brings forth a new set of important phenomena specific for short-channel structures. In compound semiconductor technology, the same trend is present but, in addition, new material combinations and device structures are constantly proposed. The device modeling of TFTs is, in many respects, still in its infancy. This is especially true for devices based on a-Si where the material and electrical properties are still not sufficiently well characterized.

In this and the next chapter, we aim to give some insight into the problems encountered in modeling the various FET families mentioned above. However, although the emphasis will be on relatively simple and basic analytical models, we will also explore some of the modeling challenges of state-of-the-art technology. As examples, we will be discussing n-channel devices, but with trivial changes in signs, the results apply to p-channel devices as well.

5.2. PRINCIPLES OF OPERATION

5.2.1. MOSFET Operation

As discussed in Section 5.1, the gate region of an FET essentially consists of an MOS capacitor or a reverse-biased rectifying diode where the conducting channel constitutes one of the electrodes. Two diffused or implanted regions serve as ohmic contacts and as reservoirs of charge carrier for the channel (see Fig. 5.1.1).

In the MOSFET, an inversion layer at the semiconductor–oxide interface acts as the conducting channel. For example, in an n-channel MOSFET, the substrate is p-type silicon and the inversion charge consists of electrons that form a conducting channel between two n^+ ohmic contacts called source and drain. The depletion regions between the n-regions (the source, the drain, and the channel) and the p-type substrate provide isolation from other devices fabricated on the same substrate. A schematic cross section of the n-channel MOSFET is shown in Fig. 5.2.1.

In Section 3.5, we discussed how channel inversion charge can be induced by applying a suitable gate voltage relative to other terminals. The onset of so-called strong inversion is defined in terms of a threshold voltage V_T being applied between the gate and the source. In order to assure that the induced inversion channel extends all the way from source to drain, it is essential that the MOSFET gate structure either overlaps slightly or, at least, aligns with the edges of these contacts (the latter is achieved by a so-called self-aligned process). Self-alignment is often preferable since it contributes to a reduction of important parasitic capacitances, especially the feedback capacitance between gate and drain.

When a drain–source bias V_{DS} is applied on a MOSFET in the above-threshold conducting state, induced charge carriers move in the channel inversion layer from source to drain. A change in the gate–source voltage V_{GS} alters the electron sheet density in the channel and, hence, modulates the channel conductance and the device current. For $V_{GS} > V_T$ in an n-channel device, the application of a positive V_{DS} gives a steady voltage increase along the channel that causes a corresponding reduction in the local gate–channel bias V_{GX} (here X signifies a position x within the channel). This reduction is greatest near drain where V_{GX} equals the gate–drain bias V_{GD}.

In a somewhat simplified picture, we may say that when V_{GD} reaches the value of

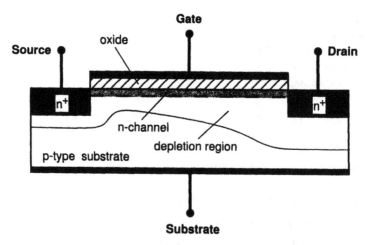

FIG. 5.2.1. Schematic cross section of an n-channel MOSFET.

the threshold voltage V_T, the inversion channel is pinched off near the drain and the drain current I_d saturates. The corresponding drain–source voltage, $V_{DS} = V_{SAT}$, is called the saturation voltage. The dependence of V_{SAT} on the gate–source voltage is $V_{SAT} = V_{GS} - V_T$.

When $V_{DS} > V_{SAT}$, the pinched-off region near the drain expands only slightly in the direction of the source, leaving the remaining inversion channel almost intact. The point of transition between the two regions, $x = x_p$, is characterized by $V_{XS}(x_p) = V_{SAT}$, where $V_{XS}(x_p)$ is the channel voltage relative to the source at the transition point. Hence, the drain current in saturation remains approximately constant, given by the voltage drop V_{SAT} across the part of the channel that remains in inversion. The voltage $V_{DS} - V_{SAT}$ across the pinched-off region creates a strong electric field that efficiently transports the electrons from the strongly inverted region to the drain.

Typical current–voltage characteristics of a so-called long-channel MOSFET, where pinch-off is the predominant saturation mechanism, are shown in Fig. 5.2.2. However, with shorter MOSFET gate lengths—typically in the submicrometer range—some degree of velocity saturation occurs in the channel near drain, leading to more evenly spaced saturation characteristics than those shown in Fig. 5.2.2. Also, phenomena such as a finite channel conductance in saturation, a drain-bias-induced shift in the threshold voltage, and an increased subthreshold current are important consequences of short gate lengths.

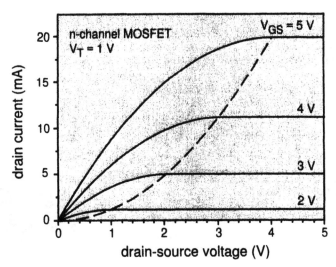

FIG. 5.2.2. Current–voltage characteristics of an n-channel MOSFET with current saturation caused by pinch-off (long-channel case). The intersections with the dotted line indicate the onset of saturation for each characteristic. The threshold voltage is assumed to be $V_T = 1$ V.

5.2.2. MESFET Operation

A schematic representation of a GaAs MESFET is shown in Fig. 5.2.3. The gate electrode forms a Schottky barrier contact (see Section 3.4) with a doped semiconductor region, usually *n*-type in order to take advantage of the excellent transport properties of electrons in GaAs. The neutral part of the semiconductor serves as a conducting channel between the source and the drain ohmic contacts. The gate bias modulates the depletion region under the gate and, hence, the cross section of the neutral conducting channel. In turn, this governs the drain current passing through the channel. The GaAs substrates used for MESFETs are semi-insulating and therefore provide excellent device isolation. Also, the parasitic substrate capacitances associated with the device are very much reduced compared to those of silicon MOS-FETs. The doped channel region is normally fabricated by ion implantation or by epitaxial growth.

Usually, GaAs MESFETs are constructed to be normally on, that is, with a doping profile such that the depletion width of the Schottky barrier does not extend all the way through the doped region at zero gate–source bias. Hence, the transistor is gradually turned off by applying a progressively more negative gate–source bias. Normally off MESFETs can also be constructed. However, the forward bias is limited by the built-in voltage of the Schottky barrier, allowing normally off MESFETs to be used only in circuits operating with low supply voltages.

Since the charge carriers of the channel are effectively removed from the gate–semiconductor interface by the gate depletion layer, problems related to interface traps are largely avoided in GaAs MESFETs.

As for the Si MOSFET, a drain–source bias applied to the MESFET creates a steadily increasing voltage along the channel in the direction from source to drain. Hence, both the gate–channel reverse bias and the depletion width under the gate also increase in the same direction. However, drain current saturation in GaAs MESFETs is primarily a result of velocity saturation in the channel near the drain

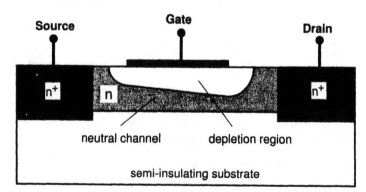

FIG. 5.2.3. Schematic representation of a GaAs MESFET. The asymmetry indicated in the depletion region under the gate is the result of an applied drain–source voltage.

instead of channel pinch-off. The reason is the relatively low field needed to achieve electron velocity saturation in gallium arsenide, a field that is normally exceeded in the conducting channel of GaAs FETs.

We also note that the conduction band structure of GaAs, with its low-mass central valley, promotes significant nonstationary phenomena such as velocity overshoot. Velocity overshoot allows the electrons to be accelerated well above the steady-state saturation velocity in a large portion of the FET channel, thus boosting both the drain current and the device speed beyond the values predicted by drift-diffusion-based transport theories. These features, together with a well-matured processing base, has made GaAs MESFET technology very popular for both high-speed digital and analog applications.

5.2.3. HFET Operation

The HFET (or HEMT) exploits the rich potential for band gap engineering in III–V semiconductors, by utilizing a heterostructure in the gate junction of FETs. In most cases, the conducting channel is formed as a narrow triangular energy well for electrons at the interface between materials with a large and a smaller band gap, for example, n-type AlGaAs and undoped GaAs. This is indicated in Fig. 5.2.4, which shows the conduction band at the heterointerface. (See Section 3.3 for further details on heterojunctions.)

Using modern crystal-growing techniques such as molecular beam epitaxy (MBE) or metal-organic chemical vapor deposition (MOCVD), the heterointerface can be grown quite free of defects. Moreover, a low doping of the material where the channel is located assures a drastic reduction in ionized impurity scattering in the channel. Hence, such a channel is characterized by a very high electron mobility, especially at low temperatures, where also phonon scattering is low. Owing to the reduced scattering rate in the conducting channel, velocity overshoot is even more pronounced in the HFET than in the MESFET, contributing to an exceptionally high speed performance of the device.

Other material systems in addition to AlGaAs/GaAs, with even better properties, are also being used in HFETs, for example, InGaAsP/InP and AlGaAs/

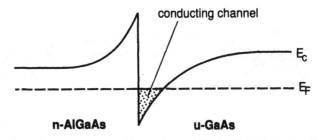

FIG. 5.2.4. Conducting channel formed at an AlGaAs/GaAs heterointerface.

InGaAs/GaAs. The latter is an example of a so-called pseudomorphic layer structure where the channel is located in a thin layer of nonlattice matched InGaAs sandwiched between the GaAs substrate and the AlGaAs layer. In this system, InGaAs provides both a reduced band gap and a higher electron velocity compared to GaAs. A consequence of the reduced band gap in InGaAs is an increase in the conduction band offset at the interface. Part of this offset can be sacrificed by lowering the aluminum content of the AlGaAs layer, resulting in a lowered density of deep-level defects (known as DX centers) that tend to trap electrons and degrade device performance.

The material with the wide band gap in Fig. 5.2.4 is contacted by a gate metal forming a Schottky barrier. It is essential that the wide-band-gap material is sufficiently thin to be fully depleted, allowing the channel charge to be controlled by the gate bias. In fact, the wide-band-gap material in HFETs plays the same role as the gate insulator of MOSFETs, blocking the electrons from passing between the gate electrode and the channel under normal operation. Figure 5.2.5 shows the cross section of a self-aligned HFET.

Note that the wide-band-gap material is usually undoped in a thin layer adjacent to the interface in order to remove the ionic scatterers from the vicinity of the channel. Also, the doping profile in the remaining part of this material may vary from a uniform doping to only a narrow doping layer—so-called delta doping. In fact, the wide-band-gap material may even be undoped since the channel charge carriers can be drawn in from the source and drain contacts by applying an appropriate gate bias. Hence, by using n-type or p-type contacts, we may obtain n-channel and p-channel devices on the same chip, allowing for the construction of CMOS-type logic.

In conclusion, the main advantage of the HFET is its ability to form a narrow conducting channel of high electron density in close proximity to the gate in a region of high material perfection and reduced electron scattering. This translates into an exceptional device performance in terms of frequency bandwidth and switching time. In fact, cutoff frequencies in excess of 300 GHz have been reported for deep submicrometer HFETs, higher than for any other transistor type.

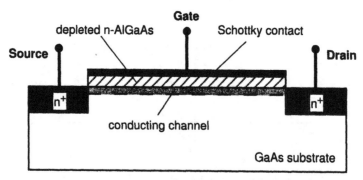

FIG. 5.2.5. Cross section of a self-aligned AlGaAs/GaAs HFET.

5.2.4. Amorphous Silicon TFT Operation

A schematic diagram of an a-Si TFT is shown in Fig. 5.2.6. The a-Si TFT operates by inducing carriers via field effect near the amorphous silicon–gate insulator interface, similar to what takes place in a conventional MOSFET. However, it is known that amorphous silicon, even with a high concentration of hydrogen, has a large number of localized states distributed throughout the forbidden gap, causing the operation of a-Si TFTs to be very different from that of its crystalline counterparts. One effect of the disorder is that the low-field mobility in a-Si TFTs is very small, typically on the order of 1 cm²/V s.

In addition to a dramatic reduction of field-effect mobility, the dominant role of traps leads to other important differences between the characteristics of a-Si TFTs and crystalline MOSFETs. Some of these differences will be described next.

Subthreshold Regime. In what may be termed the "below-threshold" regime, nearly all induced charge goes into localized acceptor-like states, corresponding to an off state of the transistor. At higher gate voltages, an increasing fraction of the induced charge populates the conduction band as mobile carriers and we enter the "above-threshold" regime—the on state of the device. Only at voltages too large for most practical applications, may we reach a mode of operation that corresponds to that of a crystalline MOSFET, with mobilities estimated to be on the order of 10 cm²/V s.

Frequency Response. In conventional MOSFETs or compound semiconductor FETs, the frequency response is primarily determined by the carrier transit time t in the channel. Ideally, for such devices, the transistor cutoff frequency $f_T = 1/2\pi t$. This cutoff frequency may also be expressed as $f_T = g_m/(2\pi C_{GS})$, where g_m is the transconductance and C_{GS} is the gate–source capacitance. In contrast, the frequency response of traps is very important in a-Si TFTs (and also in poly-Si TFTs; see below).

From a physics point of view, the electron transport in the channel of these devices can be visualized as electrons traveling a short distance while in the conduc-

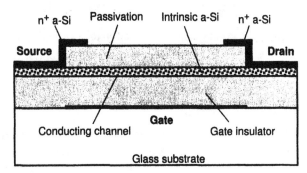

FIG. 5.2.6. Schematic diagram of an a-Si TFT. (After Lee et al., 1993.)

tion band and then being trapped, then reemitted into the conduction band, and so on. In this case, when the frequency becomes comparable to or larger than the inverse characteristic trapping time, the transistor response tapers off. This effect can be described in terms of an RC time constant, similar to the one we introduced for the MIS capacitor in Section 3.5. To a first-order approximation, this time constant becomes $R_{ch}C_{GS}$, where R_{ch} is a fraction of the channel resistance.

Stress Effects. Even for conventional FETs with a near-ideal behavior, a prolonged application of high bias voltage and/or high temperature may cause significant changes in the device characteristics. In a-Si TFTs, such effects are very pronounced, and the device characteristics may literally change during the measurement itself. The reason for this behavior is that high electric fields or temperature stress create (and sometimes remove) localized states (traps) in the energy gap. This is observed as changes in the threshold voltage, in the subthreshold behavior, and in the device leakage current.

Nonlinear Contact Resistance. Note that with the a-Si TFT design indicated in Fig. 5.2.6, the n^+ contact regions are separated from the inversion channel by a layer of intrinsic a-Si. Therefore, we rely on charge injection to the channel through the intrinsic layer, which means that the contact resistance for this device is, in fact, nonlinear and should decrease with increasing device current. Also, at high drain voltages, the drain–gate overlap acts as a second gate that induces electrons at the interface between the passivating dielectric and the intrinsic a-Si layer under the drain. Hence, the effective gate length may be drain voltage dependent.

Typically, a-Si TFT integrated circuits are fabricated on glass substrates. First, metal gates are deposited followed by a layer of silicon nitride gate dielectric, a layer of intrinsic a-Si, and a layer of a passivating insulator (silicon nitride). The gate insulator thickness may usually be on the order of 3000 Å. The intrinsic amorphous silicon layer is quite thin (about 500 Å) in order to minimize the off current. The passivating insulator thickness is of the order of 1 μm and is patterned to be self-aligned with the gate metal. Finally, a thin (~100-Å) layer of highly doped n^+ a-Si with a contact metal on top is deposited, and the source and drain contacts are patterned. Typically, the length of the gate metal strip is about 10–15 μm. The overlaps of the source and drain contact metal over the passivating dielectric are on the order of 2 μm.

Amorphous silicon is also known to be a highly photosensitive material with applications in printers, copiers, imagers, and consumer electronics. Hence, illumination may also affect the a-Si TFT characteristics.

5.2.5. Polysilicon TFT Operation

Polysilicon TFTs are fabricated using crystallized amorphous silicon films. Typically, the substrates, which may be quartz or glass, are coated with a silicon dioxide film onto which an amorphous silicon film with a thickness of typically 1000 Å is deposited. This film is then heat treated, resulting in a polycrystalline film with many grains. The grain boundaries contain traps and impede the charge carrier transport.

The room temperature electron mobility of such poly-Si films typically varies between 30 cm^2/V s for films with small grain sizes and several hundred square centimeters per volt seconds for large grain sizes. Otherwise, the poly-Si TFT structure and operation are very similar to those of conventional MOSFETs. However, the gate oxide of poly-Si TFTs is deposited instead of being thermally grown. Highly doped polysilicon is used as gate material, and the source and drain regions are formed by ion implantation. A final exposure of the devices to a hydrogen plasma contributes to a passivation of the grain boundaries. A schematic cross section of a poly-Si TFT is shown in Fig. 5.2.7.

The differences between crystalline MOSFETs and poly-Si TFTs are related to the presence of grains in the polysilicon material. The grain boundaries represent an important impediment to current flow. They act as barriers for the charge carriers and also contain dangling bonds and other types of traps.

As a first approximation, the grain boundaries can be accounted for by using the "effective medium approach," that is, by treating, somewhat superficially, poly-Si as a uniform material with a certain density of localized states (traps) in the forbidden gap. In this model, poly-Si TFTs are treated as devices with an intermediate behavior between that of crystalline MOSFETs and a-Si TFTs. Indeed, many features specific for a-Si TFTs are also typical for poly-Si TFTs, albeit much less pronounced since the effective density of localized states is much less in polysilicon. However, poly-Si TFTs have their own specific features that are somewhat different from those of both crystalline MOSFETs and a-Si transistors. These features include the following:

1. The "kink" effect, which is an increase in drain current at high drain bias, related to impact ionization. This effect may also occur in crystalline devices but at much higher electric fields and at smaller gate lengths than for polysilicon devices.

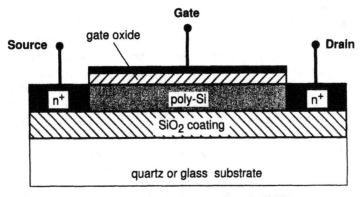

FIG. 5.2.7. Schematic diagram of a poly-Si TFT.

2. Frequency dependence of the channel capacitance, which is related to the trapping time of charge carriers. This effect is somewhat similar to that of a-Si devices but occurs at much higher frequencies in poly-Si devices.

3. Gate bias dependence of the field-effect mobility, which is distinctly different from those of both crystalline Si and a-Si transistors. From the typical mobility data given above, we see that poly-Si TFTs have superior current drive and device transconductance compared to a-Si TFTs.

4. Stochastic nature of leakage current, due to the different positions of the grains between device channels. This effect may vary drastically from device to device.

5.3. BASIC MOSFET MODELS

5.3.1. Gradual Channel Approximation

Analytical or semianalytical modeling of FETs are usually based on the so-called gradual channel approximation (GCA). In contrast to the situation in an ideal two-terminal device, where the charge density profile is determined from a one-dimensional Poisson equation (see Chapter 3), the FETs generally pose a two-dimensional electrostatic problem. The reason for this is the geometric effects and the application of a drain–source bias, which create an electric field component in the longitudinal direction of the channel, perpendicular to the vertical field associated with the ideal gate structure.

The GCA states that under certain conditions the electrostatic problem of the gate region can be expressed in terms of two coupled one-dimensional equations: a Poisson equation for determining the vertical charge density profile under the gate and a charge transport equation for the channel. This allows us to determine self-consistently both the channel potential and the charge profile at any position along the gate. A direct inspection of the two-dimensional Poisson equation for the channel region shows that the GCA is valid if we can assume that the electric field gradient in the longitudinal direction of the channel is much less than that in the vertical direction perpendicular to the channel (see, e.g., Lee et al., 1993).

The validity of the GCA can be checked by making estimates of the variation in the longitudinal and vertical components of the electric field in the gate region. Typically, we find that the GCA is valid for long-channel FETs where the ratio between the gate length and the vertical distance of the space charge region from the gate electrode (the so-called aspect ratio) is large. However, if the FET is biased in saturation, the GCA always becomes invalid near the drain as a result of the large longitudinal field gradient that develops in this region. In Fig. 5.3.1, this is schematically illustrated for a MOSFET in saturation.

We now investigate a MOSFET operating in the above-threshold regime, that is, when the gate voltage is large enough to cause inversion in the entire length of the channel at zero drain–source bias. We start by considering the inverted part of the channel of a long-channel device by assuming that the GCA is applicable and that

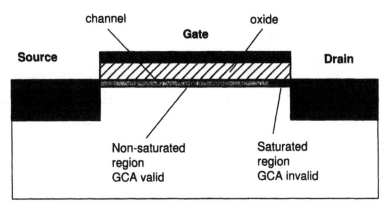

FIG. 5.3.1. Schematic representation of a MOSFET in saturation, where the channel is divided into a nonsaturated region where GCA is valid and a saturated region where GCA is invalid.

the carrier mobility can be taken to be constant. By also including the effects of the depletion charge, we arrive at the well-known Meyer I–V model for MOSFETs, which describes current saturation by the pinch-off mechanism.

As a limiting case of the Meyer model for small substrate doping, we obtain the simple long-channel charge control model, also known as the Shichman–Hodges model. For devices with channel lengths of a few micrometer and below, short-channel effects become noticeable. Therefore, in a modified version of the simple charge control model, we include the effects of velocity saturation by introducing a suitable velocity–field relationship. Finally, the implementation of these models in SPICE is discussed.

5.3.2. Meyer *I*–*V* Model

Applying the GCA, the total induced charge q_s per unit area in the semiconductor of an n-channel device can be expressed in terms of Gauss's law as follows (see Section 3.5):

$$q_s = -c_i[V_{GS} - V_{FB} - 2\varphi_b - V(x)] \tag{5.3.1}$$

Here the contents of the bracket expresses the voltage drop across the insulator layer, V_{GS} is the gate–source voltage, V_{FB} is the flat-band voltage, $V(x) \equiv V_{XS}$ is the channel voltage relative to source, $c_i = \epsilon_i/d_i$ is the gate insulator capacitance per unit area, d_i is the insulator thickness, $\varphi_b = V_{th} \ln(N_a/n_i)$, N_a is the acceptor concentration in the substrate (assumed to be constant), and n_i is the intrinsic carrier density. In Eq. (5.3.1), we assumed that the source and the semiconductor substrate are both connected to ground.

The induced sheet charge density includes both the inversion charge density $q_i = -qn_s$ and the depletion charge density q_{dep}, that is, $q_s = q_i + q_{dep}$. Using Eq. (3.5.15) and including the added channel–substrate bias caused by the channel voltage, the depletion charge per unit area can be expressed as

$$q_{dep} = -qN_a d_{dep} = -\sqrt{2\epsilon_s qN_a[2\varphi_b + V(x)]} \tag{5.3.2}$$

where d_{dep} is the local depletion layer width at position x. Hence, the inversion sheet charge density becomes

$$q_i = -qn_s = -c_i[V_{GS} - 2\varphi_b - V_{FB} - V(x)] + \sqrt{2\epsilon_s qN_a[2\varphi_b + V(x)]} \tag{5.3.3}$$

I–V Model. Assuming a constant electron mobility μ_n, the electron velocity can be written as $v_n = \mu_n \, dV/dx$. Neglecting diffusion current (only important in the subthreshold regime), the absolute value of the drain current (independent of position x) can be written as

$$I_d = Wq \, \mu_n \frac{dV}{dx} n_s \tag{5.3.4}$$

where W is the device width. Substituting for qn_s from Eq. (5.3.3) and integrating Eq. (5.3.4) over the channel length L, which corresponds to a change in V from zero at $x = 0$ to V_{DS} at $x = L$, we obtain the following expression for the current–voltage characteristics (see Meyer, 1971):

$$I_d = \frac{W\mu_n c_i}{L}\left\{\left(V_{GS} - V_{FB} - 2\varphi_b - \frac{V_{DS}}{2}\right)V_{DS}\right.$$

$$\left. - \frac{2\sqrt{2\epsilon_s qN_a}}{3c_i}\left[(V_{DS} + 2\varphi_b)^{3/2} - (2\varphi_b)^{3/2}\right]\right\} \tag{5.3.5}$$

We note that Eq. (5.3.5) is valid only for values of V_{DS} such that the inversion layer still exists even at the drain side of the gate, that is, $n_s(x = L) \geq 0$. Pinch-off is conveniently defined by the condition $n_s = 0$ at the drain. From Eq. (5.3.3), we find that pinch-off occurs when the drain–source voltage reaches the saturation value given by

$$V_{SAT} = V_{GS} - 2\varphi_b - V_{FB} + \frac{\epsilon_s qN_a}{c_i^2}\left[1 - \sqrt{1 + \frac{2c_i^2(V_{GS} - V_{FB})}{\epsilon_s qN_a}}\right] \tag{5.3.6}$$

At low doping levels, we see that $V_{SAT} \rightarrow V_{GS} - V_T \equiv V_{GT}$ when the gate voltage approaches threshold.

The pinch-off condition implies a vanishing carrier concentration at the drain side of the channel. Hence, at a first glance, one might think that the drain current should also vanish. However, as discussed in Section 5.2, the drain current instead saturates and becomes nearly independent of the drain–source bias for $V_{DS} > V_{SAT}$. The saturation current I_{sat} may be determined by substituting $V_{DS} = V_{SAT}$ from Eq. (5.3.6) into Eq. (5.3.5). In reality, of course, the electron concentration never vanishes and the electric field never becomes infinite. This simply indicates that the GCA becomes invalid near the drain in saturation, pointing to the need for a more accurate and detailed analysis of the saturation regime. Such an analysis was given, for example, by Lee et al. (1993).

Threshold Voltage. From Section 3.5, we have the following expression for the threshold voltage for the MIS capacitor [see Eq. (3.5.17)]:

$$V_T = V_{FB} + 2\varphi_b + \frac{\sqrt{4\epsilon_s q N_a \varphi_b}}{c_i} \tag{5.3.7}$$

Most MOSFETs have a fourth contact attached to the substrate, as indicated in Fig. 5.2.1, to allow the application of a substrate bias. Here, V_{BS} denotes such a bias relative to source. The difference between the interface potential and V_{BS} defines the potential drop across the gate depletion region and, therefore, also the depletion charge. Hence, V_T in Eq. (5.3.7) should be modified as follows to account for the substrate bias:

$$V_T = V_{FB} + 2\varphi_b + \frac{\sqrt{2\epsilon_s q N_a (2\varphi_b - V_{BS})}}{c_i} \tag{5.3.8}$$

We note that this expression is only valid for negative or slightly positive values of V_{BS}, when the junction between the source contact and the p-substrate is either reverse biased or slightly forward biased. For $V_{BS} > 2\varphi_b$, large leakage currents will take place.

An example of a calculated dependence of the threshold voltage on substrate bias is shown in Fig. 5.3.2 for different values of gate insulator thickness. As can be seen from this figure and from Eq. (5.3.8), the threshold voltage decreases with decreasing insulator thickness and is quite sensitive to the substrate bias. This so-called body effect is essential for device characterization and in threshold voltage engineering. In fact, for real devices it is important to be able to carefully adjust the threshold voltage to match specific application requirements.

From Eq. (5.3.8), we notice that, in addition to the application of a substrate–source bias, V_T can be permanently shifted by changing the doping or by using a different gate metal (including heavily doped polysilicon). As discussed in Section 3.5, the gate metallization affects the flat-band voltage through the work function difference between the metal and the semiconductor. Threshold voltage adjustment by means of doping is often performed with an additional ion implantation through the gate oxide. A more comprehensive discussion of threshold voltage engineering in MOSFETs can be found in Lee et al. (1993).

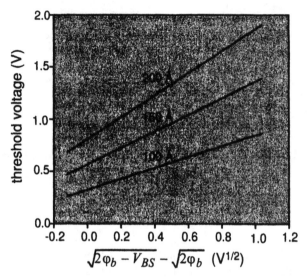

FIG. 5.3.2. Body plot: dependence of the threshold voltage on substrate bias in MOSFETs with different insulator thicknesses. Parameters used in the calculation: flat-band voltage, −1 V; substrate doping density, 10^{22} m^{-3}; temperature, 300 K. The slope of the plots are given in terms of the body effect parameter $\gamma = \sqrt{2\epsilon_s q N_a}/c_i$.

5.3.3. Simple Charge Control Model

I–V Model. For very small drain–source voltages, the terms in the curly brackets of the Meyer expression in Eq. (5.3.5) may be expanded into a Taylor series, leading to the following simplified current–voltage characteristics in the linear region:

$$I_d \approx \frac{W \mu_n c_i}{L} V_{GT} V_{DS} \tag{5.3.9}$$

This equation indicates that, at very small drain–source bias, the mobile charge induced into the channel does not depend on the channel voltage. Hence, we find for the inversion charge, $q n_s \approx c_i (V_{GS} - V_T)$, which corresponds to the charge of a simple parallel-plate capacitor. In this case, the electric field along the channel is nearly constant and given by $F \approx V_{DS}/L$. Multiplying the inversion sheet charge density $q n_s$ by the electron velocity $v_n = \mu_n F$ and by the gate width W, we obtain Eq. (5.3.9).

Based on so-called charge control models, a similar approach may be applied to arrive at simplified, analytical descriptions of the complete current–voltage characteristics of MOSFETs. In the simplest such model, we assume that the concentration of free carriers induced into the channel is still given by the parallel plate capacitor expression above, adjusted only for the channel voltage $V(x)$, that is,

$$q n_s = c_i [V_{GT} - V(x)] \tag{5.3.10}$$

where $V_{GT} \equiv V_{GS} - V_T$. The use of this charge control expression is equivalent to neglecting the variation of the depletion layer charge along the channel. The drain current can then be written as

$$I_d = W\mu_n q n_s F = W\mu_n c_i (V_{GT} - V) \frac{dV}{dx} \qquad (5.3.11)$$

Integrating Eq. (5.3.11) over the gate length, we obtain the following expression for the current–voltage characteristics (Shichman and Hodges, 1968):

$$I_d = W\mu_n c_i / L \times \begin{cases} [V_{GT} V_{DS} - \frac{1}{2} V_{DS}^2] & \text{for } V_{DS} \leq V_{SAT} \\ \frac{1}{2} V_{GT}^2 & \text{for } V_{DS} > V_{SAT} \end{cases} \qquad (5.3.12)$$

where the saturation voltage now becomes $V_{SAT} = V_{GT}$. The MOSFET current–voltage characteristics shown in Fig. 5.2.2 were calculated using this simple charge control model.

An important device characteristic is the transconductance, defined as

$$g_m = \frac{dI_d}{dV_{GS}}\bigg|_{V_{DS}} \qquad (5.3.13)$$

From Eq. (5.3.12), we find

$$g_m = \begin{cases} \beta V_{DS} & \text{for } V_{DS} \leq V_{SAT} \\ \beta V_{GT} & \text{for } V_{DS} > V_{SAT} \end{cases} \qquad (5.3.14)$$

where $\beta = W\mu_n c_i / L$ is called the transconductance parameter.

As can be seen from Eq. (5.3.14), a high transconductance is obtained with a high value of the electron mobility, a large gate insulator capacitance (i.e., thin gate insulator layer), and/or a large gate-width-to-gate-length ratio.

C–V Model. For the simulation of dynamic events in MOSFET circuits, we also have to account for variations in the stored charges of the devices. In a MOSFET, we have, for example, stored charges in the gate electrode, in the conducting channel, and in the depletion layers. Electrically, the variation in the stored charges can be expressed through different capacitive elements, as indicated in Fig. 5.3.3.

We distinguish between the so-called parasitic capacitive elements and the capacitive elements of the intrinsic transistor. The parasitics include the overlap capacitances between the gate electrode and the highly doped source and drain regions (C_{os} and C_{od} in Fig. 5.3.3), the junction capacitances between the substrate and the source and drain regions (C_{js} and C_{jd} in Fig. 5.3.3), and the capacitances between the metallic electrodes of the source, the drain, and the gate.

As indicated in Fig. 5.3.3, the semiconductor charges of the intrinsic gate region of the MOSFET are divided between the mobile inversion charge and the de-

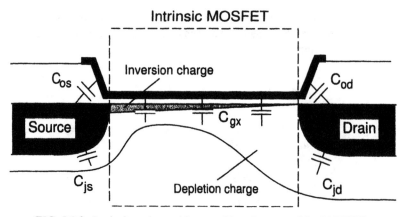

FIG. 5.3.3. Intrinsic and parasitic capacitive elements of the MOSFET.

pletion charge. In addition, these charges are nonuniformly distributed along the channel when drain–source bias is applied. Hence, the capacitive coupling between the gate electrode and the semiconductor is also distributed, making the channel resemble an *RC* transmission line. In practice, however, because of the short gate lengths and limited bandwidths of FETs, the distributed capacitance of the intrinsic device is usually very well represented in terms of a lumped capacitance model, that is, by capacitive elements connecting the various (imaginary) intrinsic device terminals.

Nonetheless, accurate modeling of the intrinsic device capacitance requires an analysis of the distribution from source to drain of the mobile charge and of the depletion charge versus terminal bias voltages. As discussed by Ward and Dutton (1978), such an analysis leads to a set of charge-conserving and nonreciprocal capacitances between the different intrinsic terminals (nonreciprocity means $C_{ij} \neq C_{ji}$, where i and j denote source, drain, gate, and substrate).

In a straightforward analysis by Meyer (1971) based on the simple charge control model, a set of reciprocal capacitances ($C_{ij} = C_{ji}$) were obtained as derivatives of the total gate charge with respect to the various terminal voltages. (In reality, the Meyer capacitances represent a subset of the Ward–Dutton capacitances in the so-called quasi-static approximation.) Although charge conservation is not strictly enforced in this case, the resulting errors in circuit simulations are usually small, except in some cases of transient analyses of especially demanding circuits. Here, we first consider Meyer's capacitance model for the long-channel case but return with comments of the applicability of the model for short-channel MOSFETs in Section 5.3.4. In Chapter 6, we present additional modifications of this model and comments on charge-conserving capacitance models in our discussion of the universal FET models.

In Meyer's capacitance model, the distributed intrinsic MOSFET capacitance is split into the following three lumped capacitances between the intrinsic terminals:

$$C_{GS} = \frac{\partial Q_G}{\partial V_{GS}}\bigg|_{V_{GD}, V_{GB}} \qquad C_{GD} = \frac{\partial Q_G}{\partial V_{GD}}\bigg|_{V_{GS}, V_{GB}} \qquad C_{GB} = \frac{\partial Q_G}{\partial V_{GB}}\bigg|_{V_{GS}, V_{GD}} \qquad (5.3.15)$$

where Q_G is the total intrinsic gate charge. The intrinsic MOSFET equivalent circuit corresponding to this model is shown in Fig. 5.3.4.

In strong inversion, Q_G is dominated by the inversion charge and is determined by integrating the sheet charge density qn_s, given by Eq. (5.3.10), over the gate area, that is,

$$Q_G \approx qW \int_0^L n_s \, dx = Wc_i \int_0^L [V_{GT} - V(x)] \, dx \qquad (5.3.16)$$

From Eq. (5.3.11), we notice that $dx = W\mu_n c_i(V_{GT} - V) \, dV/I_d$, which allows us to make a change of integration variable from x to V in Eq. (5.3.16). Hence, we obtain, for the nonsaturated regime,

$$Q_G = \frac{W^2 c_i^2}{I_d} \mu_n \int_0^{V_{DS}} (V_{GT} - V)^2 \, dV = \tfrac{2}{3} C_i \frac{(V_{GS} - V_T)^3 - (V_{GD} - V_T)^3}{(V_{GS} - V_T)^2 - (V_{GD} - V_T)^2} \qquad (5.3.17)$$

where $C_i = WLc_i$ and where we expressed I_d using Eq. (5.3.12) and replaced V_{DS} by $V_{GS} - V_{GD}$ everywhere.

By performing the differentiations in Eq. (5.3.15), the following strong-inversion, long-channel Meyer capacitances are obtained:

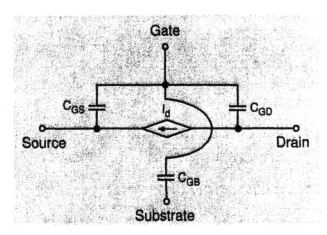

FIG. 5.3.4. Large-signal equivalent circuit of intrinsic MOSFET based on Meyer's capacitance model.

$$C_{GS} = \tfrac{2}{3}C_i\left[1 - \left(\frac{V_{GT} - V_{DS}}{2V_{GT} - V_{DS}}\right)^2\right]$$ (5.3.18)

$$C_{GD} = \tfrac{2}{3}C_i\left[1 - \left(\frac{V_{GT}}{2V_{GT} - V_{DS}}\right)^2\right]$$ (5.3.19)

$$C_{GB} = 0$$ (5.3.20)

Note that in the simple charge control model, $V_{SAT} = V_{GT}$.

The Meyer capacitances in saturation are found by replacing V_{DS} by V_{SAT} in the above expressions, that is,

$$C_{GSs} = \tfrac{2}{3}C_i \qquad C_{GDs} = C_{GBs} = 0$$ (5.3.21)

This result indicates that in saturation, a small change in the applied drain–source voltage does not contribute to the gate or the channel charge, since the channel is pinched off. Instead, the entire channel charge is "assigned" to the source terminal, giving a maximum value of the capacitance C_{GS}. Normalized dependencies of the Meyer capacitances C_{GS} and C_{GD} on bias conditions are shown in Fig. 5.3.5.

In the subthreshold regime, the inversion charge becomes negligible compared to the depletion charge, and the MOSFET gate–substrate capacitance will be the same as that of an MOS capacitor in depletion, that is, a series connection of the gate oxide capacitance C_i and the depletion capacitance C_d [see Eqs. (3.5.19)–(3.5.23)]. According to the discussion in Section 3.5, the applied gate–substrate voltage V_{GB} can be subdivided as follows [see Eqs (3.5.12)–(3.5.14)]:

$$V_{GB} = V_{FB} + \psi_s - \frac{q_{dep}}{c_i}$$ (5.3.22)

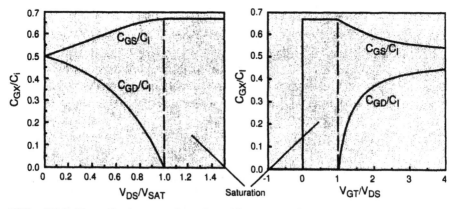

FIG. 5.3.5. Normalized strong-inversion Meyer capacitances according to Eqs. (5.3.18)–(5.3.21) versus (*a*) drain–source bias and (*b*) gate–source bias. Note that $V_{SAT} = V_{GT}$ in this model.

where V_{FB} is the flat-band voltage, ψ_s is the potential across the semiconductor depletion layer (i.e., the surface potential relative to the substrate interior), and $-q_{dep}/c_i$ is the voltage drop across the oxide. In the depletion approximation, the depletion charge per unit area q_{dep} is related to ψ_s by $q_{dep} = -\gamma c_i \sqrt{\psi_s}$ (see Section 3.5), where $\gamma = \sqrt{2\epsilon_s q N_a}/c_i$ is the body-effect parameter. Using this relationship to substitute for ψ_s in Eq. (5.3.22), we find

$$Q_G = -WLq_{dep} = \gamma C_i \left(\sqrt{\tfrac{1}{4}\gamma^2 + V_{GB} - V_{FB}} - \tfrac{1}{2}\gamma \right) \qquad (5.3.23)$$

from which we obtain the following subthreshold Meyer capacitances:

$$C_{GS} = C_{GD} = 0 \qquad C_{GB} = \frac{C_i}{\sqrt{1 + 4(V_{GB} - V_{FB})/\gamma^2}} \qquad (5.3.24)$$

Note that Eq. (5.3.24) gives $C_{GB} = C_i$ at the flat-band condition, which is somewhat different from the flat-band capacitance of Eq. (3.5.33). This discrepancy arises from neglecting the effects of the free carriers in the subthreshold regime (see Section 3.5). For the same reason, we observe the presence of discontinuities in the Meyer capacitances at threshold. Discontinuities in the derivatives of the Meyer capacitances occur at the onset of saturation as a result of additional approximations. Such discontinuities should be avoided in the device models since they give rise to increased simulation time and conversion problems in SPICE. These problems are addressed in the universal FET models discussed in Chapter 6.

5.3.4. Velocity Saturation Model

The simple MOSFET models discussed above using a linear velocity–field relationship work reasonably well for long-channel devices. However, the implicit notion of a diverging carrier velocity as we approach pinch-off is, of course, unphysical. Instead, current saturation is better described in terms of a saturation of the carrier drift velocity when the electric field near drain becomes sufficiently high.

The following two-piece model is a simple, first approximation to a realistic velocity–field relationship:

$$v(F) = \begin{cases} \mu F & F < F_s \\ v_s & F \geq F_s \end{cases} \qquad (5.3.25)$$

Here v_s is the saturation velocity, μ is the low-field mobility, and $F_s = v_s/\mu$ is the saturation field. In this picture, current saturation in FETs occurs when the field at the drain side of the gate reaches the saturation field. More realistic velocity–field relationships for MOSFETs are obtained from

$$v(F) = \frac{\mu F}{[1 + (\mu F/v_s)^m]^{1/m}} \qquad (5.3.26)$$

where $m = 2$ and $m = 1$ are reasonable choices for n-channel and p-channel MOSFETs, respectively. Figure 5.3.6 shows different velocity–field models for electrons and holes in silicon MOSFETs.

We now consider the simplest possible extension of the MOSFET charge control model discussed above using the two-piece linear approximation in Eq. (5.3.25) for the carrier velocity.

I–V Model. In the strong inversion regime, the surface carrier concentration of electrons in the channel of an n-channel MOSFET is obtained from the charge control model in Eq. (5.3.10). The current–voltage characteristics described by Eq. (5.3.12) are therefore still valid for $F(L) \leq F_s$. The saturation voltage V_{SAT} in this case is defined as the drain–source voltage at the onset of velocity saturation, that is, when $F(L) = F_s$ and $V(L) = V_{SAT}$. Hence, we find the following expressions for the drain current and the saturation voltage:

$$I_d = \frac{W\mu_n c_i}{L} \times \begin{cases} V_{GT}V_{DS} - \tfrac{1}{2}V_{DS}^2 & \text{for } V_{DS} \leq V_{SAT} \\ V_L^2\left[\sqrt{1 + \left(\dfrac{V_{GT}}{V_L}\right)^2} - 1\right] & \text{for } V_{DS} > V_{SAT} \end{cases} \quad (5.3.27)$$

$$V_{SAT} = V_{GT} - V_L\left[\sqrt{1 + \left(\frac{V_{GT}}{V_L}\right)^2} - 1\right] \quad (5.3.28)$$

where $V_L \equiv F_s L$.

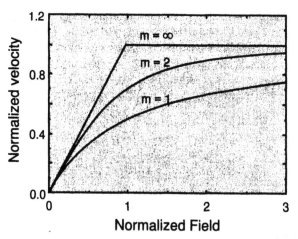

FIG. 5.3.6. Velocity–field relationships for charge carriers in silicon MOSFETs. The electric field and the velocity are normalized to F_s and v_s, respectively. The two lower curves are calculated from Eq. (5.3.26) using $m = 1$ for holes and $m = 2$ for electrons. The curve marked $m = \infty$ corresponds to the linear two-piece model in Eq. (5.3.25).

For large values of V_L such that $V_L \gg V_{GT}$, the square-root terms in Eqs. (5.3.27) and (5.3.28) may be expanded into a Taylor series, yielding the long-channel results derived earlier in the simple charge control model without velocity saturation. Assuming, as an example, that $V_{GT} \approx 3$ V, $\mu_n = 0.08$ m^2/V s, and $v_s = 1 \times 10^5$ m/s, we find that velocity saturation effects may be neglected for L $\gg 2.4$ μm. Hence, velocity saturation is important in modern MOSFETs which typically have gate lengths in the submicrometer range.

In the opposite limit, when $V_L \ll V_{GT}$, we obtain

$$V_{SAT} \approx V_L \tag{5.3.29}$$

$$I_{sat} \approx \beta V_L V_{GT} \tag{5.3.30}$$

where $\beta = W\mu_n c_i/L$ is the transconductance parameter. Since I_{sat} is proportional to V_{GT}^2 in long-channel devices and proportional to V_{GT} in a short-channel devices, we can use this difference to identify the presence of short-channel effects on the basis of measured device characteristics.

The drain and source parasitic series resistances R_d and R_s may play an important role in limiting the device performance. These resistances can be accounted for by relating the intrinsic gate–source and drain–source voltages considered so far to their extrinsic (measured) counterparts, V_{gs} and V_{ds}, by including the voltage drops across the series resistances:

$$V_{GS} = V_{gs} - R_s I_d \tag{5.3.31}$$

$$V_{DS} = V_{ds} - (R_s + R_d)I_d \tag{5.3.32}$$

Combining Eqs. (5.3.28) with Eqs. (5.3.31) and using $I_d = I_{sat}$, we can express the saturation current as follows in terms of extrinsic gate voltage overdrive V_{gt}:

$$I_{sat} = \frac{\beta V_{gt}^2}{1 + \beta R_s V_{gt} + \sqrt{1 + 2\beta R_s V_{gt} + (V_{gt}/V_L)^2}} \tag{5.3.33}$$

The extrinsic saturation voltage V_{sat} can likewise be obtained by combining Eqs. (5.3.27), (5.3.28), (5.3.31), and (5.3.32) using $V_{DS} = V_{SAT}$, $V_{ds} = V_{sat}$, and $I_d = I_{sat}$:

$$V_{sat} = V_{gt} + \left(R_d - \frac{1}{\beta V_L}\right)I_{sat} \tag{5.3.34}$$

C–V Model. In the MOSFET velocity saturation model, the Meyer capacitances developed in Section 5.3.3 for the nonsaturated, above-threshold regime are still valid up to the velocity saturation voltage V_{SAT} given by Eq. (5.3.28). The capacitance values at the saturation point are found by using $V_{DS} = V_{SAT}$ in Eqs. (5.3.18) and (5.3.19):

$$C_{GSs} = \tfrac{2}{3} C_i \left[1 - \left(\frac{V_{SAT}}{2V_L} \right)^2 \right] \qquad (5.3.35)$$

$$C_{GDs} = \tfrac{2}{3} C_i \left[1 - \left(1 - \frac{V_{SAT}}{2V_L} \right)^2 \right] \qquad (5.3.36)$$

However, well into saturation, the intrinsic gate charge will change very little with increasing V_{DS}, similar to what takes place in the case of saturation by pinch-off (see Section 4.3.3). Hence, the capacitances have to approach the same limiting values in saturation as the Meyer capacitances, that is, $C_{GS}/C_i \to \tfrac{2}{3}$ and $C_{GD}/C_i \to 0$. In fact, since the behavior of C_{GS} and C_{GD} in the velocity saturation model and in the Meyer model coincide for $V_{DS} < V_{SAT}$ and have the same asymptotic values in saturation, the Meyer capacitance model offers a reasonable approximation for the MOSFET capacitances also in short-channel devices. This suggests a separate "saturation" voltage for the capacitances close to the long-channel pinch-off voltage, which is larger than V_{SAT} associated with the onset of velocity saturation. However, the functional form of the approach to the saturation values depend on the form of the velocity–field relationship chosen. We will return with more flexible C–V expressions in our discussion of the universal MOSFET model in Chapter 6.

5.3.5. Comparison of Basic MOSFET Models

Figure 5.3.7 shows I–V characteristics calculated using the various basic MOSFET models discussed above, that is, the Meyer I–V model (MM), the simple charge control model (SCCM), and the velocity saturation model (VSM). The same set of MOSFET parameters were used in all cases. We note that all models coincide at small drain–source voltages. However, in saturation, SCCM always gives the highest current. This is a direct consequence of omitting velocity saturation and spatial variation in the depletion charge in SCCM, resulting in an overestimation of both carrier velocity and inversion charge. The characteristics for VSM and MM clearly demonstrate how inclusion of velocity saturation and distribution of depletion charge, respectively, affect the saturation current.

The intrinsic capacitances for the same device are shown in Fig. 5.3.8. Meyer's capacitance model is applicable for all the above MOSFET models (SCCM, MM, and VSM).

In the present device example, we note that velocity saturation and depletion charge may be quite important. Therefore, we emphasize that the SCCM is usually applicable only for long-channel, low-doped devices, while the MM applies to long-channel devices with an arbitrary doping level. The VSM gives a reasonable description of short-channel devices, although some important short-channel effects such as channel length modulation and drain-induced barrier lowering (DIBL) are still unaccounted for in this model. A discussion of these mechanisms is deferred to the presentation of advanced FET models in Chapter 6.

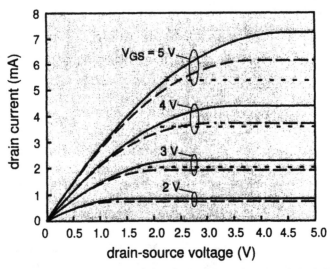

FIG. 5.3.7. Comparison of I–V characteristics obtained for a given set of MOSFET parameters using the three basic MOSFET models: SCCM (solid curves), MM (dashed curves), and VSM (dotted curves). The MOSFET device parameters are $L = 2$ μm, $W = 20$ μm, $d_i = 300$ Å, $\mu_n = 0.06$ m²/V s, $v_s = 10^5$ m/s, $N_a = 10^{22}$ m⁻³, $V_T = 0.43$ V, $V_{FB} = -0.75$ V, $\epsilon_i = 3.45 \times 10^{-11}$ F/m, $\epsilon_s = 1.05 \times 10^{-10}$ F/m, $n_i = 1.05 \times 10^{16}$ m⁻³.

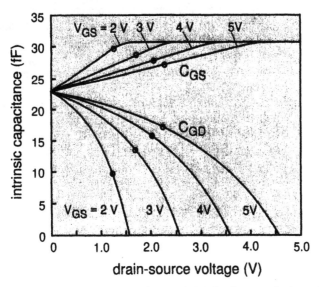

FIG. 5.3.8. Intrinsic MOSFET C–V characteristics for the same device as in Fig. 5.3.7, obtained from the Meyer capacitance model. The circles indicate the onset of saturation according to Eq. (5.3.28).

5.3.6. SPICE Implementation of Basic MOSFET *I–V* Models

A wide variety of MOSFET models have been developed and implemented in the different versions of SPICE. Here we briefly discuss the implementation of three standard models, known as MOSFET Levels 1, 2, and 3, which are related to the basic models presented above. Level 1 is a slightly modified version of the SCCM (Shichman–Hodges model), while Levels 2 and 3 may be regarded as enhanced versions of the MM and the VSM, respectively. Only the basic features of their implementation in SPICE are presented here.

   Level 1, which is the simplest model, is mostly useful for rough estimates, and is not recommended where some degree of precision is required. Levels 2 and 3 have approximately the same level of precision, but Level 3 is normally preferable because it is more efficient in terms of convergence and simulation time. For a more comprehensive account, we refer to the book by Massobrio and Antognetti (1993).

Level 1. The threshold voltage V_T for a MOSFET with a uniformly doped substrate is given by Eq. (5.3.8), which, in terms of Level 1 SPICE parameters (shown in boldface), can be rewritten as

$$V_T = \mathbf{VTO} + \mathbf{GAMMA}(\sqrt{\mathbf{PHI} - V_{BS}} - \sqrt{\mathbf{PHI}}) \qquad (5.3.37)$$

Here **VTO** is the threshold voltage at zero substrate bias ($V_{BS} = 0$), **GAMMA** $= \gamma = \sqrt{2\epsilon_s q N_a}/c_i$ is the body-effect parameter, and **PHI** $= 2\varphi_b = 2V_{th}\ln(N_a/n_i)$ is the semiconductor surface potential under the gate at threshold (for $V_{BS} = 0$). Alternatively, we may specify a set of parameters consisting of the gate oxide thickness **TOX** $= d_i$, the substrate doping **NSUB** $= N_a$, and the mobility **UO**.

   The Level 1 *I–V* characteristics are similar to those of Eq. (5.3.12), except that the effective gate length is adjusted for lateral diffusion of dopants under the gate (parameter **LD**) and for an additional effective channel length reduction in saturation (parameter **LAMBDA**). The former effect is associated with the ion implantation of source and drain contacts, and the second effect, known as channel length modulation, concerns the finite extent of the saturated part of the channel near drain when the drain bias exceeds the saturation voltage (see, e.g., Lee et al., 1993). The resulting *I–V* characteristics can be written as follows:

$$I_d = \frac{\mathbf{W} \cdot \mathbf{KP}(1 + \mathbf{LAMBDA} \cdot V_{DS})}{\mathbf{L} - 2\mathbf{LD}} \times \begin{cases} V_{GT}V_{DS} - \tfrac{1}{2}V_{DS}^2 & V_{DS} \le V_{GT} \\ \tfrac{1}{2}V_{GT}^2 & V_{DS} > V_{GT} \\ 0 & V_{GT} < 0 \end{cases} \qquad (5.3.38)$$

Here **W** and **L** are the nominal MOSFET gate width and gate length, respectively (defined in the device line of the circuit description), and **KP** $= \mu_n c_i$ (sometimes referred to as the transconductance parameter, but different from β introduced earlier).

   In addition, MOSFET Level 1 incorporates the Meyer capacitance model, slightly modified to avoid the discontinuities in C_{GS} and C_{GB} at threshold (see Massobrio and Antognietti, 1993). Level 1 also includes parameters associated with the overlap capacitances between the gate–source and gate–drain, with the source and drain series resistances, with the junctions between the substrate and the source and drain contacts, and with noise. The temperature dependencies of key parameters such as **UO (KP)** and **PHI** (V_T) are also contained in the SPICE model code.

Level 2. Equation (5.3.5) shows the Meyer expression for the I–V characteristics below saturation. This model is modified as follows in MOSFET Level 2, where the SPICE parameters have the same meaning as for Level 1:

$$I_d = \frac{W}{L - 2LD} \frac{KP}{1 - LAMBDA \cdot V_{DS}} \{(V_{GS} - V_{FB} - PHI - \tfrac{1}{2} V_{DS}) V_{DS}$$

$$- \tfrac{2}{3} GAMMA[(V_{DS} - V_{BS} + PHI)^{3/2} - (-V_{BS} + PHI)^{3/2}]\}$$

$$(5.3.39)$$

Again, the effects of lateral diffusion and channel length modulation are included. Note that the specification of the latter effect is now contained in the denominator, with an appropriate switch of sign. Strictly speaking, channel length modulation is associated with saturation, but its mathematical description has also been included in Eq. (5.3.39) for the purpose of maintaining continuity in the characteristics and in the output conductance at the onset of saturation.

The saturation voltage V_{SAT} is the same as that given in Eq. (5.3.6), that is,

$$V_{SAT} = V_{GS} - PHI - V_{FB} + \frac{GAMMA^2}{2} \left[1 - \sqrt{1 + \frac{4(V_{GS} - V_{FB})}{GAMMA^2}} \right] \quad (5.3.40)$$

and the saturation current I_{sat} is found by substituting V_{DS} by V_{SAT} in Eq. (5.3.39), setting **LAMBDA** equal to zero. In the saturation regime, the drain current is then modeled as

$$I_d = \frac{I_{sat}}{1 - LAMBDA \cdot V_{DS}} \quad (5.3.41)$$

We note that the flat-band voltage V_{FB} is determined by the work function of the gate material used (see Section 3.5). It is specified in terms of the SPICE parameter **TPG**, which has the value 1 for a poly-Si gate with the same type of doping as the substrate, -1 for a poly-Si gate with the opposite type of doping of the substrate, and 0 for an aluminum gate. The threshold voltage is the same as for Level 1, except that **VTO** can be defined by specifying parameters such as **TPG**, **NSUB**, and the semiconductor–oxide interface state density **NSS**. Likewise, **GAMMA** can be specified in terms of **TOX** and **NSUB** and **PHI** and **LAMBDA** in terms of **NSUB**.

In addition to the model expressions presented here, MOSFET Level 2 also contains refinements to express the bias dependence of the mobility (included in the parameter **KP**), the effects of velocity saturation (through the saturation velocity parameter **VMAX**), subthreshold current (using the parameter **NFS**), intrinsic capacitances (using Meyer's C–V model), junction capacitances associated with source and drain, various other parasitic effects (same as for Level 1), noise, and the temperature dependencies of key parameters.

Level 3. MOSFET Level 3 is a short-channel model containing several empirical expressions. The following below-saturation expression for the current–voltage characteristics may be viewed either as a generalization of the velocity saturation model given by Eqs. (5.3.27) and (5.3.28) or as a generalized power series expansion of Eq. (5.3.5) (the MM):

$$I_d = \frac{\mathbf{W}}{\mathbf{L} - 2\mathbf{LD}} \mathbf{KP} \left[V_{GT} V_{DS} - \frac{1 + F_B}{2} V_{DS}^2 \right] \qquad (5.3.42)$$

Here,

$$F_B = \frac{\mathbf{GAMMA} \cdot F_s}{2\sqrt{\mathbf{PHI} - V_{BS}}} + F_n \qquad (5.3.43)$$

is introduced to express the dependence of the depletion charge (neglected in the velocity saturation model) on device geometry. Short- and narrow-channel effects are contained in F_s and F_n, respectively. Also, F_s and F_n enter into the following generalized expressions for the threshold voltage:

$$V_T = V_{\mathrm{FB}} + \mathbf{PHI} + F_s \mathbf{GAMMA}(\sqrt{\mathbf{PHI} - V_{BS}} - \sqrt{\mathbf{PHI}})$$
$$+ F_n(\mathbf{PHI} - V_{BS}) - \sigma V_{DS} \qquad (5.3.44)$$

The last term in Eq. (5.3.44) is an empirical expression for the effect of static drain–channel feedback. [Today, we usually interpret the dependence of V_T on V_{DS} as resulting from DIBL that is, a drain-bias-induced lowering of the injection barrier between the source and the channel caused by the drain bias (see Chapter 6 for further details)].

The MOSFET Level 3 saturation voltage has the same form as that of the basic velocity saturation model [see Eq. (5.3.28)], the only difference being the presence of F_B and the parameter **LD**:

$$V_{\mathrm{SAT}} = \frac{V_{GT}}{1 + F_B} + \frac{\mathbf{VMAX} \cdot (\mathbf{L} - 2\mathbf{LD})}{\mathbf{UO}}$$
$$- \sqrt{\left(\frac{V_{GT}}{1 + F_B} \right)^2 + \left(\frac{\mathbf{VMAX} \cdot (\mathbf{L} - 2\mathbf{LD})}{\mathbf{UO}} \right)^2} \qquad (5.3.45)$$

Here **UO** is the mobility at threshold. The dependence of the low field mobility on gate bias is modeled according to

$$\mu = \frac{\mathbf{UO}}{1 + \mathbf{THETA} \cdot V_{GT}} \tag{5.3.46}$$

where **THETA** is an additional SPICE parameter.

MOSFET Level 3 essentially has the same additional features as mentioned for the previous model. In addition, Level 3 incorporates both a Meyer-type capacitance model and the charge-based capacitance model by Ward and Dutton. For a more comprehensive treatment of the subthreshold current, we refer to the discussion of advanced FET models in Chapter 6.

5.3.7. SPICE Example: MOSFET I–V Characteristics

AIM-Spice can easily be used as a curve tracer to calculate the current–voltage characteristics of transistors. A typical arrangement is shown in Fig. 5.3.9. The two independent voltage sources V_{ds} and V_{gs} are varied and the drain current of the transistor is calculated.

To be able to plot the drain current during the simulation in AIM-Spice, an additional independent voltage source with zero value is inserted between V_{ds} and the drain node of transistor M_1 to act as an ampere meter. Let us, for example, investigate how the simple Level 1 model and the more comprehensive Level 3 model fare in simulating the I–V characteristics of a modern submicrometer n-channel MOSFET with a channel length of 0.6 μm. The circuit description for the simulation with Level 1 is shown in Fig. 5.3.10. Note that except for the transistor gate dimensions defined in the `m1` device line (line 4), all nondefault device parameters are included in the `.model` statement.

To calculate the current–voltage characteristics, we select a DC Transfer Curve Analysis with two sweep sources: `vds` sweeps from 0 to 3 V in steps of 50 mV and `vgs` sweeps from 1 to 3 V in steps of 1 V. Before starting the simulation, we select to plot the current through the voltage source `vids`.

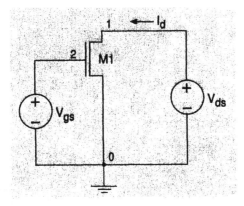

FIG. 5.3.9. AIM-Spice curve-tracer arrangement for calculating the current–voltage characteristics of an FET.

```
MOSFET I-V characteristics
vds 1 0 dc 0
vids 1 100 dc 0
m1 100 2 0 0 mn l=0.6u w=50u
vgs 2 0 dc 0
.model mn nmos level=1 vto=0.51 tox=100E-10
+ u0=610 ld=0.07u rs=20 rd=20
```

FIG. 5.3.10. Circuit description for tracing MOSFET current–voltage characteristics using the simple charge control MOSFET model (Level 1).

Typically, we would like to compare simulated *I–V* characteristics with measured data. AIM-Postprocessor has a command called Load Experimental Data that enables us to plot the measurements in the same graph as the simulated results. However, before we can generate such a plot in AIM-Postprocessor, we have to save the simulation results to an output file and format the measured data in a text file with a special format—the so-called circuit file (see Appendix A1.2).

The plot in Fig. 5.3.11 clearly shows that the simulated *I–V* characteristics ob-

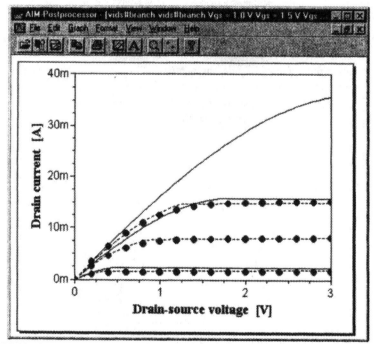

FIG. 5.3.11. Comparison of experimental short-channel MOSFET characteristics (symbols) with simulations based on the Level 1 Shichman–Hodges model (solid curves) and the Level 3 MOSFET model (dashed curves). Gate bias: 1, 2, and 3 V.

tained using the simple charge control (Level 1) MOSFET model agree very poorly with the measured data from the 0.6-μm device, especially in saturation. The reason for this poor fit is that MOSFET Level 1 does not include velocity saturation and other short-channel effects. For comparison, we include in Fig. 5.3.11 a more successful simulation of the same device using the MOSFET Level 3 model where short-channel effects are taken into account. To apply this model, we replace the model statement in the circuit description of Fig. 5.3.10 by

```
.model mn nmos level=3 vmax=1e5 vto=0.51 tox=100E-10
+ u0=610 ld=0.07u rs=20 rd=20 theta=0.03 delta=0
+ eta=0.01 nfs=1.3e12
```

5.4. BASIC MESFET MODELS

5.4.1. Shockley Model

The basic Shockley model for the GaAs MESFET is comparable to the SCCM for MOSFETs: Both rely on the gradual channel approximation (GCA) and on the assumption of a constant mobility. However, ignoring velocity saturation is a more severe assumption for GaAs FETs than for Si MOSFETs because of the low saturation field in GaAs. Hence, the applicability of the Shockley model is restricted to fairly long-channel devices, that is, with gate lengths of several micrometers. For modern-day MESFETs, with gate lengths on the order of 1 μm or less, it certainly gives a poor description. But we include the Shockley model here since it offers a suitable introduction to the more advanced models to be discussed later, in which velocity saturation and other refinements are included.

Figure 5.2.3 shows a schematic illustration of the MESFET. Here, however, we will concentrate on the intrinsic gate region of the MESFET, shown in Fig. 5.4.1.

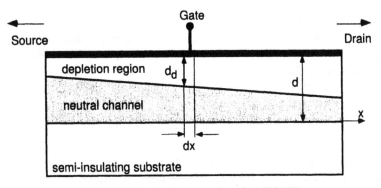

FIG. 5.4.1. Intrinsic gate region of a MESFET.

Assuming that GCA is valid, the depletion width d_d at position x in the channel can be expressed as follows:

$$d_d(x) = \sqrt{\frac{2\epsilon_s}{qN_d}[V_{bi} - V_{GS} + V(x)]} \tag{5.4.1}$$

where N_d is the doping of the active layer (assumed to be uniform), V_{bi} is the built-in voltage of the gate Schottky contact, V is the potential relative to the source side of the channel, V_{GS} is the intrinsic gate–source voltage, and ϵ_s is the semiconductor dielectric permittivity.

The threshold voltage V_T for the MESFET corresponds to the gate–source voltage at which the depletion width at zero drain–source bias $[V(x) = 0]$ equals the channel width, or in terms of Eq. (5.4.1),

$$V_T = V_{bi} - V_{po} \tag{5.4.2}$$

where V_{po} is the so-called pinch-off voltage that, for a uniformly doped active layer of thickness d, is given by

$$V_{po} = \frac{qN_d d^2}{2\epsilon_s} \tag{5.4.3}$$

At gate–source voltages above the threshold voltage, a neutral, conducting channel exists, which allows a significant drain current to pass when applying a drain–source bias. When $V_{GS} < V_T$, the neutral channel disappears and the drain current drops to a low value characteristic of the subthreshold regime.

From Eq. (5.4.1), it is obvious that the depletion width under the gate increases from source to drain when a positive drain–source bias is applied. The depletion width $d_d(L)$ at the drain side of the gate is obtained by replacing the channel potential by the intrinsic drain–source voltage V_{DS} in this expression. Without velocity saturation, $d_d(L)$ increases with increasing drain–source voltage until the channel is pinched off. Pinch-off occurs when $d_d(L) = d$, corresponding to $V_{DS} = V_{GS} - V_T \equiv V_{GT}$. As for the simple charge control MOSFET model, this defines the saturation drain–source voltage $V_{SAT} = V_{GT}$ for long-channel MESFETs.

I–V Model. In the Shockley model, we assume that the electron drift velocity is proportional to the longitudinal electric field, that is, $v_n = \mu_n\, dV/dx$. Following the derivation used for the current–voltage characteristics of the basic MOSFET models, the absolute value of the channel current of the MESFET can be written as

$$I_d = qN_d W[d - d_d(x)]\mu_n \frac{dV}{dx} \tag{5.4.4}$$

where W is the gate width and $d - d_d(x)$ is the thickness of the conducting (neutral) channel at position x. Integrating Eq. (5.4.4) over the entire gate length L leads to the following expression for the drain current:

$$I_d = g_o\{V_{DS} - \tfrac{2}{3}[(V_{DS} + V_{bi} - V_{GS})^{3/2} - (V_{bi} - V_{GS})^{3/2}]/V_{po}^{1/2} \qquad (5.4.5)$$

Here $g_o = qN_d\mu_n\, dW/L$ is the conductance of the full, undepleted active layer. Note that Eq. (5.4.5) is valid only for drain–source voltages below pinch-off, that is, for $V_{DS} \leq V_{SAT}$. The saturation drain current is obtained by using the condition $V_{DS} = V_{SAT} = V_{GT}$ in this expression.

C–V Model. As for the MOSFET, the dynamic variation in the stored charges of the MESFET is expressed in terms of capacitive elements. Again, we distinguish between the parasitic capacitive elements and those of the intrinsic device. Here we concentrate on the intrinsic capacitive elements associated with the depletion charge in the MESFET intrinsic gate region.

The simplest approach is to define an intrinsic gate–source capacitance C_{GS} and an intrinsic gate–drain capacitance C_{GD} in terms of two separate Schottky barrier diodes connecting the gate to source and drain, respectively. Furthermore, each of the two diodes is assumed to occupy half the gate area and to have a constant channel potential equal to the source and drain potentials, respectively. The large-signal equivalent circuit for this model, ignoring the gate leakage current, is illustrated in Fig. 5.4.2. This equivalent circuit is similar to that of the Meyer model for MOSFETs (see Fig. 5.3.4), except that the substrate terminal is removed

From Eqs. (5.4.1)–(5.4.3), we find the following gate charges associated with the two diodes:

$$Q_{GS} = -\tfrac{1}{2}qN_dWLd_d(x = 0) = -\tfrac{1}{2}qN_dWLd\sqrt{1 - V_{GT}/V_{po}} \qquad (5.4.6)$$

$$Q_{GD} = -\tfrac{1}{2}qN_dWLd_d(x = L) = -\tfrac{1}{2}qN_dWLd\sqrt{1 - (V_{GD} - V_T)/V_{po}} \qquad (5.4.7)$$

from which we obtain the following capacitances of the "two-diode" model in the nonsaturated regime, that is, $V_{DS} < V_{SAT}$,

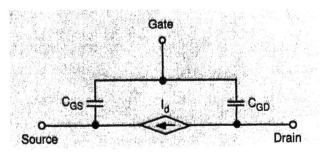

FIG. 5.4.2. Large-signal equivalent circuit corresponding to the intrinsic "two-diode" and Meyer-type MESFET models. The gate leakage current is ignored.

$$C_{GS} = \frac{dQ_{GS}}{dV_{GS}} = \frac{C_o/2}{\sqrt{1 - V_{GT}/V_{po}}} \tag{5.4.8}$$

$$C_{GD} = \frac{dQ_{GD}}{dV_{GD}} = \frac{C_o/2}{\sqrt{1 - (V_{GT} - V_{DS})/V_{po}}} \tag{5.4.9}$$

Here $C_o = \epsilon_s WL/d$ corresponds to the gate channel capacitance at threshold (i.e., fully depleted channel) using the above-threshold formalism.

As for the MOSFET, the gate charge becomes very insensitive to the drain bias once we enter saturation. Hence, in this regime, the total gate charge will be solely determined by V_{GS}, and we find that C_{GS} remains at the same value as in Eq. (5.4.8) while $C_{GD} \rightarrow 0$ in saturation.

Note that this model is only valid above threshold, and it does not account for the charges associated with the depletion zone extensions beyond the gate region, toward the source and drain (see Fig. 5.2.3). In fact, the capacitances of the depletion zone extensions, together with other parasitic capacitances, will become dominant in the subthreshold regime.

An alternative and somewhat more precise C–V model for MESFETs is obtained from an adaptation of the Meyer MOSFET capacitance model. As was discussed in Section 5.3, the analysis by Meyer is based on the SCCM applicable for MOSFETs, which yields the total gate charge Q_G as a function of the bias voltages. However, applying the same procedure to MESFETs results in quite complicated expressions for C_{GS} and C_{GD}. Instead, we recognize that the dependencies of C_{GS} and C_{GD} on the drain–source bias are basically quite similar for long-channel MOSFETs and MESFETs. Hence, a reasonable approximation for the MESFET is to adopt the MOSFET expressions of Eqs. (5.3.18) and (5.3.19), but replacing the MOSFET oxide capacitance C_i (which is the total above-threshold gate channel capacitance for $V_{DS} = 0$) by the corresponding gate channel capacitance C_{ch} for the MESFET.

The total MESFET gate charge at zero drain–source bias is simply

$$Q_G = 2Q_{GS} = -qN_d WLd\sqrt{1 - V_{GT}/V_{po}} \tag{5.4.10}$$

from which we obtain for the gate channel capacitance

$$C_{ch} = \frac{dQ_G}{dV_{GS}} = \frac{C_o}{\sqrt{1 - V_{GT}/V_{bi}}} \tag{5.4.11}$$

Hence, the expressions for C_{GS} and C_{GD} in this model, in the nonsaturated above-threshold regime, are as follows:

$$C_{GS} = \tfrac{2}{3} C_{ch}\left[1 - \left(\frac{V_{GT} - V_{DS}}{2V_{GT} - V_{DS}}\right)^2\right] \tag{5.4.12}$$

$$C_{GS} = \tfrac{2}{3} C_{ch}\left[1 - \left(\frac{V_{GT}}{2V_{GT} - V_{DS}}\right)^2\right] \tag{5.4.13}$$

In saturation, we find

$$C_{GS} = \tfrac{2}{3}C_{\text{ch}} \qquad C_{GD} = 0 \qquad\qquad (5.4.14)$$

Note that $V_{\text{SAT}} = V_{GT}$ in the Shockley model.

This Meyer-type $C–V$ model does not include short-channel effects and a description of the subthreshold regime. However, effects of velocity saturation are considered in Section 5.4.2, and the subthreshold regime is considered in Chapter 6, where this model is further developed within the framework of the universal modeling concept.

Figure 5.4.3 shows a comparison of the predictions of the two-diode and the Meyer-type capacitance models for MESFETs. The capacitances are normalized to C_{ch}. We note that both models give capacitance values of the same order of magnitude, but their dependencies on drain–source bias are quite different. Especially the two-diode model is quite crude.

As discussed previously, precise FET capacitance modeling requires a careful and consistent analysis of bias voltage dependence of the spatial distribution of the intrinsic device charges, preferably using the charge-conserving modeling approach similar to that developed for MOSFETs by Ward and Dutton (1978). Such a model for MESFETs, which is also valid for short gate lengths, is discussed in Chapter 6.

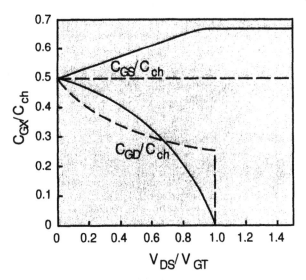

FIG. 5.4.3. Normalized intrinsic MESFET capacitances versus V_{DS}/V_{GT} for the Meyer-type model (solid curves) and for the two-diode model (dashed curves). In the latter case, we have assumed $V_{GT} = 0.75V_{\text{po}}$.

5.4.2. Velocity Saturation Models

In the Shockley model, the electron drift velocity was assumed to be proportional to the longitudinal electric field in the channel. The simplest way of dealing with velocity saturation is to apply the two-piece linear velocity–field relationship defined in Eq. (5.3.25), as was done for MOSFETs in Section 5.3.4. A comparison of this model with a realistic velocity–field relationship for GaAs is shown in Fig. 5.4.4. Although the peak in the real drift velocity is not reproduced by the two-piece model, it still represents the important low- and high-field regions of the velocity–field relationship quite well. In addition, transient effects such as velocity overshoot play an important role in modern submicrometer GaAs FETs, reducing the significance of a precise modeling of the stationary velocity–field relationship. In fact, much is gained by using a simple model combined with a judicial choice of parameters.

I–V Model. Following the arguments used for the MOSFET, we note that the results from the Shockley model are still valid as long as the maximum electric field in the channel $F(L)$ does not exceed the saturation field F_s. Hence, the saturation voltage for this case, defined as the drain–source voltage at the onset of velocity saturation, can be determined implicitly from Eq. (5.4.5) in combination with Eqs. (5.4.1) and (5.4.2) using $F(L) = F_s$ and $V(L) \leq V_{SAT}$, as discussed by Shur (1987) (see also Lee et al., (1993). Moreover, it can be shown that the Shockley saturation voltage $V_{SAT} = V_{GT}$ is recovered when $F_s L \equiv V_L >> V_{po}$. In the opposite limit, when $V_L << V_{po}$, which corresponds to near velocity saturation in the entire channel, we find $V_{SAT} = V_L$. For intermediate cases, a simple interpolation formula for the saturation voltage can be established by combining the results for the two limiting cases:

$$V_{SAT} \approx \left(\frac{1}{V_L} + \frac{1}{V_{GT}} \right)^{-1} \tag{5.4.15}$$

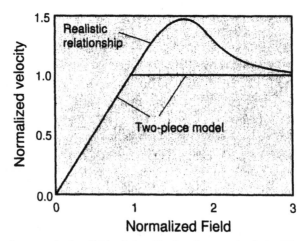

FIG. 5.4.4. Stationary velocity–field relationship for electrons in GaAs. The electric field and the velocity are normalized to F_s and v_s, respectively.

Likewise, the following approximation for the saturation current is valid for devices with relatively low pinch-off voltages (Shur, 1987):

$$I_{\text{sat}} \approx \beta V_{GT}^2 \tag{5.4.16}$$

where

$$\beta = \frac{2\epsilon_s v_s W}{d(V_{po} + 3V_L)} \tag{5.4.17}$$

is called the transconductance parameter. We note that the "square-law" expression of Eq. (5.4.16) is of the same form as that used in SPICE modeling of the saturation current in JFETs, and it has also been used to describe the saturation characteristics in SPICE simulation of GaAs MESFETs.

Statz et al. (1987) proposed the following more general version of Eq. (5.4.16) to also cover devices with higher pinch-off voltages:

$$I_{\text{sat}} \approx \frac{\beta V_{GT}^2}{1 + \alpha V_{GT}} \tag{5.4.18}$$

This expression approaches a linear behavior in I_{sat} versus V_{GT} at large V_{GT}, in better accordance with experimental observations.

The velocity saturation model allows us to make rough estimates of the intrinsic high-speed performance of GaAs MESFETs. Such estimates show that high cut-off frequencies can be obtained by using a small gate length and a small pinch-off voltage. Normally, it is also desirable to have devices with a high current drive, which favors a large doping density per unit area in the channel. The best trade-off is therefore to use a thin and highly doped channel. A thin channel is also desirable for the purpose of reducing short-channel effects in submicrometer devices.

Source and drain series resistances R_s and R_d may play an important role in determining the current–voltage characteristics of GaAs MESFETs. These resistances can be taken into account by using the relationships between intrinsic and extrinsic voltages of Eqs. (5.3.31) and (5.3.32). The saturation current in terms of the extrinsic gate voltage overdrive V_{gt} is readily obtained by combining Eqs. (5.3.31) and (5.4.16):

$$I_{\text{sat}} = \frac{2\beta V_{gt}^2}{1 + 2\beta V_{gt} R_s + \sqrt{1 + 4\beta V_{gt} R_s}} \tag{5.4.19}$$

For device modeling suitable for CAD, one has to model the current–voltage characteristics in the entire range of drain–source voltages, not only in the saturation regime. An empirical interpolation expression for the full, extrinsic MESFET I–V characteristics was proposed by Curtice (1980) in terms of a hyperbolic tangent function:

$$I_d = I_{\text{sat}}(1 + \lambda V_{ds}) \tanh\left(\frac{g_{\text{ch}} V_{ds}}{I_{\text{sat}}}\right) \tag{5.4.20}$$

Here, λ is an empirical constant to account for the finite output conductance in saturation, and g_{ch} is the extrinsic channel conductance of the linear region given by

$$g_{ch} = \frac{g_{chi}}{1 + g_{chi}(R_s + R_d)} \tag{5.4.21}$$

where g_{chi} is the intrinsic channel conductance at very low drain–source voltage. For a uniformly doped channel, we find, from Eq. (5.4.5),

$$g_{chi} = \left.\frac{\partial I_d(V_{DS} \to 0)}{\partial V_{DS}}\right|_{V_{GS}} = \frac{q N_d \mu_n W d}{L}\left(1 - \sqrt{1 - \frac{V_{GT}}{V_{po}}}\right) \tag{5.4.22}$$

The finite output conductance in saturation, described in terms of the parameter λ in Eq. (5.4.20), can be related to short-channel effects (channel length modulation) and to parasitic currents through the substrate.

The analytical I–V models discussed above are suitable for design of GaAs MESFET devices and circuits. However, they do not explicitly take into account important effects such as subthreshold current, drain-voltage-induced shift in the threshold voltage (DIBL), and gate leakage. These effects will be addressed in Chapter 6, where more comprehensive FET models are discussed.

C–V Model. In the velocity saturation model for MESFETs, the Meyer-type capacitance model developed in Section 5.4.1 for the nonsaturated, above-threshold regime is still valid up to the new saturation voltage given by Eq. (5.4.15). The capacitance at the onset of saturation is found by using $V_{DS} = V_{SAT}$ in Eqs. (5.4.12) and (5.4.13):

$$C_{GSs} = \tfrac{2}{3}C_{ch}\left[1 - \left(\frac{V_{SAT}}{V_{SAT} + V_L}\right)^2\right] \tag{5.4.23}$$

$$C_{GDs} = \tfrac{2}{3}C_{ch}\left[1 - \left(\frac{V_L}{V_{SAT} + V_L}\right)^2\right] \tag{5.4.24}$$

Once we enter velocity saturation, the gate charge in MESFETs changes very little with V_{DS}, similar to what takes place in MOSFETs (see Section 4.3.4). Hence, the capacitances have to approach the same limiting values in saturation as the Meyer-type capacitances, that is, $C_{GS}/C_{ch} \to \tfrac{2}{3}$ and $C_{GD}/C_{ch} \to 0$. But since the behavior of C_{GS} and C_{GD} in the velocity saturation model and in the Meyer-type model coincide for $V_{DS} < V_{SAT}$ and have the same asymptotic values in saturation, the Meyer-type capacitance model offers a fairly good approximation for the MESFET capacitances also in short-channel devices. However, the functional form of the approach toward the saturation values depends on the form of the velocity–field relationship chosen. We will return with more flexible C–V expressions in our discussion of the universal MESFET model in Chapter 6.

5.4.3. SPICE Implementation of Basic MESFET Model

Several of the models discussed above have been implemented in various versions of SPICE. PSpice, for example, incorporates three models: the Curtice model of Eq. (5.4.20) (as Level 1); the Statz model of Eq. (5.4.18) (as Level 2), including a polynomial representation of the hyperbolic tangent function used by Curtice; and the TriQuint model (as Level 3), which is a generalized version of the Statz model.

In AIM-Spice, the Statz model is implemented as Level 1. The following expressions are used for the nonsaturated regime of the $I–V$ characteristics ($V_{DS} \leq$ 3/**ALPHA**):

$$I_d = (1 + \textbf{LAMBDA} \cdot V_{DS}) \frac{\textbf{BETA} \cdot V_{GT}^2}{1 + \textbf{B} \cdot V_{GT}} \left[1 - \left(1 - \frac{\textbf{ALPHA} \cdot V_{DS}}{3} \right)^3 \right] \quad (5.4.25)$$

and for the saturation regime ($V_{DS} > 3/\textbf{ALPHA}$):

$$I_d = (1 + \textbf{LAMBDA} \cdot V_{DS}) \frac{\textbf{BETA} \cdot V_{GT}^2}{1 + \textbf{B} \cdot V_{GT}} \quad (5.4.26)$$

Here, **ALPHA** = g_{ch}/I_{sat} is a saturation voltage parameter, **LAMBDA** = λ is the channel length modulation parameter, **BETA** = $2v_s\epsilon_s(W/L)(V_{po} + 3V_L)$ is the transconductance parameter [see Eq. (5.4.17)], and **B** is called the doping tail extension parameter [α in Eq. (5.4.18)]. The above expressions also contain the threshold voltage **VTO**.

The present model also includes a capacitance model for the intrinsic MESFET (Statz et al., 1987), with features similar to the Meyer-type MESFET capacitance model presented above. A noise model and temperature dependencies of key parameters are also included. However, important effects such as subthreshold current, DIBL, and gate leakage are ignored.

5.4.4. SPICE Example: MESFET Subthreshold Characteristics

Ideally, the drain current of an FET device should reduce to zero when the device is biased below threshold. However, a more careful analysis (see Chapter 6) shows that a residual subthreshold current exists in this regime. Although the current in the subthreshold regime falls off exponentially with decreasing gate voltage (n-channel device), it is still of great concern since it has consequences for logic levels and for power dissipation in digital operations, as well as for the holding time in dynamic circuits.

Here we investigate the turn-off properties of the basic MESFET Level 1 (Statz) model by comparing simulated and measured subthreshold characteristics for a 1-μm MESFET device. The nominal threshold voltage of the device is −1.3 V. The circuit description used for calculating the subthreshold current is shown in Fig. 5.4.5. Note that the voltage source vids is inserted in order to measure the transistor drain current in AIM-Spice.

```
Mesfet subthreshold characteristics
Vds 1 0
vids 1 2 dc 0
Vgs 3 0 dc 0
z1 2 3 0 mesmod a=1.4
.model mesmod nmf level=1 rd=46 rs=46 vt0=-1.3
+ lambda=0.03 alpha=3 beta=1.4e-3
* z1 2 3 0 lev2 l=1u w=20u
* .model lev2 nmf level=2 d=0.12u mu=0.23 vs=1.8e5
*+ m=3.3 vto=-1.3 eta=1.82 lambda=0.044 sigma0=0.09
*+ vsigma=0.1 vsigmat=0.9 rdi=46 rsi=46 delta=5
*+ nd=2.1e23
```

FIG. 5.4.5. Circuit description for simulation of subthreshold characteristics of a 1-μm MESFET using the basic (Level 1) MESFET model of AIM-Spice. The two statements on the comment lines refer to the universal (Level 2) MESFET model and are activated by shifting the comment symbols to the corresponding Level 1 statements.

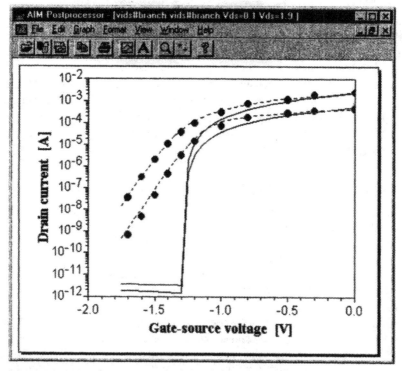

FIG. 5.4.6. Comparison of experimental 1-μm MESFET characteristics (symbols) with AIM-Spice simulations based on the Level 1 Statz model (solid curves) and the Level 2 universal MESFET model (dashed curves).

A DC Transfer Curve Analysis was performed with Vgs as the first sweep source ranging from −1.75 to 0 V in steps of 50 mV and with Vds as the second sweep source ranging from 0.1 to 1.9 V in steps of 1.8 V. From the comparison of the simulation results with the measured data shown in Fig. 5.4.6, we conclude that the simple Level 1 MESFET model does not offer a satisfactory description of the subthreshold condition. However, a more successful simulation of the subthreshold characteristics was performed using our universal MESFET model (Level 2 in AIM-Spice; see Chapter 6) where the subthreshold regime is properly included. The results of this simulation are also shown in Fig. 5.4.6. In order to use this model, the comment symbols (∗) in the circuit description should be moved to lines 5–7.

5.5. BASIC HFET MODEL

From the simple charge control concept, we can also derive a basic HFET model. Such a model was developed by Delagebeaudeauf and Linh (1982) and by Lee et al. (1983). A schematic cross section of the HFET was shown in Fig. 5.2.5. We now start by considering the band diagram in Fig. 5.5.1 of a conventional AlGaAs/GaAs HFET structure with flat bands in the GaAs buffer. As can be seen from this figure, the flat-band voltage is given by

$$V_{FB} = \phi_b - V_N - \frac{\Delta E_C + \Delta E_F}{q} \qquad (5.5.1)$$

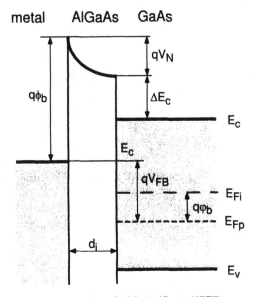

FIG. 5.5.1. Band diagram of a conventional AlGaAs/GaAs HFET structure with flat bands in the GaAs buffer.

where $q\phi_b$ is the metal–semiconductor energy barrier, V_N is the voltage drop across the AlGaAs layer at flat-band condition, ΔE_c is the conduction band discontinuity, and $\Delta E_F = E_c - E_{Fp}$. We assume that the GaAs buffer layer has a p-type doping corresponding to a Fermi energy E_{Fp}. The voltage V_N can be found by integrating Poisson's equation twice to give

$$V_N = q \int_0^{d_i} \frac{N_d(x)}{\epsilon_i(x)} x \, dx \qquad (5.5.2)$$

Here d_i is the thickness, N_d is the doping density, and ϵ_i is the dielectric permeability of the wide-band-gap AlGaAs layer. Note that Eq. (5.5.2) has been generalized to account for cases where both the doping density and the composition of the AlGaAs layer can vary with distance x.

The HFET threshold voltage is given by a similar expression as that used for the MOSFET in Eq. (5.3.7), except that c_i is now interpreted as the capacitance per unit area of the AlGaAs layer. However, in order to find a more manageable expression for V_T, we assume that the doping level in the GaAs layer is relatively low, which gives $V_T \approx V_{FB}$. In addition, ΔE_F at the interface can be estimated to be close to zero for the AlGaAs/GaAs system at threshold (see Lee et al., 1993). With these approximations, the threshold voltage simplifies to

$$V_T \approx \phi_b - \frac{qN_d d_i^2}{2\epsilon_i} - \frac{\Delta E_c}{q} \qquad (5.5.3)$$

for a uniform doping in the AlGaAs layer and

$$V_T \approx \phi_b - \frac{qn_\delta d_\delta}{\epsilon_i} - \frac{\Delta E_c}{q} \qquad (5.5.4)$$

for a delta-doped structure where d_δ is the distance between the metal gate and the doped plane and n_δ is the sheet concentration of donors in the doped plane. The calculated dependencies of V_T on d_i for uniformly doped and d_δ for planar doped devices are shown in Fig. 5.5.2.

The above-threshold sheet density of induced electrons in HFETs can be described in terms of the simple MOSFET parallel-plate capacitor model of Eq. (5.3.10). However, at large gate–source voltages, this model becomes invalid for HFETs owing to the transfer of electrons into the AlGaAs layer, and to gate leakage current (see Chapter 6). But if we neglect these effects, we obtain basic HFET models that are identical to those of the MOSFET, allowing us to use the MOSFET models implemented in SPICE for simulation of HFET circuits.

5.5.1. SPICE Example: HFET Saturation Characteristics

Normally, separate HFET device models are not included in SPICE simulators. Instead, as already indicated, MOSFET models can be used to simulate HFET devices

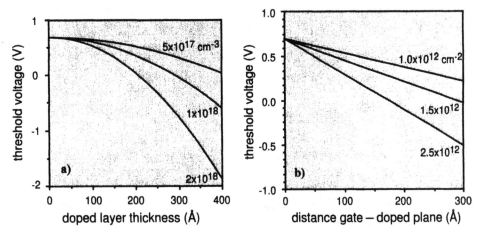

FIG. 5.5.2. Threshold voltage of (*a*) conventional and (*b*) delta-doped HFETs. (After Shur, 1990.)

and circuits. This approach can be reasonably accurate in some cases, but important effects related to gate leakage and to the transfer of carriers into the wide-gap material are, of course, not included in the MOSFET models. For example, the interlayer transfer of carriers causes a saturation of the carrier density in the conducting channel and in the current level (transconductance compression). In the present example, we illustrate this effect by comparing transfer characteristics (I_d versus V_{gs}) using the Level 3 MOSFET model and our universal HFET model (see Chapter 6). The circuit description for the simulation using the MOSFET model is shown in Fig. 5.5.3.

In order to isolate the HFET specific carrier density saturation effect, we turn off all other effects that contribute to a reduction of the slope of the transfer characteristics by setting all parasitic resistances (`rs`, `rd`) and the dependence of mobility on gate bias (`theta`) to zero in the `.model` statement of Fig. 5.5.3. We want to run a DC Transfer Curve Analysis, sweeping the voltage source `vgs` from 0 to 1 V in steps of 10 mV.

```
HFET Id versus Vgs characteristic
m1 3 2 0 0 nlev3 l=1u w=10u
vgs 2 0 dc 0
vds 1 0 dc 0.1
vids 1 3 dc 0
.model nlev3 nmos level=3 vto=0.13 rd=0 rs=0
+ tox=250E-10 ld=0 vmax=1.5e5 kp=1e-3 theta=0
+ delta=1.5 eta=0.1 kappa=1
```

FIG. 5.5.3. Circuit description for simulating HFET transfer characteristics using the Level 3 MOSFET model.

The simulation is repeated using the universal HFET model included in AIM-Spice. To select this HFET model, we have to replace the second line in the circuit description by

```
a1 3 2 0 hfet l=1u w=10u
```

and the .model statement by

```
.model hfet nhfet rdi=0 rsi=0 m=2.57 lambda=0.17
+ vs=1.5e5 mu=0.385 vt0=0.13 eta=1.32 sigma0=0.04
+ vsigma=0.1 Vsigmat=0.3 js1s=0 js1d=0 nmax=6e15
```

Here we have turned off the gate leakage current by setting the parameters js1s and js1d to zero.

The results of the simulation shown in Fig. 5.5.4 clearly illustrate the importance of carrier density saturation in the HFET channel and the shortcomings of the

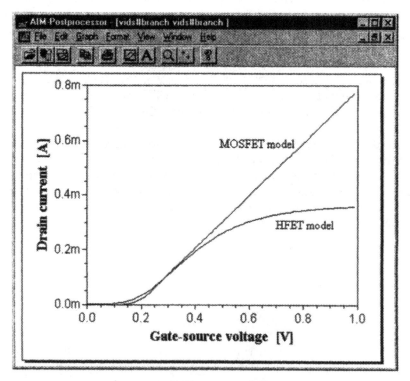

FIG. 5.5.4. AIM-Spice simulations of HFET transfer characteristics using the Level 3 MOS-FET model and the universal HFET model.

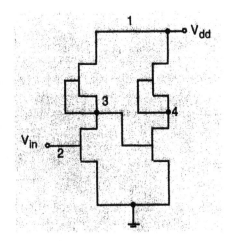

FIG. 5.5.5. Two stage HFET DCFL Inverter.

```
DCFL inverter circuit

.subckt inv 1 2 3
*             | | |
*           Vdd | |
*             Vin |
*               Vout
a1 1 3 3 aload l=1u w=10u
a2 3 2 0 adrv l=1u w=10u
.ends

vdd 1 0 dc 2
vin 2 0 dc 0 pwl(0,0V 1ns,0V 1.005ns,1V 2ns,1V)
x1 1 2 3 inv
x2 1 3 4 inv

.model adrv nhfet rd=60 rs=60 m=2.57 lambda=0.17
+ vs=1.5e5 mu=0.385 vt0=0.3 eta=1.32 sigma0=0.04
+ vsigma=0.1 vsigmat=0.3 js1s=1e-12 js1d=1e-12
+ nmax=6e15

.model aload nhfet rd=60 rs=60 m=2.57 lambda=0.17
+ vs=1.5e5 mu=0.385 vt0=-0.3 eta=1.32 sigma0=0.04
+ vsigma=0.1 vsigmat=0.3 js1s=1e-12 js1d=1e-12
+ nmax=6e15
```

FIG. 5.5.6. Circuit description of the HFET inverter circuit shown in Fig. 5.5.5.

MOSFET model in cases of strong forward gate bias. In addition to the current saturation effect demonstrated here, we normally also have a significant gate leakage current at comparable forward gate biases (see Chapter 6).

The saturation of the drain current gives rise to a reduction of the transconductance, the so-called transconductance compression effect, which has an adverse affect on the speed of HFET digital circuits. To illustrate this, we simulate an HFET DCFL inverter with a second inverter connected to the output node and compare the delay times with and without transconductance compression. The purpose of the second inverter is to simulate a realistic load capacitance. The circuit is shown in Fig. 5.5.5 and the corresponding circuit description is shown in Fig. 5.5.6.

A DCFL inverter utilizes two HFET devices. The driver transistor is an enhancement mode device and the load transistor is a depletion mode device having a negative threshold voltage. Hence, we need two HFET model definitions in Fig. 5.5.6 with different values of the threshold voltage parameter VTO.

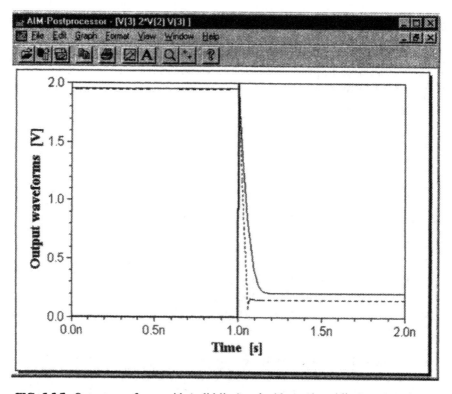

FIG. 5.5.7. Output waveforms with (solid line) and without (dotted line) transconductance compression. The thick vertical line indicates where the rising edge of the input voltage occurs.

In the circuit description, we use a subcircuit definition to define the inverter as a circuit element. This is a very convenient feature of SPICE that helps to simplify the netlist. The subcircuit feature is described in detail in Appendix A2. The inverter subcircuit is defined in lines 2–10 and is referenced in the main circuit in lines 14 and 15. Note also that we have added source and drain series resistances (rd=rs=60) to make the simulation more realistic. The parameter NMAX gives rise to the transconductance compression, and its value is the same as that used in the previous simulation. To turn the transconductance compression off, we simply increase the value of NMAX by a factor of 100.

To compare the inverter delay times for the first inverter, we apply a voltage step to the input node (node 2) as defined in line 13 and run a transient analysis using a step size of 0.1 ns and a final time of 2 ns. We then plot the output waveform at node 3 and use the cursor feature in Postprocessor to calculate the delay times. Output waveforms are shown in Fig. 5.5.7. The delay time with the transconductance compression turned on is found to be 47.1 ps compared to 27.8 ps with the effect turned off, an increase of almost 70%.

REFERENCES

W. R. Curtice, *IEEE Transactions on Microwave Theory and Techniques*, MTT-28, p. 448 (1980).

D. Delagebeaudeuf and N. T. Linh, "Metal-(*n*)AlGaAs-GaAs Two-Dimensional Electron Gas FET," *IEEE Transactions on Electron Devices*, vol. ED-29, no. 6, pp. 955–960 (1982).

K. Lee, M. Shur, T. J. Drummond, and H. Morkoç, "Electron Density of the Two-Dimensional Electron Gas in Modulation Doped Layers", *Journal of Applied Physics*, vol. 54, no. 4, pp. 2093–2096 (1983).

K. Lee, M. Shur, T. A. Fjeldly, and T. Ytterdal, *Semiconductor Device Modeling for VLSI*, Series in Electronics and VLSI, Prentice-Hall, Englewood Cliffs, NJ (1993).

G. Massobrio and P. Antognetti, *Semiconductor Device Modeling with SPICE*, 2nd ed., Mc-Graw-Hill, New York (1993).

J. E. Meyer, "MOS Models and Circuit Simulation," *RCA Review*, vol. 32, pp. 42–63 (1971).

H. Schichman and D. A. Hodges, "Modeling and Simulation of Insulated-Gate Field Effect Transistor Switching Circuits", *IEEE Journal of Solid-State Circuits*, vol. SC-3, pp. 285–289 (1968).

M. Shur, *GaAs Devices and Circuits*, Plenum, New York (1987).

M. Shur, *Physics of Semiconductor Devices*, Prentice-Hall, Englewood Cliffs, NJ (1990).

H. Statz, P. Newman, I. W. Smith, R. A. Pucel, and H. A. Haus, *IEEE Transactions on Electron Devices*, vol. ED-34, p. 160 (1987).

D. E. Ward and R. W. Dutton, "A Charge-Oriented Model for MOS Transistor Capacitances," *IEEE Journal of Solid-State Circuits*, vol. SC-13, pp. 703–708 (1978).

PROBLEMS

5.2.1. Consider the circuit in Fig. P5.2.1.

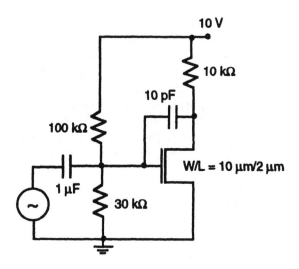

FIG. P5.2.1

Use default parameters for the n-channel MOSFET and compute the voltage gain as a function of frequency from 10 kHz to 1 GHz with and without the capacitance of 10 pF between the gate node and the drain node of the transistor. Comment on the role of this gate-to-drain capacitance.

5.3.1. Plot the threshold voltage as a function of the substrate bias using the MOSFET parameters specified in the caption of Fig. 5.3.2.

5.3.2. Derive the expressions for the long-channel MOSFET drain current and the three Meyer capacitances using the simple charge control model of Section 5.3.3. Compare the transconductance and the channel conductance for this model.

5.3.3. Derive the intrinsic expressions of Eqs. (5.3.27) and (5.3.28) for the drain current and the saturation voltage in the velocity saturation model of Section 5.3.4. Accounting for the source and drain series resistances, derive the corresponding extrinsic expressions of Eqs. (5.3.33) and (5.3.34). Use these results to find the extrinsic transconductance and saturation voltage for the case when the parasitic resistances are very large. Comment on the result.

5.3.4. (a) Review the example in Section 5.3.7.
(b) Modify the example to calculate the drain current versus gate–source voltage with the drain–source voltage as a parameter. Use drain–source volt-

age values of 0.1 and 1.1 V for both Level 1 and Level 3. Show the resulting drain currents versus V_{GS} in a semilog plot. Comment on the results.

5.3.5. Use SPICE to show graphically how the *I–V* characteristics of a MOSFET changes as R_s increases and R_d increases. What are the implications of an increase in R_s and R_d for the noise margin of an inverter?

5.3.6. Figure P5.3.6 shows an NMOS inverter utilizing an enhancement mode switching transistor and a depletion mode load transistor.

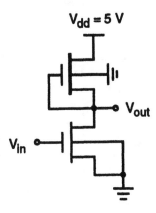

FIG. P5.3.6

(a) Use the SPICE model descriptions below to calculate and plot the dc transfer curve of the inverter. Assume $W = 20$ μm and $L = 1$ μm for both transistors.

```
.model load nmos level=3 vmax=1.2e5 vto=-0.7 tox=180E-10
+ rs=30 rd=30 eta=0.03 delta=1.5 kappa=0.45 gamma=0

.model driver nmos level=3 vmax=1.2e5 vto=0.7 tox=180E-10
+ rs=30 rd=30 eta=0.03 delta=1.5 kappa=0.45 gamma=0
```

(b) Find a *W/L* for the depletion load such that the logic inversion voltage is 2.5 V.
(c) Discuss the body effect by setting gamma=0.4 and repeat the simulation.

5.3.7. Simulate the response to the input voltage pulse indicated for the three MOSFET circuits in Fig. P5.3.7. Use $W = 20$ μm and $L = 1$ μm for all transistors and default SPICE parameters for Level 1 and Level 3. Compare the results.

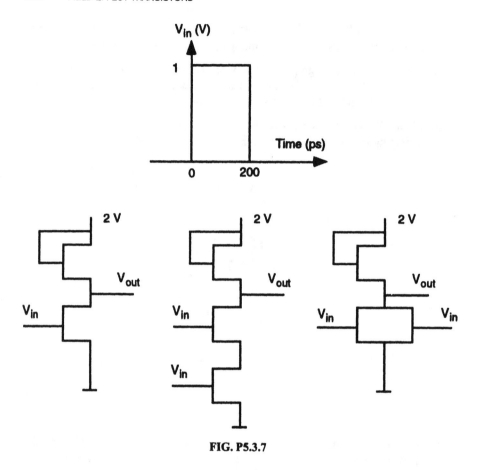

FIG. P5.3.7

5.4.1. Derive the extrinsic expression for the MESFET drain saturation current in Eq. (5.4.19). Find the corresponding expression based on the intrinsic Statz model of Eq. (5.4.18).

5.4.2. Consider a Si n-channel MOSFET with gate length and width of 1 μm and 20 μm, respectively, and with a threshold voltage of 0.2 V. Other parameters of this device are default SPICE parameters. Design a GaAs MESFET with the same dimensions and threshold voltage (using a reasonable channel thickness and Schottky barrier built-in voltage). Use SPICE to simulate and compare the output and transfer characteristics of these two devices (Level 1 in both cases).

5.4.3. (a) Review the example in Section 5.4.4.
(b) Modify the example to calculate the I–V characteristics in the above-threshold regime for both Level 1 and Level 2. Use gate–source voltages from –1.2 to 0 V in steps of 0.4 V. Assume that the Level 2 results can be taken as measured data and adjust the Level 1 parameters to obtain a best possi-

ble fit to the data. In this process, consider which parameters may be the most important in the various parts of the characteristics (saturation, linear, near threshold).

5.4.4. Repeat the simulations of Problem 5.3.7 using MESFETs but this time with a negative voltage pulse of –1 V at the input and with the load gates connected to the output. Use MESFET Level 1 with vto=-0.7 and otherwise default parameters.

CHAPTER 6

ADVANCED FET MODELING

6.1. INTRODUCTION

The rapid evolution of semiconductor electronics technology is fueled by a never-ending demand for better performance, combined with a fierce global competition. For silicon CMOS technology, this evolution is often measured in generations of three years, the time it takes for manufactured memory capacity on a chip to be increased by a factor of 4 and for logic circuit density to increase by a factor of between 2 and 3. Technologically, this long-term trend is made possible by a steady downscaling of transistor feature size (i.e., MOSFET gate length), by about a factor of 2 per two generations (Bohr, 1995).

At present, MOSFETs in high-volume manufacturing have progressed to 0.35 μm feature size. Following the evolutionary trend, the critical dimension in MOSFET VLSIs is expected to decrease below 0.1 μm within the next decade, as indicated in Fig. 6.1.1 (Hu, 1994; Geppert, 1996). Simultaneously, the performance of CMOS ICs rises steeply, packing up to 60 million transistors on a chip and operating with clock rates of more than 1 GHz by the turn of the century.

Very important issues in this development are the increasing level of complexity of the fabrication process and the many subtle mechanisms that govern the properties of submicrometer FETs. These mechanisms, dictated by the device physics, have to be described and implemented into process modeling and circuit design tools to empower the circuit designers with tools to fully utilize the potential of existing technology.

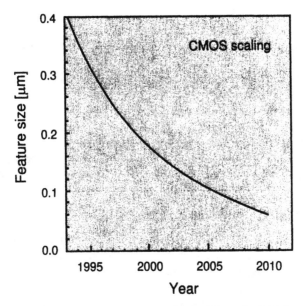

FIG. 6.1.1. Scaling of MOSFET feature size. (After Fjeldly et al., 1997. Data from Geppert, 1996.)

6.1.1. Challenges in Advanced FET Modeling

The scaling of FETs into the submicrometer regime tends to augment two sets of nonideal phenomena, both of which have to be incorporated into any viable device model for use in circuit simulation and device design based on submicrometer technology. On the one hand, well-known effects from earlier FET generations become magnified at short gate lengths owing to enhanced electric fields. Examples of such effects are channel length modulation (CLM), bias dependence of the field-effect mobility, and phenomena related to hot electron induced impact ionization near the drain.

On the other hand, we have a set of "new" phenomena—the so-called short-channel effects—that basically are related to a weakening of gate control over the channel charge. Typical manifestations of short-channel phenomena are serious leakage currents related to shifts in threshold voltage and to punchthrough and an increased output conductance in saturation caused by drain-bias-induced lowering of the injection barrier at the source. The loss of gate control usually results from an improper collective scaling of dimensions, doping levels, and voltages in the device, since an ideal scaling scheme is difficult to enforce in practice. In fact, with proper scaling, the FET would always possess good gate control and have ensured long-channel behavior.

Additional challenges in compound semiconductor FET modeling are gate leakage, temperature dependence and self-heating, side and back gating, frequency-dependent parameters, kink effect (caused by impact ionization and substrate charging), and the increasing relative importance of series resistances with downscaling.

In amorphous and polysilicon TFT electronics, the emphasis is less on small feature size than on the fabrication of large-area circuits for use in flat-panel display devices, printers, scanners, and three-dimensional LSI circuits. The important challenges in the modeling of such devices is to describe the effects of grain boundaries, traps, and impact ionization (kink effect) in poly-Si TFTs and on the very high density of defect states in the energy gap of a-Si TFTs.

For application of the models in circuit simulators such as SPICE, it is also necessary to emphasize the use of unified device models, that is, models that are continuous in their functional values and in all their derivatives through all regimes of operation. This allows us to avoid discontinuities that, when imbedded in device models, are well-known sources of convergence problems in circuit simulations.

6.1.2. Advanced FET Modeling Approach

For any FET, the threshold gate voltage V_T is a key parameter. It separates the on (above-threshold) and the off (subthreshold) states of operation. As indicated in Fig. 6.1.2, the potential energy of the channel electrons in the off state is high relative to the source, creating an effective barrier against electron transport from source to drain, while in the on state, this barrier is significantly lowered. For long-channel devices with gate lengths of several micrometers and operating with large power supply voltages, the behavior in the transition region near threshold is not important in digital applications. However, for FETs with submicrometer feature size and reduced power supply voltages (especially in low-power operation), the transition region becomes increasingly important, and the distinction between on and off states becomes blurred. Accordingly, a precise modeling of all regimes of device operation, including the near-threshold regime, is needed for short-channel devices.

In the basic FET models considered in Chapter 5, the subthreshold regime was simply considered an off state of the device, ideally blocking all drain current (although the SPICE implementations in some cases contain descriptions of the subthreshold regime). In practice, however, there will always be some leakage current

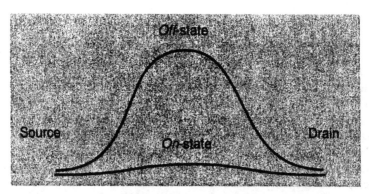

FIG. 6.1.2. Schematic conduction band profile through the channel region of a short-channel FET in the on state and the off state. (After Fjeldly et al., 1997.)

in the off state owing to a finite amount of mobile charge in the channel and a finite injection rate of carriers from the source into the channel.

This effect is enhanced in modern-day submicrometer FETs owing to short-channel phenomena such as DIBL. Drain-induced barrier lowering is a mechanism whereby the application of a drain–source bias causes a lowering of the source–channel junction barrier. In a long-channel device biased in the subthreshold regime, the applied drain–source voltage drop will be confined to the drain depletion zone. The remaining part of the channel is essentially at a constant potential (flat energy bands), where diffusion is the primary mode of charge transport (see Section 2.2).

However, in a short-channel device, the effect of the applied voltage will be distributed over the length of the channel and gives rise to a shift of the conduction band edge near the source end of the channel, as illustrated in Fig. 6.1.3. Such a shift represents an effective lowering of the injection barrier between the source and the channel, and since the dominant injection mechanism is thermionic emission, it translates into a significant increase in the injected current. This phenomenon can be described in terms of a shift in the threshold voltage [see, e.g., Fjeldly and Shur (1993) or Lee et al. (1993)]. Well above threshold, the injection barrier is much reduced, and the DIBL effect eventually disappears.

Clearly, the subthreshold current is very important since it has consequences for the bias and logic levels needed to achieve a satisfactory off state in digital operations. Hence, it affects the power dissipation in logic circuits. Likewise, the holding

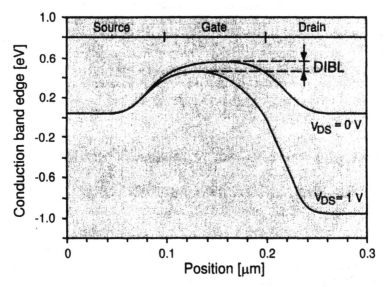

FIG. 6.1.3. Conduction band profile at the semiconductor–insulator interface of a 0.1-μm *n*-channel MOSFET with and without drain bias. The figure indicates the origin of DIBL. (After Fjeldly et al., 1997.)

time in dynamic memory circuits is controlled by the magnitude of the subthreshold current.

To correctly model the subthreshold operation of FETs, we need a charge control model for this regime. Also, in order to avoid convergence problems when using the model in circuit simulators, it is preferable to use a unified charge control model (UCCM) that covers both the above- and below-threshold regimes in one continuous expression. Such a model, by Byun et al. (1990), was introduced in Section 3.5 for the purpose of accurately describing the inversion charge density in MIS structures. This and related models have since been successfully applied to various FETs, including MOSFETs, MESFETs, HFETs, poly-Si TFTs, and a-Si TFTs (see, e.g., Lee et al., 1993).

In Section 6.2, we consider more closely how UCCM can be applied to describe the charge density in FET channels. Subsequently, we present our universal FET model, an advanced FET model based on UCCM, whose basic structure is "universal" in the sense that it applies to several types of FETs. In this upgraded model, we also include advanced features to account for effects such as velocity saturation, gate bias dependence of mobility, impact ionization, drain and source series resistances (extrinsic modeling), channel length modulation, and DIBL. In addition, this model uses continuous expressions for the current–voltage (I–V) and capacitance–voltage (C–V) characteristics, with no discontinuities in derivatives to any order and applicable at all bias conditions. All this makes the universal model very suitable for describing a variety of FETs ranging from submicrometer MOSFETs, HEMTs, and HFETs to thin-film amorphous silicon and polysilicon FETs.

The most advanced MOSFET models supplied with Berkeley SPICE and adopted with some modifications in commercial simulators, including AIM-Spice, are of the BSIM (Berkeley short-channel IGFET model) family. These models also include a number of advanced features relating to small feature size and scaling of device dimensions. Although the BSIM models are characterized by a large number of SPICE parameters (more than a hundred in some cases), they have gained a wide popularity for use in professional circuit simulation and design. The latest of these models, BSIM3, and its implementation in SPICE are briefly discussed at the end of this chapter.

6.2. UNIVERSAL FET MODELING APPROACH

6.2.1. Unified Charge Control Model for FETs

The UCCM equation discussed in Section 3.5 [see Eq. (3.5.34)] describes the two-dimensional interface charge density, $-qn_s$, encountered in MOS capacitors and in heterostructure potential wells. In order to account for the additional channel potential resulting from a drain–source bias in a MOSFET, this equation has to be modified as follows [see, e.g., Lee et al., 1993]:

$$V_{GT} - \alpha V_F = \eta V_{\text{th}} \ln\left(\frac{n_s}{n_o}\right) + a(n_s - n_o) \tag{6.2.1}$$

Here $V_{GT} = V_{GS} - V_T$ is the gate voltage overdrive, V_F is the quasi–Fermi potential measured relative to the Fermi potential at the source side of the channel, α is the bulk effect parameter, n_o is the sheet density of carriers at threshold, η is the ideality factor in the subthreshold regime, V_{th} is the thermal voltage, and $a \approx q/c_i$, where $c_i = \epsilon_i/d_i$ is the effective gate capacitance per unit area, ϵ_i is the dielectric permittivity of the gate insulator, and d_i is the thickness of the gate insulator [including a quantum correction to the insulator thickness; see Lee et al. (1993)]. Equation (6.2.1) also applies to other FETs with a two-dimensional interface charge such as HFETs and TFTs.

We note that in strong inversion the linear term will dominate on the right-hand side of Eq. (6.2.1), and the charge transport in the channel will be dominated by drift current. Hence, according to our discussion of charge transport in Section 2.2, V_F in Eq. (6.2.1) can be replaced by the channel potential V [see Eq. (2.2.27)]. This shows that, except for the parameter α, the UCCM reduces to the simple parallel-plate capacitor model of Eq. (5.3.10) above threshold.

In order to understand the origin of the parameter α, we consider the expression for the above-threshold inversion sheet charge density in a MOSFET given by Eq. (5.3.3), which can be rewritten as

$$qn_s = c_i(V_{GS} - V_{TX} - V) \tag{6.2.2}$$

where, assuming a constant substrate doping N_a,

$$V_{TX} = V_{FB} + 2\varphi_b + \frac{\sqrt{2\epsilon_s qN_a(2\varphi_b + V - V_{BS})}}{c_i} \tag{6.2.3}$$

Here V_{TX} may be regarded as a generalization of the threshold voltage V_T of Eq. (5.3.8), where the dependence of the depletion charge density on the local channel potential V is included. By linearizing Eq. (6.2.3) with respect to V, we obtain $V_{TX} \approx V_T + (\alpha - 1)V$, where $\alpha = 1 + \sqrt{2\epsilon_s qN_a/(2\varphi_b - V_{BS})}/2c_i$. Hence, the strong-inversion charge control equation, including the effect of the variation in the depletion charge along the channel, can be written as $qn_s \approx c_i(V_{GS} - V_T - \alpha V)$. In practical modeling, α may be regarded as an adjustable parameter with a value close to unity.

Below threshold, the linear term on the right-hand side of Eq. (6.2.1) can be neglected. This allows us to estimate the subthreshold drain current as follows, using Eq. (2.2.27) to express the current density in terms of the gradient of the Fermi potential:

$$I_d = qn_s W\mu_n \frac{dV_F}{dx} \approx -\frac{q\eta WD_n}{\alpha}\frac{dn_s}{dx} \tag{6.2.4}$$

Here we also used the Einstein relation $D_n = \mu_n V_{th}$ to relate the mobility to the diffusion coefficient. Equation (6.2.4) clearly shows that the subthreshold current is a diffusion current.

In the general case, the dependence of n_s on $V_{GT} - \alpha V_F$ can be expressed by approximate analytical solutions of Eq. (6.2.1) (see Fjeldly et al., 1991).

As mentioned, the UCCM formalism presented above is valid for FETs with a narrow channel confined to a semiconductor–oxide interface (MOSFETs and TFTs) or a heterointerface (HFETs). On the other hand, MESFETs have an above-threshold conducting channel consisting of a relatively thick neutral layer, as indicated in Fig. 5.2.3. However, near threshold and in the subthreshold regime, this channel also becomes narrow. Hence, a unified charge control expression for the MESFET can be established by a suitable combination of the straightforward above-threshold behavior and the subthreshold formalism implicit in the UCCM. This will be discussed in more detail in Section 6.4.

6.2.2. Basics of the Universal FET Model

The basic analytical models discussed in Section 5.3 need improvements in several areas, especially for application to modern submicrometer devices. This issue is addressed in the universal FET model presented here. Most importantly, the subthreshold regime is considered in terms of the UCCM approach. Furthermore, the universal model also accounts for series drain and source resistances, velocity saturation in the channel, gate bias dependence of mobility, impact ionization, CLM, and the threshold voltage shift caused by DIBL. In order to avoid convergence problems and excessive computation time in circuit simulations, continuous (unified) expressions are used for the current–voltage (I–V) and capacitance–voltage (C–V) characteristics, covering all regimes of operation.

Another important feature of the model is the use of relatively few, physically based parameters, all extractable from experimental data. Details of the parameter extraction procedures for the universal FET model are found in Lee et al. (1993).

I–V Model. In order to establish a continuous expression for the I–V characteristics, valid in all regimes of operation, we start by expressing the drain current in each regime and then make smooth transitions between them. Moreover, we want the model to be extrinsic, that is, to include the effects of the parasitic source and drain resistances (the corresponding intrinsic model is recovered simply by setting all parasitic resistances to zero).

For drain–source voltages in the linear regime, well below the saturation voltage, the drain current can be written as

$$I_d \approx g_{\text{chi}} V_{DS} \approx g_{\text{ch}} V_{ds} \tag{6.2.5}$$

where V_{DS} and V_{ds} are the intrinsic and extrinsic (applied) drain–source voltages, respectively [see Eq. (5.3.32)], and g_{chi} and g_{ch} are the intrinsic and extrinsic channel conductances in the linear region, respectively. The linear conductances are related by (see Appendix A4)

$$g_{\text{ch}} = \frac{g_{\text{chi}}}{1 + g_{\text{chi}} R_t} \tag{6.2.6}$$

where $R_t = R_s + R_d$ is the sum of the source and the drain series resistances. In deriving Eq. (6.2.6), we assumed that the drain current is so small that the transconductance is much less than the channel conductance. The linear intrinsic channel conductance can be written as

$$g_{chi} \approx \frac{q n_s W \mu_n}{L} \qquad (6.2.7)$$

where n_s is the unified carrier sheet density at zero drain bias and μ_n is the low-field carrier mobility. We note that near threshold and in the subthreshold regime, the channel resistance will normally be much larger than the parasitic source and drain resistances, allowing us to make the approximations $g_{ch} \approx g_{chi}$, $V_{ds} \approx V_{DS}$, and $V_{gt} \approx V_{GT}$.

The drain saturation current for the above-threshold regime was calculated in Section 5.3 for MOSFETs and MESFETs. The results in terms of extrinsic voltages are given in Eqs. (5.3.33) and (5.4.19), respectively, for the two-piece velocity saturation model of Eq. (5.3.25). In the subthreshold regime, we can calculate the saturation current from Eq. (6.2.4) by making use of the subthreshold approximation of the UCCM expression, that is, $n_s \approx n_o \exp[(V_{gt} - \alpha V_F)/\eta V_{th}]$. Straightforward integration of Eq. (6.2.4) over the channel length (corresponding to a variation in V_F from 0 to V_{ds}) gives the subthreshold current

$$I_{sub} \approx \frac{q n_o \eta D_n W}{\alpha L} \exp\left(\frac{V_{gt}}{\eta V_{th}}\right)\left[1 - \exp\left(\frac{\alpha V_{ds}}{\eta V_{th}}\right)\right] \qquad (6.2.8)$$

From this expression, the subthreshold saturation current is obtained for $V_{ds} > 2V_{th}$. According to the discussion in Section 6.2.1, this result also applies to MESFETs in the subthreshold regime. As will be shown later, the above- and below-threshold expressions for the saturation current can easily be combined into unified expressions.

The next task is to bridge the transition between the linear and the saturation regimes with one single expression. For this purpose, we propose to use the following extrinsic, universal interpolation formula:

$$I_d = \frac{g_{ch} V_{ds}(1 + \lambda V_{ds})}{[1 + (g_{ch} V_{ds}/I_{sat})^m]^{1/m}} \qquad (6.2.9)$$

Here, m is a parameter that determines the shape of the characteristics in the knee region. The factor $1 + \lambda V_{ds}$ is the same as that used in the basic models of Chapter 5, describing the finite output conductance in saturation, mainly caused by channel length modulation, that is, the effect of the finite extent of the saturated region of the channel (illustrated in Fig. 5.3.1). We note that Eq. (6.2.9) has the correct asymptotic behavior in the linear regime, in agreement with Eq. (6.2.5). In addition, when $\lambda = 0$, I_d asymptotically approaches I_{sat} in saturation, as required.

In Fig. 6.2.1, we relate a typical FET experimental I-V characteristic to the saturation voltage I_{sat}, the linear extrinsic channel conductance g_{ch}, and a finite-output conductance g_{chs} in saturation. When channel length modulation is the cause of the

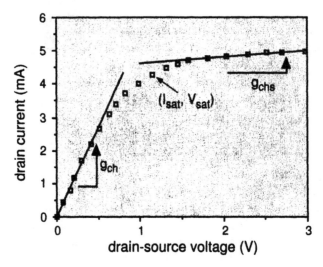

FIG. 6.2.1. Typical experimental FET *I–V* characteristic with finite-output conductance in saturation.

finite g_{chs}, we obtain $g_{chs} \approx \lambda I_{sat}$ in deep saturation. This result can be used to determine λ as part of the device characterization (see Lee et al., 1993). A more detailed discussion of the effect of channel length modulation is given below.

C–V Model. As was discussed in Section 5.3, accurate modeling of the intrinsic capacitances associated with the gate region of FETs requires an analysis of the charge distribution in the channel versus the terminal bias voltages. Normally, the problem is simplified by assigning the distributed charges to the various "intrinsic" terminals. Hence, the mobile charge Q_I of a MOSFET is divided into a source charge $Q_S = F_p Q_I$ and a drain charge $Q_D = (1 - F_p)Q_I$, where F_p is a partitioning factor. The depletion charge Q_B under the gate is assigned to the MOSFET substrate terminal. The total gate charge Q_G is the negative sum of these charges, that is, $Q_G = -Q_I - Q_B = -Q_S - Q_D - Q_B$. Note that by assigning the charges this way, charge conservation is always assured. A similar procedure can be applied for the HFET, except that the substrate depletion charge and the substrate terminal are normally missing. For the MESFET, it is the depletion charge under the gate that has to be partitioned between the source and drain terminals.

The net current flowing into terminal X can now be written as

$$I_X = \frac{dQ_X}{dt} = \sum_Y \frac{\partial Q_X}{\partial V_Y} \frac{\partial V_Y}{\partial t} = \sum_Y \chi_{XY} C_{XY} \frac{\partial V_Y}{\partial t} \qquad (6.2.10)$$

where the indices X and Y run over the terminals G, S, D (and B when applicable). In this expression, we have introduced a set of intrinsic capacitance elements C_{XY}—so-called transcapacitances—defined by

$$C_{XY} = \chi_{XY} \frac{\partial Q_X}{\partial V_Y} \quad \text{where } \chi_{XY} = \begin{cases} -1 & \text{for } X \neq Y \\ 1 & \text{for } X = Y \end{cases} \tag{6.2.11}$$

These are identical to the charge-based nonreciprocal capacitances introduced by Ward (1981) and by Ward and Dutton (1978). The term *nonreciprocal* means that, in general, we have $C_{XY} \neq C_{YX}$ when $X \neq Y$. The elements C_{XX} are called self-capacitances. This is an unfamiliar concept if we think in terms of a parallel-plate capacitor. However, C_{XY} only tells us how much the charge Q_X assigned to terminal X changes by a small variation in the voltage V_Y at terminal Y. The following example illustrates clearly why C_{XY} may be different from C_{YX}: Assume that we have an FET in saturation. Then the gate charge changes little if the drain voltage is slightly perturbed, making C_{GD} very small. On the other hand, if V_G changes, Q_D may be significantly affected (depending on the partitioning scheme used), making C_{DG} large.

In the case of a four-terminal MOSFET, the Ward–Dutton description leads to a total of 16 transcapacitances. This set of 16 elements can be organized as follows in a 4 × 4 matrix, a so-called indefinite admittance matrix:

$$\mathbf{C} = \begin{bmatrix} C_{GG} & C_{GS} & C_{GD} & C_{GB} \\ C_{SG} & C_{SS} & C_{SD} & C_{SB} \\ C_{DG} & C_{DS} & C_{DD} & C_{DB} \\ C_{BG} & C_{BS} & C_{BD} & C_{BB} \end{bmatrix} \tag{6.2.12}$$

where the elements in each column and each row must sum to zero owing to the constraints imposed by charge conservation (which is equivalent to obeying Kirchhoff's current law) and for the matrix to be reference independent, respectively [see Arora (1993)]. This means that some of the transconductances will be negative, and of the 16 MOSFET elements, only 9 are independent. The complete MOSFET large-signal equivalent circuit, including the 16 transcapacitances, is shown in Fig. 6.2.2. This compares with the simple Meyer model in Fig. 5.3.4, which contains 3 capacitances.

In three-terminal FETs, such as HFETs and MESFETs, we have a total of 9 transcapacitances, of which 4 are independent (Nawaz and Fjeldly, 1997). The equivalent circuit for this case is obtained from Fig. 6.2.2 by removing the substrate terminal B and all elements connected to it. Also, the 4 × 4 matrix in Eq. (6.2.12) reduces to a 3 × 3 matrix.

As was explained in Section 5.3, the simplified $C-V$ model by Meyer (1971) can be obtained simply as derivatives of the total gate charge with respect to the various terminal voltages [see Eqs. (5.3.18)–(5.3.24)]. In reality, the Meyer capacitances represent a subset of the Ward–Dutton capacitances in the so-called quasi-static approximation. Although charge conservation is not assured in this case, the resulting errors in circuit simulations are usually small (except in some cases of transient analyses involving especially demanding circuits). We next consider a unified version of the Meyer FET capacitances discussed in Section 5.3. [A set of unified, charge-conserving capacitances for short-channel MOSFETs was also discussed by Lee et al. (1993).]

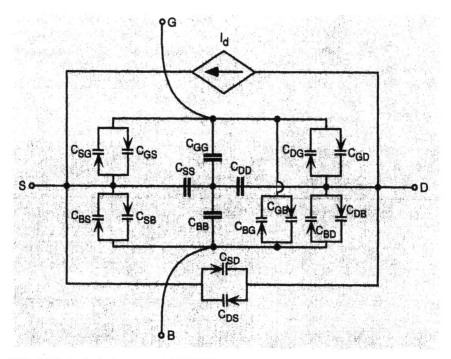

FIG. 6.2.2. Intrinsic large-signal MOSFET equivalent circuit including a complete set of nonreciprocal and charge-conserving transcapacitances. The transcapacitances C_{XY} are defined in the text. (Compare with the Meyer capacitance model of Fig. 5.3.4.).

A unified expression for the gate–channel capacitance C_{ch} of an FET at zero drain–source bias can, for example, be approximated by the following combination of the above-threshold capacitance C_a and the below-threshold capacitance C_b (for details, see the discussion in the subsequent sections).

$$C_{ch} = \frac{C_a C_b}{C_a + C_b} \tag{6.2.13}$$

By using this unified gate–channel capacitance in conjunction with Meyer's capacitance model, we obtain the following continuous expressions for the intrinsic gate–source capacitance C_{GS} and the gate–drain capacitance C_{GD}, valid for all regimes of operation:

$$C_{GS} = \tfrac{2}{3} C_{ch} \left[1 - \left(\frac{V_{GT} - V_{DSe}}{2V_{GT} - V_{DSe}} \right)^2 \right] \tag{6.2.14}$$

$$C_{GD} = \tfrac{2}{3} C_{ch} \left[1 - \left(\frac{V_{GT}}{2V_{GT} - V_{DSe}} \right)^2 \right] \tag{6.2.15}$$

Here V_{DSe} is an effective intrinsic drain–source voltage that is equal to V_{DS} for $V_{DS} < V_{GTe}$ and is equal to V_{GTe} for $V_{DS} > V_{GTe}$. Moreover, V_{GTe} is an effective gate voltage overdrive that equals V_{GT} above threshold and is of the order of the thermal voltage in the subthreshold regime. In order to make a smooth transition between the non-saturated and the saturated regimes, we interpolate V_{DSe} by the following type of expression:

$$V_{DSe} = \tfrac{1}{2}\left[V_{DS} + V_{GTe} - \sqrt{V_8^2 + (V_{DS} - V_{GTe})^2}\right] \qquad (6.2.16)$$

where V_8 is a constant voltage that determines the width of the transition region. This parameter may be treated as an adjustable parameter to be extracted from experiments. Here, V_{GTe} is needed in order to obtain a smooth transition between the correct limiting I–V and C–V expressions above and below threshold. Further details on suitable functional forms of V_{GTe} are provided in the subsequent sections.

Figure 6.2.3 shows a comparison of the normalized dependencies of C_{GS} and C_{GD} on V_{DS} for $V_8/V_{GTe} = 0$, corresponding to the Meyer capacitances, and for a more realistic value, $V_8/V_{GTe} = 0.2$. By a simple extrapolation of the discussion in Sections 5.3.4, we can state that the present unified version of the Meyer capacitances are applicable also for short-channel devices.

In practice, still more flexible expressions for the capacitances are obtained by substituting V_{GT} by χV_{GT} in Eqs. (6.2.14) and (6.2.15), where χ is an adjustable parameter close to unity.

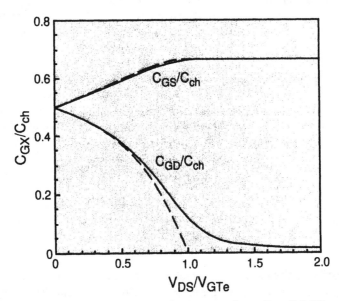

FIG. 6.2.3. Normalized standard Meyer capacitances according to Eqs. (5.3.18) and (5.3.19) (dashed lines) and unified Meyer capacitances according to Eqs. (6.2.14)–(6.2.16) (solid lines) using a transition width parameter $V_8 = 0.2V_{GTe}$.

6.2.3. High-Field Effects

As discussed in Section 6.1, it is convenient to divide the nonideal phenomena associated with submicrometer scaling into two categories: the high-field effects and the short-channel effects. The former are related to the increased electrical field in the channel owing to reduced dimensions, and the latter concerns the loss of gate control in FETs as a result of improper device scaling.

Channel Length Modulation. When the drain–source bias of an FET approaches the drain saturation voltage, a region of high electric field forms near the drain and the electron velocity in this region saturates (in long devices, we instead have pinch-off where n_s becomes very small near drain). In saturation, the length ΔL of the high-field region expands with increasing V_{DS} in the direction of the source, and the MOSFET behaves as if the effective channel length has been reduced by ΔL. This phenomenon is called channel length modulation (CLM). The following simplified expression links V_{DS} to the length of the saturated region (see Lee et al., 1993):

$$V_{DS} = V_p + V_\alpha \left[\exp\left(\frac{\Delta L}{l}\right) - 1 \right] \qquad (6.2.17)$$

where V_p, V_α, and l are constants related to the electron saturation velocity, the field effect mobility, and the drain conductance in the saturation regime. In fact, V_p is the potential at the point of saturation in the channel and is sometimes approximated by V_{SAT}. Good agreement has been obtained between the potential profile described by Eq. (6.2.17) and that obtained from a two-dimensional simulation for the saturated region of an n-channel MOSFET (see Lee et al., 1993).

Since CLM causes a reduction of the effective length of the conducting channel, its effect can be observed as a finite output conductance in saturation, as indicated in the I–V characteristic in Fig. 6.2.1. We notice that the slope of the characteristic in saturation is fairly constant over a wide range of drain biases. Also, we find that the output conductance increases steadily with increasing gate bias. This observation suggests an even simpler, first-order model for describing CLM, where the basic expression for the drain current is simply multiplied by the factor $1+\lambda V_{DS}$, as was done in Eq. (6.2.9). In this case, λ can easily be extracted from the output conductance in the saturation regime, well above threshold.

Channel length modulation is described in terms of the first-order approximation in several FET models used in AIM-Spice, while expressions similar to Eq. (6.2.17) are used in the MOSFET model BSIM3.

Hot-Carrier Effects. Hot-carrier effects are among the main concerns when shrinking FET dimensions into the deep submicrometer regime. Reducing the channel length while retaining the power supply level, known as constant-voltage scaling, results in increased electric field strength near the drain end of the channel that accelerates and heats the channel charge carriers. The following are some of the manifestations of hot electrons on device operation:

- Impact ionization → substrate current, breakdown
- Interface damage → interface charge → device degradation
- Hot-electron emission → gate current
- Light emission → photocurrent

Impact ionization results in electron–hole pair generation and initially gives rise to a substrate current in MOSFETs (see Iñiguez and Fjeldly, 1996), that may overload substrate bias generators, introduce snap-back breakdown and CMOS latchup, and generate a significant increase in the subthreshold drain current. A complete model for the substrate current is too complex to be used in circuit level simulation. Instead approximations are used to obtain an analytical expression for the substrate current as a function of applied voltages. The following expression is a widely used analytical model for the substrate current:

$$
I_{\text{substr}} = I_d \frac{A_i}{B_i}(V_{DS} - V_{\text{SAT}}) \exp\left(-\frac{l_d B_i}{V_{DS} - V_{\text{SAT}}}\right) \tag{6.2.18}
$$

Here I_d is the channel current, A_i and B_i are ionization constants, V_{SAT} is the saturation voltage, and l_d is the effective ionization length. This expression can also be used in the subthreshold regime by choosing $V_{\text{SAT}} = 0$.

In FETs fabricated on an insulating layer, such as fully depleted SOI/MOSFETs, the impact ionization may give rise to a charging of the transistor body, causing an onset of increased drain current in saturation (floating-body effect). Related mechanisms where holes are trapped in a GaAs semi-insulating substrate or in traps associated with grain boundaries in poly-Si TFTs produce the so-called kink effect in the saturation characteristics of these FETs.

At higher rates of impact ionization, we have a breakdown in the drain current of all types of FETs. In MOSFETs, a substantial amount of the majority carriers created by impact ionization near the drain will flow toward the source and forward bias the source–substrate junction, causing injection of minority carriers into the substrate. This effect can be modeled in terms of conduction in a parasitic bipolar transistor, as described by Sze (1981). In MESFETs, the breakdown usually takes place in the high-field depletion extension toward the drain.

Electron trapping in the oxide and generation of interface traps caused by hot-electron emission induce degradation of the MOSFET channel near the drain in conventional MOSFETs or cause changes in the parasitic drain resistances in low-doped drain (LDD) MOSFETs (Ytterdal et al., 1995a). Reduced current drive capability and transconductance degradation are manifestations of interface traps in n-MOSFET characteristics. Reduced current also leads to circuit speed degradation, such that the circuits may fail to meet speed specifications after aging.

Photon emission and subsequent absorption in a different location of the device may cause unwanted photocurrent that, for example, may degrade the performance of memory circuits.

Temperature Dependence and Self-Heating. Since electronic devices and circuits have to operate in different environments, including a wide range of temperatures, it is imperative to be able to perform reliable modeling for such eventualities. Heat generated from power dissipation in an integrated circuit chip can be considerable, and the associated temperature rise must be accounted for in both device and circuit design. In conventional silicon substrates, the thermal conductivity is high enough and a well-designed chip placed on a good heat sink may achieve a relatively uniform and tolerable operating temperature. However, such design becomes increasingly difficult as the device dimensions are scaled down and power dissipation increases.

The thermal behavior of MOSFETs has been extensively studied in the past, and the temperature dependencies of major model parameters have been incorporated in SPICE models such as BSIM3.

For circuits fabricated on GaAs substrates or on insulating silicon dioxide layers, individual devices are more susceptible to a significant self-heating effect (SHE). In thin-film SOI CMOS, the buried SiO_2 layer inhibits an effective heat dissipation, and the self-heating manifests itself as a reduced drain current and even as a negative differential conductance at high-power inputs. Hence, for a reliable design of SOI circuits, accurate and self-consistent device models that account for SHE are needed for use in circuit simulation.

The influence of SHE on the electrical characteristics of SOI MOSFETs can be evaluated using a two-dimensional device simulator incorporating heat flow or by combining a temperature rise model with an I–V expression through an iteration procedure. But the effect can also be described in terms of a temperature dependent model for the device I–V characteristics, in combination with the following simplified relationship between temperature rise and power dissipation:

$$T - T_o = R_{th} I_d V_{ds} \qquad (6.2.19)$$

Here T is the actual temperature, T_o is the ambient (substrate) temperature, and R_{th} is a thermal resistance that contains information on thermal conductivity and geometry. The equations can be solved self-consistently, either numerically or analytically (Cheng and Fjeldly, 1996). Once the temperature dependence of the device parameters are established, the same procedure can also be used for describing self-heating in other types of devices, such as GaAs MESFETs and HFETs. Recently, temperature dependencies of GaAs MESFET model parameters were reported by Ytterdal et al. (1995b).

Gate Bias-Dependent Mobility. In submicrometer MOSFETs, scaling dictates that the gate dielectric must be made very thin. In a sub-0.1-μm device, the gate dielectric may be as thin as 70 Å. With a gate–source voltage of 1 V, this corresponds to a transverse electric field of nearly 500 kV/cm. In this case, electrons are confined to a very narrow region at the silicon–silicon dioxide interface, and their motion in the direction perpendicular to the gate oxide is quantized. Also, the proximity of the carriers to the interface enhances the scattering rate by surface nonuniformities, drastically reducing the field-effect mobility in comparison to that of bulk silicon.

To first-order approximation, the following simple expression accurately describes the dependence of the field-effect mobility on the gate bias (Park et al., 1991):

$$\mu_n + \mu_{on} - \kappa_n(V_{GS} + V_T) \tag{6.2.20}$$

The experimental MOSFET mobility data in Fig. 6.2.4 shows that this expression can be applied with the same set of parameters for different values of substrate bias. The parameter values are fairly close even for devices with quite different gate lengths. All in all, this leads to a reduction in the number of parameters needed for accurate modeling of the MOSFET characteristics.

A more complete expression for the MOSFET field-effect mobility that takes into account both temperature variations and scaling is used in BSIM3 (see Section 6.8).

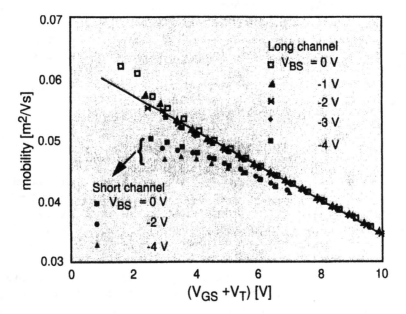

FIG. 6.2.4. Electron mobility versus $V_{GS} + V_T$ for a long-channel ($L = 20$ μm) and a short-channel ($L = 1$ μm) NMOS for different values of substrate bias. The solid line corresponds to the linear approximation used in Eq. (6.2.20). (From Park et al., 1991. Copyright 1991 by IEEE. See also Lee et al., 1993.)

6.2.4. Short-Channel Effects

Aspect Ratio. To first order, FET dimensions are scaled by preserving the device aspect ratio: the ratio between the gate length and the active vertical dimension of the device. In MOSFETs, the vertical dimension accounts for the oxide thickness d_i, the source and drain junction depths r_j, and the junction depletion depths W_s and W_d. A low aspect ratio is synonymous with short-channel behavior. The following empirical relationship indicates the transition from long-channel to short-channel behavior (Brews et al., 1980):

$$L < L_{min}(\mu m) = 0.4[r_j(\mu m)d_i(\text{Å})(W_d + W_s)^2(\mu m^2)]^{1/3} \qquad (6.2.21)$$

When $L < L_{min}$, the MOSFET threshold voltage V_T will be affected in several ways as a result of reduced gate control. First, the *depletion* charges near the source and drain are under the shared control of these contacts and the gate. In a short-channel device, the shared charge will constitute a relatively large fraction of the total gate depletion charge and can be shown to give rise to an increasingly large shift in V_T with decreasing L. Also, the shared depletion charge near the drain expands with increasing drain–source bias, resulting in an additional V_{DS}-dependent shift in V_T.

Drain-Induced Barrier Lowering. The threshold voltage is a measure of the strength of the barrier against carrier injection from the source to the channel. In the short-channel regime ($L < L_{min}$), this barrier may be significantly modified by the application of a drain bias, as was shown schematically in Fig. 6.1.3. In n-channel FETs, this DIBL translates into a lowering of the threshold voltage and a concomitant rise in the subthreshold current with increasing V_{DS}. The combined scaling and DIBL effect on the threshold voltage may be expressed as follows:

$$V_T(L) = V_{To}(L) - \sigma(L)V_{DS} \qquad (6.2.22)$$

where $V_{To}(L)$ describes the scaling of V_T at zero drain bias resulting from charge sharing and $\sigma(L)$ is the channel-length-dependent DIBL parameter. In the long-channel case, where $L > L_{min}$, V_T should become independent of L and V_{DS}, which can be modeled by letting both $V_{To}(L)$ and $\sigma(L)$ scale approximately as $\exp(-L/L_{min})$. In BSIM3, somewhat more detailed scaling functions and also a dependence on substrate bias are used.

In Fig. 6.2.5a, we show experimental data of V_T versus V_{DS} for two short-channel n-MOSFETs that demonstrate good agreement with the linear relationship of Eq. (6.2.22). Also, the exponential scaling for V_T versus L is confirmed by experiments, except for a deviation at the shortest gate lengths, as shown in Fig. 6.2.5b (Fjeldly and Shur, 1993).

As discussed in Section 6.1, DIBL vanishes well above threshold. Hence, for

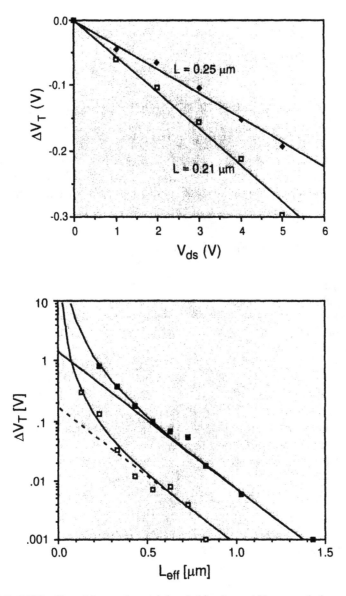

FIG. 6.2.5. DIBL-effect: (*a*) experimental threshold voltage shift versus drain–source voltage for two *n*-MOSFETs with different gate lengths; (*b*) experimental threshold voltage shifts versus gate length compared with exponential scaling. (From Fjeldly and Shur, 1993. Copyright 1993 by IEEE.)

modeling purposes, we adopt the following empirical expression for σ (see Lee et al., 1993):

$$\sigma = \frac{\sigma_o}{1 + \exp[(V_{gto} - V_{\sigma t})/V_\sigma]} \tag{6.2.23}$$

where V_{gto} is the gate voltage overdrive at zero drain–source bias and the parameters $V_{\sigma t}$ and V_σ determine the voltage and the width of the DIBL fade-out, respectively. We note that $\sigma \rightarrow \sigma_o$ for $V_{gto} < V_{\sigma t}$ and $\sigma \rightarrow 0$ for $V_{gto} > V_{\sigma t}$.

The effects of DIBL can be accounted for in our I–V models by adjusting the threshold voltage according to Eq. (6.2.22) in the expressions for the saturation current and the linear channel conductance. Likewise, UCCM has to be modified as follows:

$$V_{GTo} + \sigma V_{DS} - \alpha V_F \approx \eta V_{th} \ln\left(\frac{n_s}{n_o}\right) + a(n_s - n_o) \tag{6.2.24}$$

where V_{GTo} is the intrinsic threshold voltage overdrive at zero drain–source bias. All this will result in a contribution to the finite-output conductance in saturation and to an increase in the subthreshold current. It is important to emphasize that the parameters λ and σ_o are strongly dependent on the gate length, the aspect ratio, the contact depth, and other factors.

At a sufficiently large gate voltage overdrive, typically $V_{gto} > V_{\sigma t} + 3V_\sigma$, the effects of DIBL vanish and we have $g_{chs} \approx \lambda I_{sat}$ in deep saturation, as discussed above.

A related effect of device miniaturization is observed in narrow-channel FETs. In such devices, charges associated with the extension of the gate depletion regions beyond the width of the gate may become a significant fraction of the total gate depletion charge. In this case, a one-dimensional analysis will underestimate the total depletion charge and give a wrong prediction of the threshold voltage. In practice, the threshold voltage increases (n-MOSFET) as the channel width is reduced. A common method of modeling this effect is to add an additional term in the threshold voltage expression containing a $1/W$ term, where W is the effective width of the gate.

6.3. UNIVERSAL MOSFET MODEL

In order to establish a complete MOSFET model based on the universal modeling concept outlined above, we have to include unified expressions for the linear intrinsic channel conductance g_{chi}, the extrinsic saturation current I_{sat}, and the channel capacitance C_{ch}.

6.3.1. MOSFET I–V Model

From Eq. (6.2.7), we observe that g_{chi} is directly proportional to the sheet density of inversion charge, which in turn is given by the UCCM expression of Eq. (6.2.1). Unfortunately, UCCM cannot be solved analytically with respect to n_s. However, we can use the following generalized version of the approximate analytical solution of UCCM introduced in Section 3.5 for the MIS capacitor [see Eq. (3.5.37)]:

$$n_s = 2n_o \ln\left[1 + \tfrac{1}{2} \exp\left(\frac{V_{GT} - \alpha V_F}{\eta V_{th}}\right)\right] \tag{6.3.1}$$

where the carrier sheet density at threshold is given by

$$n_o = \frac{\eta V_{th} c_i}{2q} \tag{6.3.2}$$

We note that Eq. (6.3.1) has the correct limiting behavior above and below threshold and is in good overall agreement with UCCM.

For the purpose of calculating g_{chi}, we set $V_F = 0$ in Eq. (6.3.1) since the drain current is taken to be small. For the same reason, the intrinsic gate voltage overdrive V_{GT} can be replaced by its extrinsic counterpart V_{gt}. Hence, we obtain for the MOS-FET

$$g_{chi} = \frac{2Wq\mu_n n_o}{L} \ln\left[1 + \tfrac{1}{2} \exp\left(\frac{V_{gt}}{\eta V_{th}}\right)\right] \tag{6.3.3}$$

The extrinsic MOSFET drain saturation current for the above-threshold regime was derived in Section 5.3 for the two-piece velocity saturation model [see Eq. (5.3.33)]. The subthreshold saturation current is obtained from Eq. (6.2.8) when V_{ds} exceeds the subthreshold saturation voltage $V_{sat} \approx 2V_{th}$. We first note that in strong inversion we can write $\beta V_{gt} = g_{chi}$ since $\beta = W\mu_n c_i/L$ and $qn_s = c_i V_{gt}$ at small V_{ds}. Hence, all occurrences of βV_{gt} in Eq. (5.3.33) can be substituted by g_{chi}. Next, we replace all remaining occurrences of V_{gt} by the effective gate voltage overdrive V_{gte} that coincides with V_{gt} well above threshold and equals $2V_{th}$ below threshold, similar to what was done for the C–V model in Section 3.2.2 [see Eq. (6.2.16)]. This converts Eq. (5.3.33) to the following unified form that approximates the correct limiting behavior above and below threshold:

$$I_{sat} = \frac{g_{chi} V_{gte}}{1 + g_{chi} R_s + \sqrt{1 + 2g_{chi} R_s + (V_{gte}/V_L)^2}} \tag{6.3.4}$$

Here $V_L = F_s L$ where F_s is the saturation field [see Eq. (5.3.25)]. The following is a suitable MOSFET expression for the effective gate voltage overdrive:

$$V_{gte} = V_{th}\left[1 + \frac{V_{gt}}{2V_{th}} + \sqrt{\delta^2 + \left(\frac{V_{gt}}{2V_{th}} - 1\right)^2}\right] \tag{6.3.5}$$

were δ determines the width of the transition region. Typically, $\delta = 3$ is a good choice.

Hence, using Eq. (6.3.4) for the saturation current and Eq. (6.3.3) combined with

Eq. (6.2.6) for the channel conductance, the universal $I-V$ expression of Eq. (6.2.9) becomes truly unified, with the correct limiting behavior both above and below threshold and with a continuous transition between all regimes of operation. Furthermore, the short-channel DIBL effect is included by adjusting the threshold voltage according to Eq. (6.2.22).

The quality of the present universal MOSFET model is illustrated in Fig. 6.3.1 for a deep submicrometer n-channel MOSFET. Figure 6.3.1a shows the above-threshold $I-V$ characteristics and Fig. 6.3.1b shows the subthreshold transfer characteristics in a semilog plot. As can be seen from the figures, our model reproduces quite accurately the experimental data in the entire range of bias voltages over several decades of current variation. The parameters used in the model calculations were obtained from the parameter extraction procedure described by Shur et al. (1992b) (see also Lee et al., 1993).

6.3.2. MOSFET C–V Model

As discussed in Section 6.2, we have several ways of handling intrinsic capacitive effects in MOSFETs. One is based on a proper assignment of the intrinsic charges of the MOSFET to the various terminals and calculating the full set of charge-conserving, nonreciprocal transcapacitances according to Ward (1981) and Ward and Dutton (1978) (see Fig. 6.2.2). An equivalent method is to implement the charge expressions into the SPICE code and use them directly in the calculation of the node currents. Details of this technique are outlined in Johannessen et al. (1994). Additional models with emphasis on short-channel MOSFETs are discussed by Rho et al. (1993) and by Iñiguez and Moreno (1996).

However, here we concentrate on the unified version of the Meyer capacitances discussed in Section 6.2. The equivalent circuit is shown in Fig. 5.3.4. The MOSFET gate–source and gate–drain capacitances are given by Eqs. (6.2.14) and (6.2.15), respectively. The only device-specific part of these equations is the channel capacitance C_{ch} at zero drain bias, which for MOSFETs can be derived from the approximate UCCM expression of Eq. (6.3.1):

$$C_{\text{ch}} = WLq \frac{dn_s}{dV_{GT}} \approx C_i \left[1 + 2 \exp\left(-\frac{V_{GT}}{\eta V_{\text{th}}} \right) \right]^{-1} \tag{6.3.6}$$

Well above or below threshold, this expression has the asymptotic forms

$$C_a \approx C_i \tag{6.3.7}$$

$$C_b \approx \frac{C_i}{2} \exp\left(\frac{V_{GT}}{\eta V_{\text{th}}} \right) \tag{6.3.8}$$

respectively. In Fig. 6.3.2, we show C_{ch} versus V_{GT} in a linear and a semilog plot. From UCCM and from Eq. (6.3.6), we find that $C_{\text{ch}} = C_i/3$ at threshold, which may

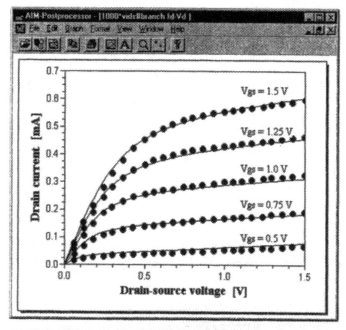

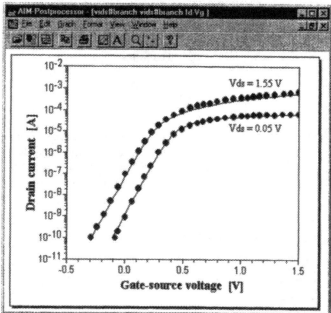

FIG. 6.3.1. Experimental (symbols) and modeled (*a*) above-threshold and (*b*) subthreshold characteristics for a deep submicrometer NMOS with effective gate length $L = 0.09$ μm. Device parameters: $W = 1$ μm, $d_i = 3.5$ nm, $\mu_n = 0.026$ m^2/V s, $v_s = 6 \times 10^4$ m/s, $m = 2.2$, $R_s = R_d = 200$ Ω, $\lambda = 0.142$ V^{-1}, $\eta = 1.7$, $V_{To} = 0.335$ V, $\sigma_o = 0.11$, $V_\sigma = 0.2$ V, $V_{\sigma t} = 0.18$ V. Substrate–source bias $V_{bs} = 0$ V. (From Ytterdal et al., 1994. Data from Mii et al., 1994.)

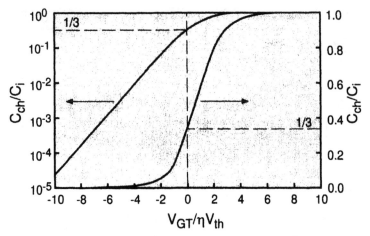

FIG. 6.3.2. Normalized channel capacitance versus $V_{GT}/\eta V_{th}$ according to Eq. (6.3.6) in a linear plot (right) and a semilog plot (left). The condition $C_{ch}/C_i = \frac{1}{3}$ at threshold is indicated.

serve as a convenient and straightforward way of determining the threshold voltage from experimental C_{ch}-versus-V_{GS} curves.

In the subthreshold regime, the gate–substrate capacitance C_{GB} is the dominant Meyer capacitance in MOSFETs [see Eq. (5.3.24)]. Above threshold, C_{GB} vanishes in the ideal long-channel case. A unified version of C_{GB} that includes a gradual phase-out above threshold can be modeled as follows:

$$C_{GB} = \frac{C_i/(1 + n_s/n_o)}{\sqrt{1 + 4(V_{GS} - V_{BS} - V_{FB})/\gamma^2}} \qquad (6.3.9)$$

Here n_s is the unified electron density given by UCCM or its approximate solution [see Eqs. (6.2.1) and (6.3.1)], n_o is the electron sheet density at threshold, and $\gamma = \sqrt{2\epsilon_s q N_a}/c_i$ is the body-effect coefficient. Equation (6.3.9) utilizes the fact that the high density of inversion charge above threshold shields the substrate from the influence of the gate electrode. A typical plot of C_{GB} versus V_{GT} is shown in Fig. 6.3.3.

6.3.3. Implementation in AIM-Spice

The universal MOSFET model implemented in AIM-Spice is labeled Level 7. The expression for the threshold voltage is similar to that of the SPICE Level 4 MOS-FET model (BSIM1):

$$V_T = \mathbf{VTO} - \sigma V_{ds} + \gamma_S(\sqrt{\mathbf{PHI} - V_{bs}} - \sqrt{\mathbf{PHI}}) + \gamma_L V_{bs} \qquad (6.3.10)$$

Here **VTO** is the threshold voltage at zero drain and substrate bias and $\mathbf{PHI} = 2\varphi_b$ is the semiconductor surface potential under the gate at threshold. The second term on

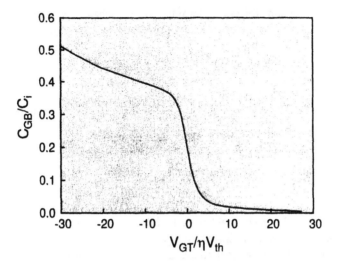

FIG. 6.3.3. Normalized and unified Meyer-type gate–substrate capacitance versus $V_{GT}/\eta V_{th}$ according to Eq. (6.3.9) for $V_{BS} = 0$. Typical values for an n-channel MOSFET with a polysilicon gate were used: $V_T \approx 0.7$ V, $V_{FB} \approx -1$ V, $\gamma \approx 1$ V$^{1/2}$, and $\eta \approx 1.33$.

the right of Eq. (6.3.10) describes the DIBL effect, and the two remaining terms contain the effects of the substrate doping profile and of geometry (body effect).

The DIBL effect coefficient of the universal FET models has the following dependence on the gate bias [see Eq. (6.2.23)]:

$$\sigma = \text{SIGMA0} \bigg/ \left[1 + \exp\left(\frac{V_{gto} - \text{VSIGMAT}}{\text{VSIGMA}}\right) \right] \qquad (6.3.11)$$

where $V_{gto} = V_{gs} - \text{VTO}$, **SIGMA0** is the subthreshold DIBL parameter, and **VSIGMAT** and **VSIGMA** are the above-threshold DIBL phase-out voltage and width, respectively.

The body-effect parameter γ_S corresponds to that of a MOSFET with a uniformly doped substrate but adjusted as follows for fringing effects of small geometries (length and width) on the depletion charge (see Lee et al., 1993):

$$\gamma_S = \text{GAMMAS0} + \text{LGAMMAS} \cdot \left(1 - \frac{\text{L0}}{\text{L}}\right) + \text{WGAMMAS} \cdot \left(1 - \frac{\text{W0}}{\text{W}}\right) \quad (6.3.12)$$

Here **GAMMAS0** is the same as the parameter **GAMMA** in Levels 1, 2, and 3, **LGAMMAS** and **WGAMMAS** are sensitivity parameters for γ_S with respect to the effective gate length **L** and the effective gate width **W** (both specified on the device line), while **L0** and **W0** are the nominal gate length and width, respectively.

The last term on the right in Eq. (6.3.10) introduces an additional body-effect parameter γ_L to account for nonuniformities in the substrate doping. The effects of small dimensions on γ_L are described in the same way as for γ_S:

$$\gamma_L = \text{GAMMAL0} + \text{LGAMMAL} \cdot \left(1 - \frac{\text{L0}}{\text{L}}\right) + \text{WGAMMAL} \cdot \left(1 - \frac{\text{W0}}{\text{W}}\right) \quad (6.3.13)$$

Here **GAMMAL0, LGAMMAL,** and **LWGAMMAL** are additional AIM-Spice parameters.

The MOSFET Level 7 I–V characteristics are given by [see Eq. (6.2.9)]

$$I_d = \frac{g_{\text{ch}} V_{ds}(1 + \text{LAMBDA} \cdot V_{ds})}{[1 + (g_{\text{ch}} V_{ds}/I_{\text{sat}})^{\text{M}}]^{1/\text{M}}} \quad (6.3.14)$$

where **LAMBDA** is the channel length modulation parameter (same as for MOSFET Levels 1 and 2) and **M** is the knee shape parameter.

The extrinsic linear channel conductance g_{ch} in Eq. (6.3.14) has the following form:

$$g_{\text{ch}} = \frac{g_{\text{chi}}}{1 + g_{\text{chi}}(\text{RSI} + \text{RDI})} \quad (6.3.15)$$

where **RSI** and **RDI** are the source and drain series resistances and

$$g_{\text{chi}} = q n_s \mu_n \frac{\text{W}}{\text{L}} \quad (6.3.16)$$

is the intrinsic linear conductance. The sheet density of channel electrons n_s at small drain–source bias is approximated by

$$n_s = \frac{V_{\text{th}} \epsilon_i \text{ETA}}{q \text{TOX}} \ln\left[1 + \tfrac{1}{2}\exp\left(\frac{V_{gt}}{V_{\text{th}}\text{ETA}}\right)\right] \quad (6.3.17)$$

where **ETA** is the subthreshold ideality factor and **TOX** is the oxide thickness. The dependence of the channel mobility on the vertical electric field is expressed through the gate voltage overdrive as follows:

$$\mu_n = \frac{\text{U0}}{1 + \text{THETA} \cdot (V_{gte} + 2V_{\text{T}})/\text{TOX}} \quad (6.3.18)$$

where **U0** is the low-field mobility, **THETA** is a mobility parameter, and V_{gte} is the effective gate voltage overdrive defined as follows in terms of the transition parameter **DELTA** to smooth the transition to the subthreshold regime:

$$V_{gte} = V_{\text{th}}\left[1 + \frac{V_{gt}}{2V_{\text{th}}} + \sqrt{\text{DELTA}^2 + \left(\frac{V_{gt}}{2V_{\text{th}}} - 1\right)^2}\right] \quad (6.3.19)$$

The saturation current I_{sat} entering into Eq. (6.3.14) is given by

$$I_{sat} = \frac{g_{chi}V_{gte}}{1 + g_{chi}\mathbf{RSI} + \sqrt{1 + 2g_{chi}\mathbf{RSI} + \left(\frac{\mu_n V_{gte}}{\mathbf{VMAX \cdot L}}\right)^2}} \qquad (6.3.20)$$

where **VMAX** is the electron saturation velocity.

The universal MOSFET model (Level 7) implemented in AIM-Spice also includes the unified version of the Meyer capacitances described in Section 6.2.2 and in Eqs. (6.3.6) and (6.3.9).

6.3.4. SPICE Example: CMOS Operational Amplifier

In this example, we investigate properties of a simple CMOS operational amplifier (op-amp) shown in Fig. 6.3.4. The corresponding circuit description is presented in Fig. 6.3.5. The model parameters were extracted for n- and p-channel devices with a nominal gate length of 2.0 μm, a gate width of 68 μm, and an oxide thickness of 250 Å.

We first simulate the large-signal differential input transfer characteristics of the amplifier using the DC Transfer Curve Analysis. We specify vin to sweep from −5 to 5 mV in steps of 0.1 mV. The results of this dc sweep is plotted in Fig. 6.3.6. Using AIM-Postprocessor, we may plot the derivative of V(8) and use the cursors to directly estimate the maximum small-signal differential gain to be 4.6 kV/V.

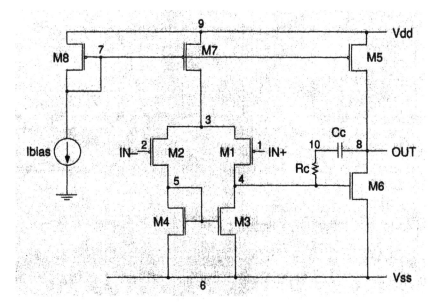

FIG. 6.3.4. Simple CMOS operational amplifier.

```
A Simple CMOS OPAMP
* Power supply and bias current source
vdd 9 0 dc 2.5
vss 6 0 dc -2.5
ibias 7 0 dc 5.19ua
* Amplifier
m1 4 1 3 3 mp l=2u w=18u
m2 5 2 3 3 mp l=2u w=18u
m3 4 5 6 6 mn l=2u w=3u
m4 5 5 6 6 mn l=2u w=3u
m5 8 7 9 9 mp l=2u w=6u
m6 8 4 6 6 mn l=2u w=18u
m7 3 7 9 9 mp l=2u w=2u
m8 7 7 9 9 mp l=2u w=2u
rc 10 8 7042
cc 4 10 200fF
* Input biasing circuit (not shown in figure)
vin 101 0 dc 0 ac 1 pwl(0,0V 100ns,0V 103ns,100mV 200ns,100mV)
rd 101 0 1
ev+ 1 0 101 0 +0.5
ev- 2 0 101 0 -0.5
.model mn nmos level=7 vto=0.736 gammas0=0.825 phi=0.686 pb=0.8
+js=3.2e-5 tox=250e-10 nsub=8e15 xj=0.25e-6 ld=0.24e-6 vmax=5e4
+ uo=418 eta=1.745 lambda=0.00969 m=2.9 sigma0=0.0065
+ cgdo=1.27E-10 cgso=1.27E-10 cgbo=7.66E-11 cj=570E-6 mj=0.3
+ cjsw=260E-12 mjsw=0.33 rs=12.87 rd=12.87
.model mp pmos level=7 vto=-0.622 gammas0=0.514 phi=0.635
+ pb=0.635 js=3.2e-5 tox=250e-10 nsub=3e15 xj=0.4e-6 ld=0.3e-6
+ vmax=2.5e4 uo=195 eta=1.745 lambda=0.0429 m=3 sigma0=0.0065
+ cgdo=1.27E-10 cgso=1.27E-10 cgbo=7.66E-11 cj=190E-6 mj=0.3
+ cjsw=260E-12 mjsw=0.33 rs=76.45 rd=76.45
```

FIG. 6.3.5. Circuit description for the operational amplifier in Fig. 6.3.4.

Another important property of op-amps is the slew rate, which affects the high-frequency operation of op-amp circuits. The worst-case slew rate is obtained by connecting the output node to the inverting input and applying a step input of 100 mV to the noninverting input. To illustrate this, we run a transient analysis and observe the transient response at the output node. The transient analysis parameters are 0.01 μs step size and 0.15 μs final time. The results shown in Fig. 6.3.7 indicate that the slew-rate is about 13 V/μs.

6.4. UNIVERSAL MESFET MODEL

The basic MESFET models discussed in Section 5.3 can be improved in many ways within the universal modeling concept. In addition to short-channel and high-field phenomena (DIBL, subthreshold current, channel length modulation, impact ionization, etc.) and the effects of parasitic resistances, discussed above, a

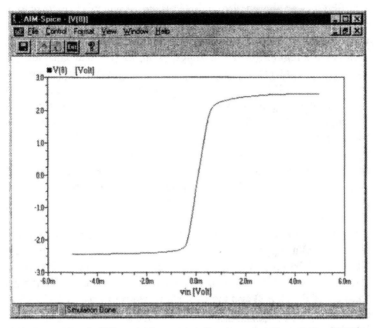

FIG. 6.3.6. Large-signal differential-input transfer characteristics of the CMOS op-amp circuit shown in Fig. 6.3.4.

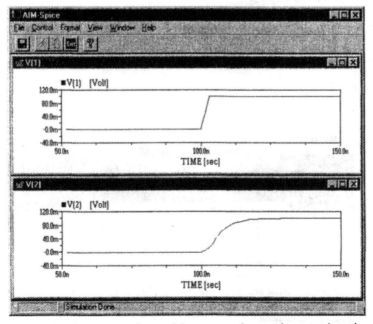

FIG. 6.3.7. Input and output waveforms of the op-amp when a voltage step input is applied.

viable MESFET model should also include effects of nonuniform channel doping, gate leakage current, and temperature dependencies.

Nonuniformities are caused, for example, by ion-implanted channel doping and may be tailored as a part of the device design. Also, we note that since the gate metal forms a Schottky contact with the channel, we can expect a significant gate leakage current when the MESFET gate is biased in the forward direction. In normal operation, electron–hole generation in the gate depletion region gives rise to a small reverse leakage current (see Chapter 3). A description of the dependence of device performance on temperatures is important because MESFET electronics is applied in a wide temperature range, from cryogenic conditions up to several hundred degrees centigrade. However, although gate leakage and temperature effects have been included in our universal MESFET model implemented in AIM-Spice, a comprehensive discussion of these effects fall outside the scope of this book, and we instead refer to Lee et al. (1993) and to Ytterdal et al. (1995b).

Here, we describe a MESFET model based on the universal modeling concept outlined earlier. Our approach follows the same basic philosophy used for MOSFETs; that is, by deriving unified expressions for the linear channel conductance, the saturation current, and the channel capacitance to be used in the universal $I-V$ expression of Eq. (6.2.9) and the $C-V$ expressions of Eqs. (6.2.13)–(6.2.16). A parameter extraction procedure for the universal MESFET model was described by Shur et al. (1992a) (see also Lee et al., 1993).

6.4.1. MESFET *I–V* Model

The linear intrinsic and extrinsic conductances are dependent on the electron sheet density n_s in the conducting channel, as indicated in Eqs. (6.2.6) and (6.2.7). Referring to the schematic MESFET cross section in Fig. 6.4.1, the above-threshold value

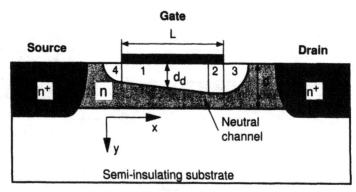

FIG. 6.4.1. Schematic cross section of a MESFET. The numbered parts of the gate depletion region correspond to (1) the nonsaturated part, (2) the saturated part, and (3, 4) the source and drain depletion extensions.

n_{sa} of n_s is given by the sheet density of doping atoms in the neutral part of the channel, that is,

$$n_{sa} = \int_{d_d}^{d} N(y)\, dy \qquad (6.4.1)$$

Here, d is the effective channel depth, d_d is the depth of the depleted region, and $N(y)$ is the position-dependent doping density at the vertical position y.

The positively charged depletion region under the gate is created by the combined effects of the built-in voltage V_{bi}, the gate–source voltage V_{GS}, and the channel voltage V. The depletion depth d_d can be determined from a one-dimensional Poisson equation, assuming that the GCA is valid in the nonsaturated region 1 of the channel (see Fig. 6.4.1). Integrating this equation in the vertical dimension, we find the following implicit expression for d_d, accounting for a nonuniform channel-doping profile:

$$V_{bi} - V_{GS} + V = \frac{q}{\epsilon_s} \int_0^{d_d} y N(y)\, dy \qquad (6.4.2)$$

Here we assumed that the doping profile does not vary too abruptly, so that the extent of the partially depleted boundary layer separating the neutral channel from the depletion region is small compared to the characteristic length of the doping profile variation. [The special but important case of delta-doped MESFETs was addressed by Lee et al. (1993).]

The threshold voltage, corresponding to the gate–source voltage that exactly depletes the entire channel at zero drain bias, is given by $V_T = V_{bi} - V_{po}$ (see Section 5.3), where the pinch-off voltage V_{po} can be expressed as

$$V_{po} = \frac{q}{\epsilon_s} \int_0^{d} y N(y)\, dy \qquad (6.4.3)$$

For the special case of a uniformly doped MESFET, Eq. (6.4.2) reduces to the simplified expression of Eq. (5.4.1) in the above-threshold regime, and the pinch-off voltage becomes $V_{po} = q N_d d^2 / 2\epsilon_i$. Hence, we obtain the following analytical expression for the above-threshold channel electron sheet density:

$$n_{sa} = N_d d \left[1 - \sqrt{1 - \frac{V_{GT} - V}{V_{po}}} \right] \qquad (6.4.4)$$

Below threshold, the electrons mainly collect near the bottom of a potential well, at the interface between the doped channel and the substrate. This vertical confinement of the electrons at a well-defined distance from the gate electrode is very similar to the situation in a MOSFET. Hence, the below-threshold approximation of the UCCM expression [see Eq. (6.2.1)] is suitable for describing the electron sheet density of a MESFET in the subthreshold regime, that is,

$$n_{sb} \approx n_o \exp\left(\frac{V_{GT} - V_F}{\eta V_{\text{th}}}\right) \tag{6.4.5}$$

Here, n_o is the MESFET electron sheet density at threshold (determined below), V_F is the Fermi potential relative to the source, and η is the ideality factor.

At low drain–source voltages, corresponding to the linear region of the I–V characteristics, we simplify the above equations by taking $V(x) = 0$ and $V_F = 0$. Also, since the drain current is assumed to be small, we can replace V_{GT} by its extrinsic counterpart V_{gt}. We now want to construct a unified expression for the electron sheet density for this condition, to be used for calculating the unified linear channel conductance. One possibility is to use the interpolation formula

$$n_s \approx \frac{n_{sa} n_{sb}}{n_{sa} + n_{sb}} \tag{6.4.6}$$

But in order to arrive at a meaningful result for $V_{gt} < 0$, we introduce the following effective gate voltage overdrive to replace V_{GT} in n_{sa}:

$$V_{gte} = \frac{V_{\text{th}}}{2}\left[1 + \frac{V_{gt}}{V_{\text{th}}} + \sqrt{\delta^2 + \left(\frac{V_{gt}}{V_{\text{th}}} - 1\right)^2}\right] \tag{6.4.7}$$

(This is similar to what was done for the universal MOSFET model in Section 6.3; see also the discussion on capacitance modeling in Section 6.2.) We note that V_{gte} approaches asymptotically V_{th} below threshold and V_{gt} above threshold, and n_s attains the correct limiting behavior in both regimes. The parameter δ determines the width of the transition region. Typically, $\delta = 3, \ldots, 5$ is a good choice.

Hence, for MESFETs with a uniformly doped channel region, we obtain

$$n_s = \left[\frac{1}{N_d d}\left(1 - \sqrt{1 - \frac{V_{gt}}{V_{\text{po}}}}\right)^{-1} + \frac{1}{n_o}\exp\left(-\frac{V_{gt}}{\eta V_{\text{th}}}\right)\right]^{-1} \tag{6.4.8}$$

The corresponding expression for a delta-doped MESFET was given by Lee et al. (1993). We note that Eq. (6.4.8) does not depend explicitly on the drain–source voltage since we assumed this voltage to be small when calculating the linear channel conductance. However, when using n_s in the universal I–V expression, it will have some implicit dependence on V_{ds} through the DIBL-induced shift in threshold voltage (see Section 6.2).

The above-threshold saturation current, hereafter denoted I_{sata}, was obtained in Section 5.4 using the basic velocity saturation model for MESFETs. The subthreshold saturation current I_{satb} is given by Eq. (6.2.8) assuming $V_{ds} \gg V_{\text{th}}$. Based on these expressions, a unified saturation current can be written in terms of the following interpolation formula:

$$I_{\text{sat}} = \frac{I_{\text{sata}} I_{\text{satb}}}{I_{\text{sata}} + I_{\text{satb}}} \tag{6.4.9}$$

Again, in order to obtain meaningful results below threshold, we replace V_{gt} in I_{sata} by the effective gate voltage swing V_{gte} of Eq. (6.4.7). A further correction of I_{sata} involves the introduction of a transconductance compression parameter t_c to account for more general doping profiles (see Statz et al., 1987). These modifications lead to

$$I_{\text{sata}} = \frac{2\beta V_{gte}^2}{(1 + 2\beta V_{gte}R_s + \sqrt{1 + 4\beta V_{gte}R_s})(1 + t_c V_{gte})} \tag{6.4.10}$$

where the transconductance parameter $\beta = 2\epsilon_s v_s W/d(V_{po} + 3V_L)$ is taken from the basic theory of Section 5.3. The value of the transconductance compression factor depends on the doping profile and may also depend on the properties of the substrate–channel interface and other factors. Typically, $t_c \approx 0$ for uniformly doped or ion-implanted devices with small pinch-off voltages (<2.5 V). For MESFETs with larger pinch-off voltages (>6 V), Statz et al. (1987) found $t_c \approx 0.1$ V^{-1}. We should point out that the nonuniformity in the doping profile does not influence the saturation current very strongly. In fact, some of this dependence is already included via the pinch-off voltage contained in β and in V_T. Accordingly, a model where the shape of the doping profile is accounted for by the additional parameter t_c is quite adequate for modeling I_{sat}. As was shown by Lee et al. (1993), this model works well even for the limiting case of delta-doped MESFETs.

In Fig. 6.4.2, the present universal MESFET model has been applied to experimental I–V characteristics obtained for a 0.5-μm GaAs MESFET. Figure 6.4.2a shows the above-threshold I–V characteristics and Fig. 6.4.2b shows the subthreshold characteristics in a semilog plot. As can be seen from the figures, our model quite accurately reproduces the experimental data in the entire range of bias voltages, over several decades of the current variation, including the residual drain current at large negative values of V_{gt}. The parameters used in this calculation were obtained using the parameter extraction procedure similar to that described in Shur et al. (1992a) (see also Lee et al., 1993).

6.4.2. MESFET C–V Model

In Sections 5.4 and 6.2, we discussed different ways of modeling MESFET capacitances, including the two-diode model, the Meyer-type model, and the charge-based model.

The two-diode model is the simplest and the least precise and will not be discussed any further here. The Meyer-type MESFET model, modified to include velocity saturation effects and the subthreshold regime, is probably adequate for most circuit simulation purposes. This model has been implemented in AIM-Spice and has the benefit of being relatively simple. The most advanced and accurate MESFET capacitance model is a charge-based model similar to that developed for MOSFETs by Ward and Dutton (1978). Although straightforward, this model results in relatively complicated expressions for the charges and capacitances (Nawaz and Fjeldly, 1997).

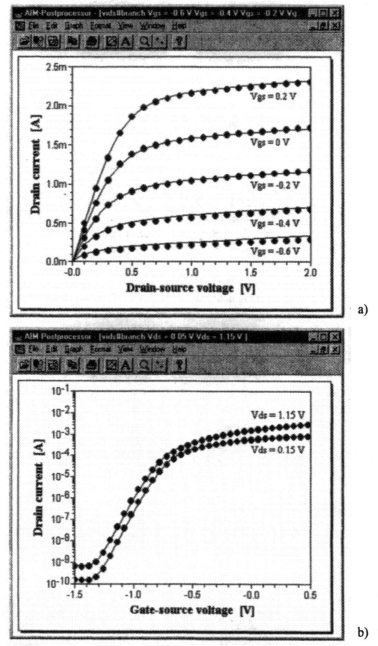

FIG. 6.4.2. (*a*) Above-threshold and (b) subthreshold I–V characteristics of a 0.5-μm ion-implanted MESFET. Experimental data (symbols) and model calculations (solid lines). Device parameters: $L = 0.5$ μm, $W = 10$ μm, $d = 0.09$ μm, $\mu_n = 0.22$ m²/V s, $v_s = 2.35 \times 10^5$ m/s, $m = 2.8$, $V_{bi} = 0.75$ V, $V_{To} = -0.874$ V, $\eta = 1.59$, $R_s = R_d = 50$ Ω, $\lambda = 0.043$ V⁻¹, $t_c = 0.4$ V⁻¹, $\zeta = 1.62$, $\sigma_o = 0.051$, $V_\sigma = 0.2$ V, $V_{\sigma t} = 0.45$ V.

Unified Meyer-Type C–V Model. The unified intrinsic gate channel capacitance C_{ch} of the uniformly doped MESFET at zero drain bias can, in principle, be obtained by differentiating the part of the unified gate charge that corresponds to the depletion charge under the gate (regions 1 and 2 in Fig. 6.2.1) with respect to V_{GT}. However, since this leads to a somewhat cumbersome expression for C_{ch}, we instead consider the limiting values C_a well above threshold and C_b well below threshold. Assuming a uniformly doped MESFET, we obtain, from Eqs. (6.4.4) and 6.4.5),

$$C_a = -qWL \frac{d(N_d d - n_{sa})}{dV_{GT}} = \frac{WL\epsilon_s}{d\sqrt{(1 - V_{GT}/V_{po})}} \qquad (6.4.11)$$

$$C_b = -qWL \frac{d(N_d d - n_{sb})}{dV_{GT}} = WL \frac{qn_o}{\eta V_{th}} \exp\left(\frac{V_{GT}}{\eta V_{th}}\right) \qquad (6.4.12)$$

The electron sheet density n_o at threshold is found by considering the effective channel capacitance at threshold and zero drain–source voltage. We define the threshold channel capacitance based on the following consideration: Above threshold, we have $C_a \ll C_b$, and below threshold, we have $C_b \ll C_a$. Hence, it is reasonable to associate threshold with the condition $C_b = C_a = \frac{1}{2}C_o$, where $C_o = WL\epsilon_s/d$. At threshold, we use $V_{GT} = 0$ in Eqs. (6.4.11) and (6.4.12), which gives

$$n_o = \frac{\epsilon_s \eta V_{th}}{qd} \qquad (6.4.13)$$

From these equations, we obtain the following approximate, unified expression for the gate channel capacitance by using the interpolation formula of Eq. (6.2.13):

$$C_{ch} \approx \frac{C_a C_b}{C_a + C_b} = C_o\left[\sqrt{1 - \frac{V_{GTe}}{V_{po}}} + \exp\left(-\frac{V_{GT}}{\eta V_{th}}\right)\right]^{-1} \qquad (6.4.14)$$

Here we replaced V_{GT} in C_a by the effective gate voltage swing V_{GTe} from Eq. (6.4.7) in order to obtain meaningful results for $V_{GT} < 0$. Figure 6.4.3 shows C_{ch} versus V_{GT}.

The capacitance C_{ch} enters in the unified Meyer-type capacitances C_{GS} and C_{GD} given by Eqs. (6.2.14) and (6.2.15). Figure 6.4.4 shows calculated values of these capacitances for the MESFET considered in Fig. 6.4.2. The large-signal MESFET equivalent according to this model is shown in Fig. 6.4.5a.

Charge-Based C–V Model. The ideas of charge-based $C–V$ modeling for FETs were outlined in Section 6.2. For a MESFET, this model leads to a set of nine transcapacitances, of which four are independent. The complete large-signal equivalent circuit according to the charge-based model is shown in Fig. 6.2.5b. Since this model leads to quite complicated expressions, we restrict ourselves to an outline of the theory here.

As shown in Fig. 6.2.1, the depletion charges in a MESFET are distributed over

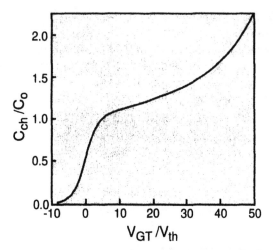

FIG. 6.4.3. Unified MESFET gate channel capacitance normalized to C_o versus V_{GT}/V_{th}. Same device as in Fig. 6.4.2. Threshold transition parameter: $\delta = 3$.

four regions: Region 1 is the nonsaturated part of the channel below the gate, region 2 is the saturated part of this channel (exists only in saturation), and regions 3 and 4 are the depletion extensions toward drain and source, respectively. So far we have ignored the contributions of the depletion extensions to the gate charge. But these contributions may be important and will be considered below.

In the Ward–Dutton scheme, the MESFET depletion charges located directly un-

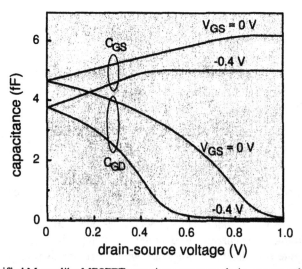

FIG. 6.4.4. Unified Meyer-like MESFET capacitances versus drain–source voltage V_{GS} of 0 V and –0.4 V. Same device as in Fig. 6.4.2. Saturation transition parameter: $V_\delta = 0.2$ V.

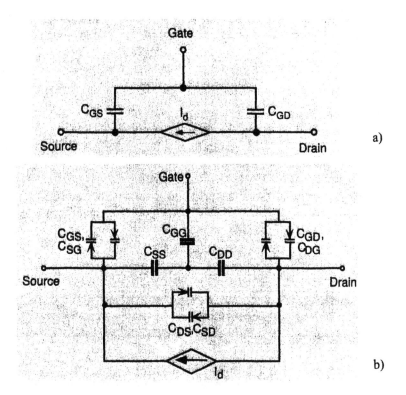

FIG. 6.4.5. Large-signal MESFET equivalent circuits corresponding to (*a*) the Meyer-type capacitance model and (*b*) the complete charge-based capacitance model. (From Nawaz and Fjeldly, 1997.)

der the gate (regions 1 and 2 in Figure 6.4.1) should be partitioned and assigned to the source and drain terminals as follows (uniform doping):

$$Q_{So} = qWN_d \int_0^L \left(1 - \frac{x}{L}\right) d_d(x)\, dx \qquad (6.4.15)$$

$$Q_{Do} = qWN_d \int_0^L \frac{x}{L} d_d(x)\, dx \qquad (6.4.16)$$

and the corresponding gate charge is

$$Q_{Go} = -(Q_{So} + Q_{Do}) = -qWN_d \int_0^L d_d(x)\, dx \qquad (6.4.17)$$

In saturation, the depletion width d_d in region 2 can be assumed to be constant for a given gate–source voltage, and the channel voltage V_c at the boundary between the regions 1 and 2 can be approximated by $V_c = V_{SAT}$. This allows the length of the saturated region to be determined as (see Shur, 1987)

$$L_s = \frac{2d}{\pi} \sinh^{-1} \left[\frac{\pi k_d (V_{DS} - V_c)}{2dF_s} \right]$$ (6.4.18)

where $F_s = v_s/\mu_n$ is the saturation field and $k_d \approx 0.1$.

Accurate capacitance modeling of MESFETs requires the inclusion of the depletion region extensions toward the drain and source, indicated as regions 3 and 4 in Fig. 6.4.1. The charges Q_{Se} and Q_{De} of these extensions depend on the bias conditions of the device. Above threshold in the nonsaturated regime, both extensions can be modeled as quarter circles to give

$$Q_{Se} = \tfrac{1}{4} \pi q N_d [d_d(0)]^2$$ (6.4.19)

$$Q_{De} = \tfrac{1}{4} \pi q N_d [d_d(L)]^2$$ (6.4.20)

In above-threshold saturation, that drain extension is modeled as a circle sector in combination with a right-angle triangle, while below threshold, both extensions are modeled this way. We always assign the charge Q_{Se} to the source terminal and the charge Q_{De} to the drain terminal. The corresponding contribution to the gate charge is $Q_{Ge} = -(Q_{Se} + Q_{De})$. Clearly, the depletion extensions will be the dominant contributors to the subthreshold transcapacitances of the intrinsic MESFET with a semi-insulating substrate.

Collecting the relevant charge contributions of the three terminals, i.e., $Q_X = Q_{Xo} + Q_{Xe}$, the terminal transcapacitances can be calculated according to Eq. (6.2.11). The total of nine transcapacitances may be organized in a 3×3 matrix as follows:

$$\mathbf{C} = \begin{bmatrix} C_{GG} & C_{GS} & C_{GD} \\ C_{SG} & C_{SS} & C_{SD} \\ C_{DG} & C_{DS} & C_{DD} \end{bmatrix}$$ (6.2.21)

where the elements in each column sum to zero because of Kirchhoff's current law and the elements in each row sum to zero for the matrix to be reference independent. As a result of these constraints, only four of the nine elements are independent. Also, some of the transcapacitances will have negative values.

Performing the calculations of the charges and the capacitances for the MESFET leads to fairly complicated expressions that are not reproduced here. We instead refer to Nawaz and Fjeldly (1997).

6.4.3. Implementation in AIM-Spice

Two versions of the universal MESFET model have been implemented in AIM-Spice: Level 2 typically corresponds to a MESFET with a uniform doping or with a doping profile created by ion implantation. Level 3 describes a delta-doped MES-FET. (Level 1 is the Statz model discussed in Section 5.4.)

In Levels 2 and 3, the MESFET threshold voltage includes the DIBL effect:

$$V_T = \mathbf{VTO} - \sigma V_{ds} \qquad (6.4.22)$$

Here $\mathbf{VTO} = V_{bi} - V_{po}$ is the threshold voltage at zero drain bias and the DIBL coefficient σ has the same form as that used for MOSFET Level 7 [(see Eq. (6.3.11)].

Also the universal form of the I–V characteristics and the expressions for the linear channel conductances g_{chi} and g_{ch} are the same as for MOSFET Level 7 [see Eqs. (6.3.14)–(6.3.16)].

In MESFET Level 2 (uniform or near-uniform doping), the channel sheet density of charge carriers n_s at low drain bias is approximated by

$$n_s = \left[\frac{1}{\mathbf{ND \cdot D}} \left(1 - \sqrt{1 - \frac{V_{gte}}{V_{po}}} \right)^{-1} + \frac{1}{n_o} \exp\left(-\frac{V_{gt}}{\mathbf{ETA} \cdot V_{th}} \right) \right]^{-1} \qquad (6.4.23)$$

Here $\mathbf{ND}$ is the doping density, $\mathbf{D}$ is the active layer thickness, $\mathbf{ETA}$ is the subthreshold ideality factor, $V_{po} = q\mathbf{ND \cdot D}^2/2\epsilon_s$ is the pinch-off voltage, $n_o = \epsilon_s \mathbf{ETA} \cdot V_{th}/q\mathbf{D}$ is the carrier sheet density at threshold, and

$$V_{gte} = \frac{V_{th}}{2}\left[1 + \frac{V_{gt}}{V_{th}} + \sqrt{\mathbf{DELTA}^2 + \left(\frac{V_{gt}}{V_{th}} - 1 \right)^2} \right] \qquad (6.4.24)$$

is the effective gate voltage overdrive.

The saturation current has the form

$$I_{sat} = \frac{I_{sata}I_{satb}}{I_{sata} + I_{satb}} \qquad (6.4.25)$$

where

$$I_{sata} = \frac{2\beta V_{gte}^2}{(1 + 2\beta V_{gte}\,\mathbf{RS} + \sqrt{1 + 4\beta V_{gte}\mathbf{RS}})(1 + \mathbf{TC} \cdot V_{gte})} \qquad (6.4.26)$$

$$I_{satb} = \frac{qn_o V_{th}\mathbf{ETA} \cdot \mathbf{MU} \cdot \mathbf{W}}{\mathbf{L}} \exp\left(\frac{V_{gt}}{\mathbf{ETA} \cdot V_{th}} \right) \qquad (6.4.27)$$

are the above- and below-threshold saturation currents, respectively. Here, $\mathbf{L}$ and $\mathbf{W}$ are the gate length and width, respectively (specified on the device line), $\mathbf{RS}$ is the source series resistance, $\mathbf{TC}$ is the transconductance compression factor, $\mathbf{MU}$ is the effective low-field mobility,

$$\beta = \frac{2\epsilon_s VS \cdot W}{D \cdot (V_{po} + 3VS \cdot L/MU)} \tag{6.4.28}$$

is the transconductance parameter, an VS is the saturation velocity.

The MESFET Level 2 I–V characteristics implemented in AIM-Spice also include expressions for the gate leakage current, temperature and frequency dependencies of key parameters, and side gating. Moreover, the intrinsic device C–V characteristics are modeled using the unified version of the Meyer capacitance model according to Eqs. (6.2.14)–(6.2.16). The gate channel capacitance C_{ch} for this case is given by

$$C_{ch} \approx \frac{\epsilon_s W \cdot L}{D} \left[\sqrt{\left(1 - \frac{V_{gte}}{V_{po}}\right)} + \exp\left(-\frac{V_{gt}}{ETA \cdot V_{th}}\right) \right]^{-1} \tag{6.4.29}$$

6.4.4. SPICE Example: MESFET Ring Oscillator

In this example, we simulate an 11-stage ring oscillator composed of MESFET inverters with ungated loads. The I–V characteristic of the ungated load can be modeled as follows:

$$I_L = V_{Lsw} \tanh\left(\frac{V_L}{V_{Lss}}\right)(1 + \lambda_L V_L) \tag{6.4.30}$$

where λ_L is the load device output conductance parameter, V_L is the drain–source voltage across the load, and V_{Lss} is an empirical parameter given by

$$V_{Lss} = I_L R_{offw} \tag{6.4.31}$$

where I_{Lsw} is the saturation current and R_{offw} is the resistance in the linear region. We used parameter values from Shur (1987), given in Table 6.4.1. The oscillator and its circuit description are shown in Figs. 6.4.6 and 6.4.7, respectively. We used load widths between 1 and 4 μm. The ungated loads were modeled directly in the circuit description with non-linear-dependent current sources. Note that the purpose of the noise voltage source in the circuit is to trigger the oscillations.

The MESFET model parameters for the switching transistors were extracted by the procedure outlined in Lee et al. (1993) using experimental data from Shur (1987). Table 6.4.2 gives a complete listing of the parameter values used.

TABLE 6.4.1. Ungated Load Parameters.

Parameter	Value
λ_L	0.027
I_{Lsw}	250 A/m
R_{offw}	2.24 Ω mm

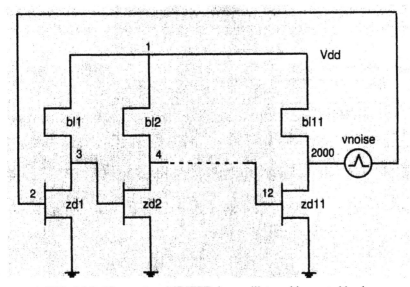

FIG. 6.4.6. Eleven-stage MESFET ring oscillator with ungated loads.

```
Mesfet Ring Oscillator with ungated load
*
.subckt mesinv 10 20 30
* Node 10: Power Supply
* Node 20: Input
* Node 30: Output
rdl 10 1 20
bl 1 2 i=0.00025*tanh(v(1,2)/0.00025/2240)*
+(1+v(1,2)*0.027)
rsl 2 30 20
zd 30 20 0 driver l=0.7u w=20u
ci 30 0 20f
.ends mesinv

.model driver nmf level=2 n=1.44 rd=20 rs=20 vs=1.9e5
+ mu=0.25 d=1e-7 vto=0.15 m=2 lambda=0.15 sigma0=0.02
+ vsigmat=0.5

vdd 1 0 dc 1.6
xinv01 1 2 3 mesinv
xinv02 1 3 4 mesinv
xinv03 1 4 5 mesinv
xinv04 1 5 6 mesinv
xinv05 1 6 7 mesinv
xinv06 1 7 8 mesinv
xinv07 1 8 9 mesinv
xinv08 1 9 10 mesinv
xinv09 1 10 11 mesinv
xinv10 1 11 12 mesinv
xinv11 1 12 2000 mesinv
vnoise 2000 2 dc 0 pwl(0 0 0.2n 0 0.3n 0.1 0.4n 0)
```

FIG. 6.4.7. Circuit description of 11-stage MESFET ring oscillator with ungated loads.

TABLE 6.4.2. Parameters Used for 11-Stage MESFET Ring Oscillator.

Parameter Description	Parameter	Value
Gate length	**L**	0.7 μm
Gate width	**W**	20 μm
Emission coefficient	**N**	1.44
Drain resistance	**RD**	20 Ω
Source resistance	**RS**	20 Ω
Saturation velocity	**VS**	1.9×10^5 m/s
Low-field mobility	**MU**	0.25 m²/V s
Channel thickness	**D**	1×10^{-7} m
Threshold voltage	**VTO**	0.15 V
Knee shape parameter	**M**	2
Output conductance parameter	**LAMBDA**	0.15 V^{-1}
DIBL parameter 1	**SIGMA0**	0.02
DIBL parameter 2	**VSIGMAT**	0.5 V
DIBL parameter 3	**VSIGMA**	0.1 V

The parameters for the transient analysis are shown in the dialog box of Fig. 6.4.8. Note that since an oscillator does not have a quiescent operating point, the UIC option is turned on, making AIM-Spice skip the bias point calculation. The output waveform of the ring oscillator is shown in Fig. 6.4.9.

6.5. UNIVERSAL HFET MODEL

We now consider an advanced HFET model based on the universal modeling principles discussed in the preceding sections. Again, we include the various physical mechanisms related to small geometries, high fields, and parasitic series resis-

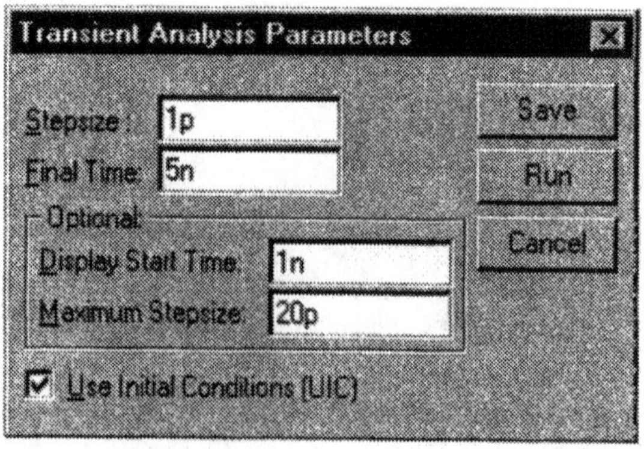

FIG. 6.4.8. Transient Analysis parameters for MESFET ring oscillator.

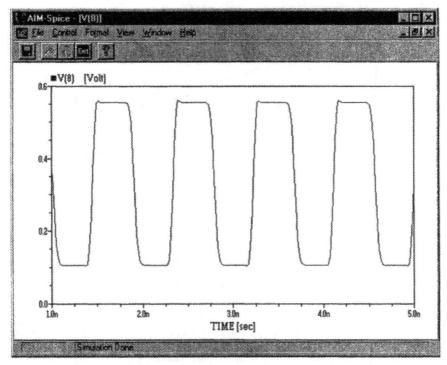

FIG. 6.4.9. MESFET oscillator output waveform.

tances. In addition, we consider the effects of electron transfer into the semiconductor barrier layer separating the HFET channel from the gate. As for the MESFET, the HFET is also subjected to gate leakage. In fact, the HFET gate structure may be viewed as a distributed back-to-back series combination of a heterojunction barrier and a Schottky barrier. Although gate leakage is included in the HFET model implemented in AIM-Spice, a detailed discussion of this subject falls outside the scope of this book. However, its effect is demonstrated in a simulation example at the end of this section.

Throughout this section, we tacitly assume that the HFETs are based on the AlGaAs/GaAs material system. However, with appropriate input parameters, our model is also applicable to HFETs fabricated from other material systems such as AlGaAs/InGaAs/GaAs or AlInAs/InGaAs/InP.

The HFET characteristics are strongly affected by the limitations imposed on the carrier sheet density in the conducting channel by the energy band discontinuities at the GaAs/AlGaAs interface. This can be understood by analyzing the energy band diagram of such structures. As an example, we plot simplified conduction band diagrams of delta-doped HFETs for different gate–source voltages in Fig. 6.5.1. In this figure, we see that the electron quasi–Fermi level E_{Fn} inside the AlGaAs layer will approach the bottom of the conduction band at large V_{GS} and cause a significant

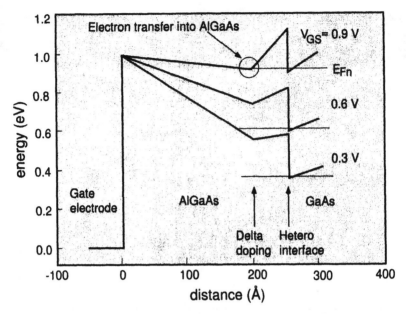

FIG. 6.5.1. Conduction band diagrams of a delta-doped HFET for different V_{GS}. At sufficiently large forward gate bias, the electron quasi–Fermi level in the semiconductor approaches the conduction band minimum in AlGaAs, causing a transfer of electrons into this layer and a reduction in the device transconductance. (After Chao et al., 1989. Copyright 1989 by IEEE.)

transfer of electrons into the AlGaAs layer. But since AlGaAs has many defects, most of these carriers are trapped once inside this layer and will not contribute much to the device current.

Computed gate bias dependencies of the sheet densities of channel electrons n_s and of electrons in the AlGaAs layer n_t are shown in Fig. 6.5.2. These results are based on a self-consistent solution of Poisson's equation and Schrödinger's equation according to Stern (1972). As can be seen from this figure, the density of electrons in the AlGaAs layer increases sharply beyond a certain threshold gate voltage, and the slope of the n_s-versus-V_{GS} dependence starts to decrease. At high gate biases, n_s tends to saturate.

6.5.1. HFET *I–V* Model

In Section 5.5, we argued that the basic velocity saturation MOSFET model can also be applied to HFETs. Likewise, as a starting point for the development of a universal HFET model, we can adopt some of the results from the universal MOSFET model. Hence, provided V_{GS} is not too large, the MOSFET expressions of Section 6.3 for the channel carrier sheet density n_s, the linear channel conductance

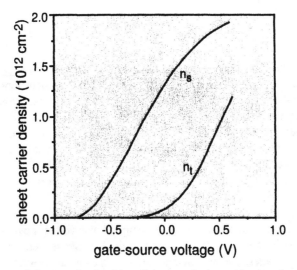

FIG. 6.5.2. Computed sheet carrier densities of electrons in the conducting channel n_s and in the AlGaAs layer n_t versus V_{GS}, indicating the transfer of electrons into the AlGaAs layer above a certain gate bias threshold. (After Chao et al., 1989. Copyright 1989 by IEEE.)

g_{chi}, the saturation current I_{sat}, and the gate channel capacitance C_{ch}, are reasonable approximations also for HFETs.

However, at sufficiently large gate bias, we also have to account for the transfer of electrons into the AlGaAs layer. This can be done, for example, by assuming that n_s cannot exceed a certain maximum value n_{max}, typically below 2×10^{12} cm^{-2}. Naturally, this maximum value is dependent on the material system used and, to a certain extent, on the doping profile in the wide-band-gap semiconductor. Here follows a modeling approach based on this idea that provides a reasonable description of the saturation phenomena in the channel sheet density and in the saturation current (Fjeldly and Shur, 1991; Lee et al., 1993).

The following is a model expression describing the saturation in n_s at large gate bias:

$$n_s = \frac{n_s'}{[1 + (n_s'/n_{max})^\gamma]^{1/\gamma}} \qquad (6.5.1)$$

Here n_s' is the sheet carrier density used in the universal MOSFET model [Eq. (6.3.1) with $V_F = 0$)] and γ is a characteristic parameter for the transition to saturation.

Once the unified expression for the surface carrier density in the HFET channel is established, the modeling of the HFET drain current becomes similar to that of the MOSFET. Equation (6.2.9), which describes the extrinsic I–V characteristics for FETs in both the linear and saturation regimes, still applies. However, note that the MOSFET expression for the drain saturation current [Eq. (6.3.4)],

$$I_{\text{sat}} = \frac{g_{\text{chi}} V_{gte}}{1 + g_{\text{chi}} R_s + \sqrt{1 + 2 g_{\text{chi}} R_s + (V_{gte}/V_L)^2}} \tag{6.5.2}$$

can still be used with the intrinsic linear channel conductance g_{chi} given by Eq. (6.2.7), except that we now use n_s given by Eq. (6.5.1) to incorporate the effect of carrier saturation. The term V_{gte} is the effective extrinsic gate voltage swing for MOSFETs, given by Eq. (6.3.5), to ensure the correct subthreshold limit of I_{sat}.

A complete, universal HFET I–V model is obtained by incorporating the above results in the universal FET I–V expression of Eq. (6.2.9). Note that the short-channel DIBL effect is included by a proper modification of the threshold voltage according to Eqs. (6.2.22) and (6.2.23). The quality of the present unified HFET model is illustrated in Fig. 6.5.3 for a submicrometer HFET.

As can be seen from this figure, the present HFET model reproduces quite accurately the experimental data in the entire range of bias voltages over several decades of current variation, even the residual drain leakage current at large negative gate–source voltages. The parameters used in this calculation were obtained using a parameter extraction procedure similar to that described for MOSFETs by Shur et al. (1992b) (see also Lee et al., 1993).

6.5.2. HFET C–V Model

For HFETs, we basically have the same choice of C–V models as for MOSFETs—the unified Meyer-type model and the charge-based model. Here we concentrate on the former, for which we only have to modify the expression for the gate-channel capacitance C_{ch} to reflect the saturation in the channel electron sheet density n_s. The rest of the model remains unchanged.

Since the expression for n_s in Eq. (6.5.1) is unified and also incorporates the saturation of n_s at large gate bias, we can write the unified HFET gate channel capacitance at zero drain–source bias as follows:

$$C_{\text{ch}} = WLq \frac{dn_s}{dV_{GS}} \approx \frac{C'_{\text{ch}}}{[1 + (n_s'/n_{\max})^\gamma]^{1+1/\gamma}} \tag{6.5.3}$$

Here C'_{ch} is the unified gate channel capacitance of an infinitely deep potential well, identical to C_{ch} of the MOSFET of Eq. (6.3.6). We note that above threshold, C'_{ch} rapidly approaches its maximum value $C_i = WL\epsilon_i/d_i$, where ϵ_i and d_i are the dielectric permeability and thickness of the wide-band-gap layer, respectively. However, when n_s' becomes comparable to or larger than $n_{\max}$, C_{ch} will decrease, as indicated in Fig. 6.5.4.

As was shown in Fig. 6.5.2, the saturation in n_s is accompanied by an increase in the sheet density n_l of electrons in the AlGaAs layer. This added charge contributes to the total differential gate capacitance C_{gtot}, which can be represented as a parallel coupling of C_{ch} and the capacitance C_{gl} associated with this added charge. To a lowest order approximation, we may assume that the carriers in the AlGaAs layer are

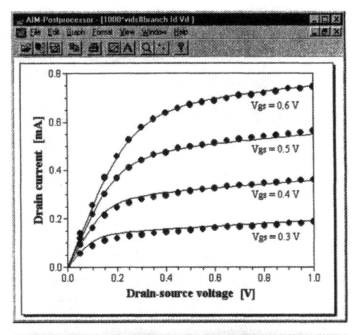

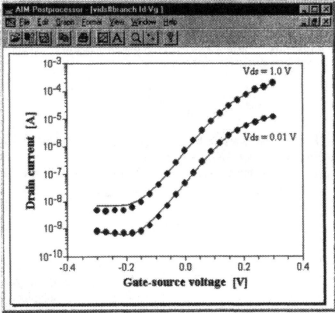

FIG. 6.5.3. (*a*) Above-threshold and (*b*) subthreshold experimental (symbols) and calculated (solid lines) HFET *I–V* characteristics. Device parameters: $L = 1$ μm, $W = 10$ μm, $\epsilon_s = 1.14 \times 10^{-10}$ F/m, $d_i = 0.04$ μm, $\Delta d = 4.5 \times 10^{-9}$ m, $\mu_n = 0.385$ m²/V s, $v_s = 1.35 \times 10^5$ m/s, $V_{T0} = 0.12$ V, $\eta = 1.32$, $R_s = R_d = 52$ Ω, $\lambda = 0.08$ V⁻¹, $n_{max} = 7 \times 10^{15}$ m⁻², $m = 2.9$, $\gamma = \delta = 3.0$, $\sigma_o = 0.04$, $V_\sigma = 0.2$ V, $V_{\sigma t} = 0.4$ V.

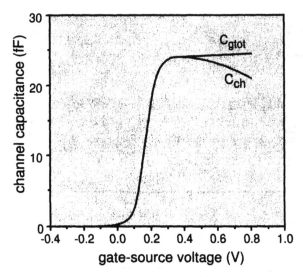

FIG. 6.5.4. Calculated gate channel capacitance C_{ch} and total gate capacitance C_{gtot} for a similar HFET as in Fig. 6.5.3. (After Fjeldly and Shur, 1991; see also Lee et al., (1993.)

located at a fixed distance d_1 from the gate, independent of V_{GS}. This assumption is quite accurate for an HFET with a delta-doped AlGaAs layer; see Fig. 6.5.1. Then, this charge can be treated in full analogy with that of a MOSFET channel, and C_{g1} can be expressed in terms of Eq. (6.3.6) as

$$C_{g1} = \frac{C_{i1}}{1 + 2 \exp[-(V_{GS} - V_{T1})/\eta_1 V_{th}]} \qquad (6.5.4)$$

Here $C_{i1} = WL\epsilon_i/d_{i1}$, V_{T1} is a threshold voltage characterizing the onset of significant charge transfer into the AlGaAs layer, and η_1 is a suitable ideality factor. The voltage V_{T1} can be estimated approximately as the gate–source voltage needed to increase the carrier density n_s to its maximum value n_{max}:

$$V_{T1} \approx V_T + WL\frac{qn_{max}}{C_i} \qquad (6.5.5)$$

where V_T is the threshold voltage of the channel at the GaAs/AlGaAs interface.

Figure 6.5.4 shows values of C_{ch} calculated from Eq. (6.5.3), and the total gate capacitance $C_{gtot} = C_{ch} + C_{g1}$ based on the above discussion for a typical HFET with nominal gate length $L = 1$ μm and gate width $W = 20$ μm. The drop in C_{ch} observed at large gate voltage is associated with the saturation of the charge density in the GaAs channel, while the slight increase in C_{gtot} in the same region is caused by the increasing carrier population in the parallel AlGaAs channel.

The present HFET capacitance model can be incorporated in the unified Meyer-type capacitance model by using C_{gtot} instead of C_{ch} in Eqs. (6.2.14) and (6.2.15).

6.5.3. Implementation in AIM-Spice

As explained above, the universal HFET and MOSFET models are intimately related. Their implementations in AIM-Spice are also quite similar. We therefore refer to the description of the implementation of the universal MOSFET model in Section 6.3. Here we only emphasize some major differences.

First, instead of using a detailed definition of the threshold voltage, we adopt the following simplified description also used for the MESFET Level 2:

$$V_T = \mathbf{VTO} - \sigma V_{ds} \tag{6.5.6}$$

where the SPICE parameter **VTO** is the threshold voltage at zero drain bias and the DIBL coefficient σ has the same form as that used for MOSFET Level 7 [(see Eq. (6.3.11)].

Second, the description of the saturation effects in n_s, I_{sat}, and C_{ch} at high gate bias requires the use of Eqs. (6.5.1)–(6.5.5). The transition to saturation is characterized in terms of the new SPICE parameter **GAMMA**.

Third, a model for the gate leakage current is included. For details on this model, we refer to Lee et al. (1993), Ytterdal et al. (1995b), and the example below.

6.5.4. SPICE Example: Gate Leakage in HFETs

Gate leakage is of great concern in HFET devices because it degrades the I–V characteristics and the transconductance. Our HFET model implemented in AIM-Spice includes a description of the gate leakage, and here we investigate how this current affects the characteristics.

We first use the circuit description in Fig. 6.5.5 to calculate drain current versus the gate–source voltage with the gate current model disabled (achieved by setting the model parameters **js1d** and **js1s** to zero). Note that the voltage sources vig, vid, and vis appearing in the circuit description are introduced to "measure" gate, drain, and source currents, respectively. The simulation was performed using the DC Transfer Curve Analysis parameters shown in the dialog box of Fig. 6.5.6, and the results of this simulation are shown as one of the traces in Fig. 6.5.7. We notice that the drain current increases steadily with increasing gate–source voltage in this case.

```
Circuit for calculating gate leakage current
a1 4 2 5 hfet l=1u w=10u
vgs 3 0 dc 0
vds 1 0 dc 0.2
vig 3 2 dc 0
vid 1 4 dc 0
vis 5 0 dc 0
.model hfet nhfet rd=60 rs=60 js1d=0 js1s=0
```

FIG. 6.5.5. Circuit description for the simulation of HFET drain current versus gate–source voltage with zero gate current.

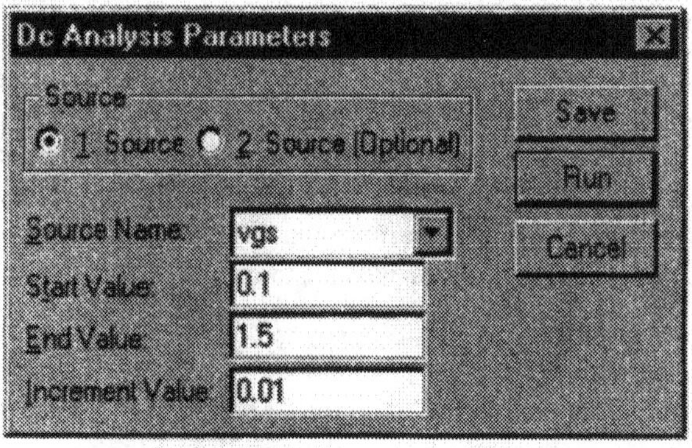

Dc Analysis Parameters

Source
◉ 1 Source ○ 2 Source (Optional)

Source Name: vgs
Start Value: 0.1
End Value: 1.5
Increment Value: 0.01

Save
Run
Cancel

FIG. 6.5.6. DC Analysis parameters used for the circuit in Fig. 6.5.5.

Gate current can be included in the simulation by using the model parameters shown in Table 6.5.1. The resulting drain and gate currents are shown in Fig. 6.5.7. As can be seen from this figure, the rapidly increasing gate current at high gate–source voltages causes a reduction in the drain current. We also notice a corresponding reduction in the device transconductance, as shown in Fig. 6.5.8.

In digital HFET circuits based on the DCFL (direct-coupled FET logic) family,

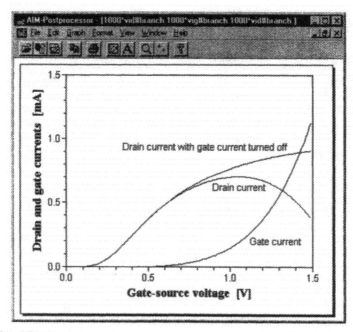

FIG. 6.5.7. HFET drain and gate current versus gate–source voltage. The drain–source bias is 0.2 V.

TABLE 6.5.1. HFET Gate Current Model Parameters.

Parameter	Value	Unit
js1d	1	A/m^2
js2d	1.15×10^6	A/m^2
m1d	1.32	—
m2d	6.9	—
js1s	1	A/m^2
js2s	1.15×10^6	A/m^2
m1s	1.32	—
m2s	6.9	—
rgd	90	Ω
rgs	90	Ω

which uses enhancement-type switching transistors, the gate current degrades the transfer characteristics. The output high voltage is reduced due to the gate current of the next inverter stage, and the output low voltage is raised because the gate current creates a voltage drop across the series source resistance. To illustrate these effects, we use the inverter circuit description shown in Fig. 6.5.9. The inverter itself is defined as a subcircuit and is referenced in the main circuit. We insert a second inverter to simulate a realistic load.

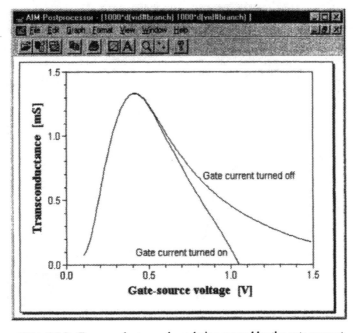

FIG. 6.5.8. Transconductance degradation caused by the gate current.

```
HFET DCFL inverter circuit (gate current turned on)
.subckt inv 1 2 3
*                | | |
*            Vdd | |
*             Vin |
*              Vout
a1 1 3 3 load l=1u w=10u
a2 3 2 0 driver l=1u w=10u
.ends

vdd 1 0 dc 2
vin 2 0 dc 0
x1 1 2 3 inv
x2 1 3 4 inv
.model load nhfet rd=60 rs=60 rgs=90 rgd=90 vto=-0.3
+ js1d=1 m1d=1.32 js2d=1.15e6 m2d=6.9 js1s=1 m1s=1.32
+ js2s=1.15e6 m2d=6.9

.model driver nhfet rd=60 rs=60 rgs=90 rgd=90 vto=0.3
+ js1d=1 m1d=1.32 js2d=1.15e6 m2d=6.9 js1s=1 m1s=1.32
+ js2s=1.15e6 m2d=6.9
```

FIG. 6.5.9. Circuit description of a HFET DCFL inverter.

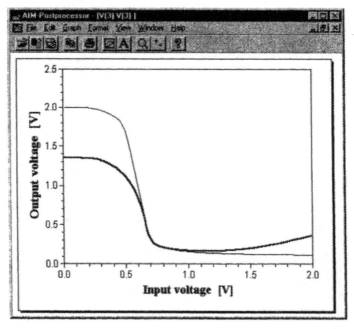

FIG. 6.5.10. Inverter transfer curves with gate current turned on (thick line) and off (thin line).

The transfer characteristics of the first inverter is obtained by sweeping vin from 0 to 2 V and plotting the voltage at node 3. The results shown in Fig. 6.5.10 clearly illustrate the effects of gate leakage current on the inverter transfer characteristics. The lower output high voltage and the larger output low voltage reduce the noise margin of the inverter.

6.6. UNIVERSAL A-Si:H TFT MODEL

As briefly discussed in Section 5.2, a large fraction of the induced carriers in the channel of a hydrogenated amorphous silicon TFT will be trapped in localized states at all relevant gate bias conditions. Hence, the carrier mobility μ_{FET} of such devices, called the field-effect mobility, will be much smaller than the mobility μ of carriers in energy band states. Also, μ_{FET} will depend on the gate bias or, equivalently, on the sheet density of induced charge n_{ind}. An analysis of the dependence of μ_{FET} on n_{ind} based on a solution of Poisson's equation in the direction perpendicular to the channel was presented by Shur et al. (1989) (see also Lee et al., 1993).

From the distribution of localized states in the forbidden gap of amorphous silicon, shown in Fig. 6.6.1, the densities of free and trapped carriers in a-Si as functions of the position of the Fermi level E_F can be calculated. The results of such a

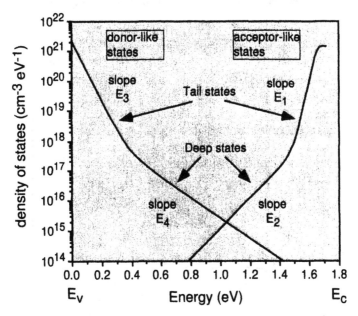

FIG. 6.6.1. Density of localized states in amorphous silicon; E_1, E_2, E_3, and E_4 are characteristic energies for the exponential variation of the density of localized states. (From Shur et al., 1989; see also Lee et al., 1993.)

calculation are shown in Fig. 6.6.2. This figure allows us to distinguish between the different regimes of a-Si TFT operation. When $E_F - E_v$ is less than approximately 1.4 eV (E_v corresponds to the top of the valence band), the device is in the below-threshold regime. In this regime, the free electron density is very small and increases exponentially with increasing gate–source voltage. When $E_{F_0} - E_v$ is less than approximately 1.7 eV and larger than 1.4 eV, the device is in the above-threshold regime where the free-carrier concentration is a significant fraction of the total induced charge. Finally, at values of $E_F - E_v$ greater than about 1.7 eV, the free-carrier density dominates and the device should operate in a crystalline-like regime. This regime, however, is only reached at gate voltages too large for most practical applications.

The analysis by Shur et al. (1989) yields the sheet density of free electrons n_s at the a-Si:H-insulator interface as a function of the gate–source voltage, shown in Fig. 6.6.3. This relationship allows us to evaluate the field-effect mobility as a function of the gate–source voltage above threshold. For modeling purposes, this dependence can be approximated quite well by the analytical expression

$$\mu_{FET} = \mu_o \left(\frac{V_{GS} - V_T}{V_{aa}} \right)^\gamma \tag{6.6.1}$$

where $V_{aa} = 10^5$ V is a typical scaling parameter and μ_o and γ are constants. Equation (6.6.1) gives an accurate fit to numerical calculations over a limited but important range of induced carrier concentrations, as indicated in Fig. 6.6.4.

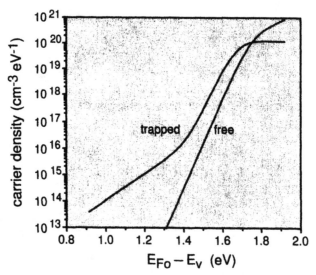

FIG. 6.6.2. Densities of free and trapped carriers in a-Si versus the position of the Fermi level relative to the valence band edge. (From Shur et al., 1989; see also Lee et al., 1993.)

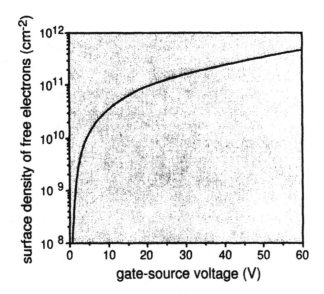

FIG. 6.6.3. Sheet density of free electrons at the a-Si-insulator interface versus the gate–source voltage. (From Shur et al., 1989; see also Lee et al., 1993.)

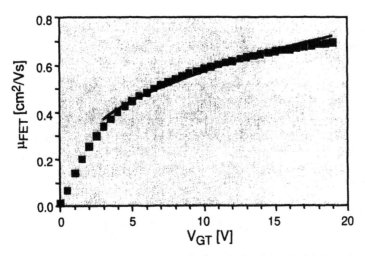

FIG. 6.6.4. Analytical (solid curve) and numerically calculated (symbols) dependencies of the field-effect mobility on V_{GT}. Parameters: $\mu_o = 10$ cm^2/V s, $\gamma = 0.363$, and $V_{aa} = 2.64 \times 10^4$ V. (From Slade, 1997.)

6.6.1. a-Si:H TFT I–V Model

Our universal a-Si:H TFT I–V model is based on the theory developed by Shur et al. (1995), combined with the universal FET modeling formalism outlined in Section 6.2. The theory uses the above results for the field-effect mobility to derive unified expressions for the linear conductance and the saturation current in a-Si:H TFTs, needed for the universal model.

The basic assumptions are that the sheet density of free electrons n_s can be written as a fraction of the total induced charge n_{ind} as

$$n_s = \frac{\mu_{FET}}{\mu} n_{ind} \tag{6.6.2}$$

and that the above-threshold sheet density of induced electrons for small drain–source voltages is given by the MOSFET expression

$$n_{ind} = \frac{\epsilon_i V_{GT}}{q d_i} \tag{6.6.3}$$

where d_i is the gate insulator thickness. By a straightforward analysis, we obtain, for the linear intrinsic channel conductance above threshold,

$$g_{chi} = \frac{W}{L} q n_{ind} \mu_{FET} = \frac{W \epsilon_i \mu_o V_{GT}}{L d_i} \left(\frac{V_{GT}}{V_{aa}} \right)^{\gamma} \tag{6.6.4}$$

In order to extend the validity of this result into the subthreshold regime, Shur et al. (1995) proposed to use the following emperical expression for n_{ind}:

$$n_{ind} = n_{so} \left[\left(\frac{t_m}{d_i} \right) \left(\frac{V_{GFB}}{V_0} \right) \left(\frac{\epsilon_i}{\epsilon_s} \right) \right]^{2 V_0 / V_e} \tag{6.6.5}$$

which is derived from a charge distribution assuming an exponential distribution of deeps states of slope $1/V_0$. In Eq. (6.6.5), t_m is the charge channel thickness, ϵ_s is the a-Si:H dielectric permittivity, $V_{GFB} = V_{GS} - V_{FB}$, V_{FB} is the flat-band voltage, V_0 is a characteristic voltage determined by the deep states, and $V_e = 2 V_{th} V_0 / (2 V_0 - V_{th})$. The characteristic sheet carrier density n_{so} is given by

$$n_{so} = N_c t_m \left(\frac{V_e}{V_0} \right) \exp \left(-\frac{dE_{Fo}}{q V_{th}} \right) \tag{6.6.6}$$

Here, N_c is the effective density of states in the conduction band and dE_{Fo} is the zero-bias dark Fermi-level position relative to the bottom of the conduction band.

The saturation current can be derived from the saturation voltage $V_{SAT} = \alpha_{sat} V_{GT}$ as

$$I_{sat} = g_{chi} \alpha_{sat} V_{GT} \tag{6.6.7}$$

where α_{sat} is a constant.

This model can be further improved by accounting for space charge injection effects that are important in the saturation regime. As was shown by Shur et al. (1989), based on two-dimensional simulations and analytical calculations, an a-Si:H TFT in the saturation regime behaves as a series combination of an "intrinsic" TFT and an n–i–n diode formed between the drain contact and the channel at a certain distance ΔL from the drain contact. To a first-order approximation, the output conductance in the saturation regime is given by the expression

$$g_{chs} = KI_{sat}/(L - \Delta L)^2 \tag{6.6.8}$$

where K is a constant.

Parasitic series source and drain resistances may have a considerable effect on the current–voltage characteristics of a-Si TFTs, and their effects can be incorporated in the above results by means of the relationships between intrinsic and extrinsic quantities in Eqs. (5.3.31), (5.3.32), and (6.2.6). However, we emphasize that the source and drain contacts of the a-Si channel are usually far from ideal ohmic contacts. These contacts behave more as very leaky Schottky diodes. Nevertheless, the theory reviewed above can give a good fit to measured current–voltage characteristics even when constant drain and source parasitic resistances are assumed. This is shown in Fig. 6.6.5 where we present typical measured and calculated current–voltage characteristics of an a-Si:H TFT.

By incorporating the above results for the linear channel conductance and the saturation current in the universal FET modeling expression of Eq. (6.2.9), we arrive at a complete universal model for a-Si:H TFTs. Note that the parameter λ of the universal model is related to the saturation channel conductance g_{chs} of a-Si:H TFTs by $\lambda = g_{chs}/I_{sat} = K/(L - \Delta L)^2$.

6.6.2. a-Si:H C–V Model

The unified Meyer-type capacitances of Eqs. (6.2.14) and (6.2.15) can also be used to describe the intrinsic a-Si:H TFT capacitances. The gate-channel capacitance can be obtained as $C_{ch} = WLq(dn_{ind}/dV_{GT})$ using Eqs. (6.6.5) and (6.6.6). However, a simpler expression is found by combining the asymptotic forms of C_{ch} above and below threshold, that is,

$$C_a = WL \frac{\epsilon_i}{d} \tag{6.6.9}$$

$$C_b = \frac{C_a}{2} \exp\left(\frac{qV_{GT}}{\eta E_2}\right) \tag{6.6.10}$$

Here, η is a factor accounting for the voltage division between the semiconductor and the gate dielectric (similar to the ideality factor in crystalline FETs). Equation (6.6.10) is obtained by assuming that the gate channel capacitance at threshold is equal to one-third of the maximum gate channel capacitance (the same condition as

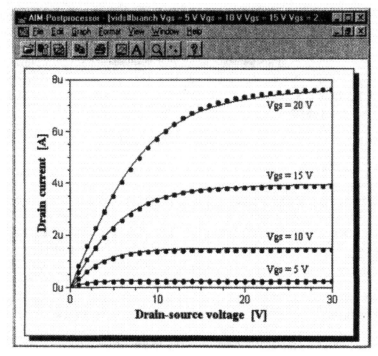

FIG. 6.6.5. Current–voltage characteristics of an a-Si TFT with W/L = 150 μm/60 μm. Symbols: experimental data; solid lines: calculated curves. Device parameters: dielectric thickness d_i = 0.3 μm, band mobility μ_o = 10 cm²/V s, power law mobility value γ = 0.187, characteristic voltage for field-effect mobility V_{aa} = 1.9 × 10⁷ V, threshold voltage V_T = 1.168 V, characteristic voltage for deep states V_0 = 0.124, and saturation voltage parameter α_{sat} = 0.54.

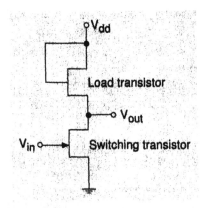

FIG. 6.6.6. Enhancement/enhancement logic inverter.

```
a-Si TFT EEL Inverter
m1 1 1 2 0 TFTL l=8u w=80u
m2 2 3 0 0 TFTD l=8u w=80u
vin 3 0 dc 0
vdd 1 0 dc 10
.model TFTL nmos level=15 vto=2.5 tox=0.1u
.model TFTD nmos level=15 vto=5 tox=0.1u
```

FIG. 6.6.7. Circuit description for simulating the a-Si:H TFT EEL inverter.

for MOSFETs) but replacing the thermal energy $k_B T$ by the characteristic energy E_2 (see Fig. 6.6.1). The parameters C_a and C_b can be combined according to Eq. (6.2.13) to give a unified expression for C_{ch}.

Note that since the resistance R_{ch} of the a-Si:H TFT channel is high, the charging time of the capacitances is significant and will in fact dominate the frequency response of the transistor. As suggested in Section 5.2.4, the charging time is typically $R_{ch} C_{GS}$.

6.6.3. SPICE Example: a-Si:H TFT EEL Inverter

We apply the a-Si:H TFT model to calculate the transfer characteristics of the enhancement/enhancement logic (EEL) inverter shown in Fig. 6.6.6. Both transistors have a gate length of 8 μm and a gate width of 80 μm. The threshold voltages are 2.5 and 5.0 V for the load and switching transistor, respectively. We use default model parameters in the circuit description of Fig. 6.6.7.

To calculate the transfer characteristics, we run a DC Analysis, sweeping vin from 0 to 20 V. Figure 6.6.8 shows the dialog box for DC Analysis with appropriate analysis parameters. The resulting transfer characteristic is displayed in Fig. 6.6.9.

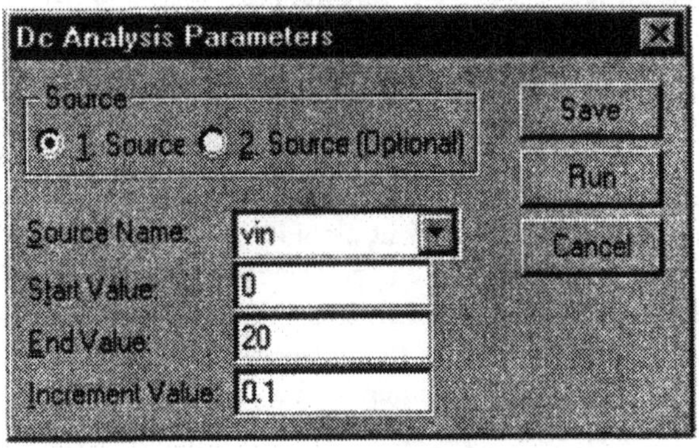

FIG. 6.6.8. DC Analysis parameters for calculating inverter transfer characteristics.

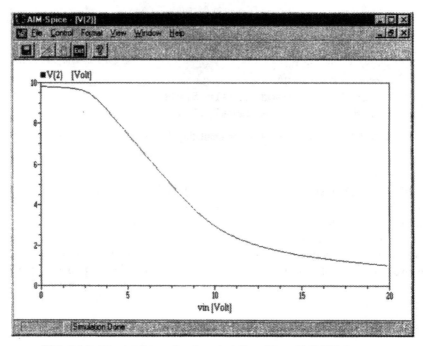

FIG. 6.6.9. Simulated transfer characteristics of the a-Si:H EEL inverter.

6.7. UNIVERSAL POLY-Si TFT MODEL

In Section 5.2, we mentioned that the room temperature mobility of polysilicon film may vary between about 30 cm²/V s and several hundred cm²/V s, depending on the grain size. This shows that poly-Si TFTs have superior current drives and device transconductances compared to a-Si TFTs.

The modeling of poly-Si TFTs is usually based on the "effective medium" approach, which treats the nonuniform polycrystalline samples as some uniform effective medium with an effective carrier mobility and an average density of states in the energy gap (see Faughan, 1987). Since poly-Si TFTs usually are long-channel devices, we can, as a first approach, utilize the simple charge control model for MOSFETs discussed in Section 5.3.3. However, we will combine this with a more realistic dependence of the field-effect mobility μ_{FET} on the induced carrier sheet density in the channel, similar to the approach used for a-Si:H TFTs. We derive expressions for the linear channel conductance, the saturation current, and the channel capacitance, which can be used in the universal modeling approach described in Section 6.2 to formulate a universal poly-Si TFT model.

In this model, however, we ignore the so-called kink-effect—an important mechanism associated with the grain structure in poly-Si TFTs. The kink effect causes a noticeable increase in the channel conductance in saturation beyond a critical onset value of the drain–source voltage. This phenomenon is thought to be related to impact ionization and can be modeled by empirical equations, as was done, for example, by Byun et al. (1992) and Owusu et al. (1996). The kink effect has been implemented in the AIM-Spice poly-Si TFT models.

6.7.1. Poly-Si TFT *I–V* Model

In the simple charge control MOSFET model, the following linear conductance and saturation current are easily obtained from Eq. (5.3.12):

$$g_{chi} = \frac{W\mu_{FET}qn_s}{L} \tag{6.7.1}$$

$$I_{sat} = \tfrac{1}{2}g_{chi}V_{GT} \tag{6.7.2}$$

where we used $V_{SAT} = V_{GT}$. For the inversion sheet charge density at low drain–source voltage or near source, we used $qn_s = \epsilon_i V_{GT}/d_i$, where d_i is the gate insulator thickness [see Eq. (5.3.10)]. As proposed by Shur et al. (1993), the dependence of the field-effect mobility in poly-Si TFTs on n_s can be modeled according to

$$\frac{1}{\mu_{FET}} = \frac{1}{\mu_o} + \frac{1}{\mu_1}\left(\frac{n_o}{n_s}\right)^\kappa \tag{6.7.3}$$

where $n_o = \epsilon_i \eta V_{th}/(2qd_i)$ is the carrier sheet density at threshold [see Eq. (6.3.2)], μ_o is the above-threshold mobility, and μ_1 and κ are adjustable parameters.

Equations (6.7.1) and (6.7.2) can now easily be converted into unified expressions, valid both above and below threshold, by using the unified form of n_s of Eq. (6.3.1) (with $V_F = 0$) and replacing V_{GT} by the effective gate overdrive voltage of Eq. (6.3.5). The resulting intrinsic and unified expressions to be used in conjunction with the universal $I–V$ model of Eq. (6.2.9) are therefore

$$g_{chi} = \frac{2W\mu_o qn_o}{L} \frac{\ln[1 + \tfrac{1}{2}\exp(V_{GT}/\eta V_{th})]}{1 + (\mu_o/\mu_1)\{2\ln[1 + \tfrac{1}{2}\exp(V_{GT}/\eta V_{th})]\}^{-\kappa}} \tag{6.7.4}$$

$$I_{sat} = \tfrac{1}{2}g_{chi}V_{GTe} = \tfrac{1}{2}g_{chi}V_{th}\left[1 + \frac{V_{GT}}{2V_{th}} + \sqrt{\delta^2 + \left(\frac{V_{GT}}{2V_{th}} - 1\right)^2}\right] \tag{6.7.5}$$

where δ is a parameter that determines the width of the transition from above to below threshold in I_{sat}.

Even though poly-Si TFTs are long-channel devices, the parasitic resistances are usually important and should not be neglected. Hence, in order to obtain an extrinsic poly-Si model, we can write $g_{ch} \approx g_{chi}/[1 + g_{chi}(R_s + R_d)]$ and replace Eq. (6.7.5) by its extrinsic counterpart given by Eq. (6.3.4).

As mentioned above, the kink effect causes a noticeable increase in the channel conductance in saturation beyond a critical onset value of the drain–source voltage, as shown in Fig. 6.7.1. This phenomenon is caused by impact ionization in the high-field region near drain, and the added drain current can be expressed by an equation similar to that used for the substrate current [Eq. (6.2.18)]. The effect is further enhanced by a positive feedback mechanism caused by the traps associated with the grain boundaries.

A continuous model for the differential output resistance, including the kink regime, was proposed for poly-Si TFTs by Byun et al. (1992) (see also Lee et al., 1993) and is implemented in AIM-Spice. Below, we illustrate how the kink-effect degrades the transfer characteristics of a poly-Si TFT inverter.

6.7.2. Poly-Si TFT C–V Model

In the present poly-Si TFT model, the expressions for the channel capacitance and the Meyer capacitances will be identical to those of the universal MOSFET model discussed in Section 6.3. The reason is that we used the same form of the universal charge control model in both cases.

As for a-Si TFTs, the traps will contribute to the frequency response of poly-Si TFTs, and the characteristic trapping time may typically be expressed as a product of the channel series resistance and the gate–source capacitance.

6.7.3. SPICE Example: Poly-Si TFT Inverter

In this example, we want to illustrate how the kink effect degrades the transfer characteristics of a poly-Si inverter with depletion load. Figure 6.7.2 shows the circuit

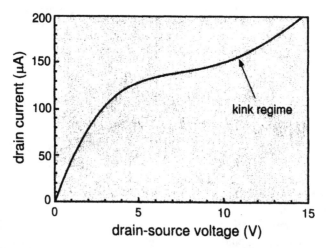

FIG. 6.7.1. Measured I–V characteristic for an n-channel poly-Si TFT with $L = 15$ μm and $W = 50$ μm. (From Byun et al., 1992.)

```
Poly-Si Inverter with depletion load
m1 1 2 2 0 TFTL l=30u w=50u
m2 2 3 0 0 TFTD l=30u w=50u
vin 3 0 dc 0
ds 1 0 dc 30
.model TFTL nmos level=12 vto=-10 uo=85.0 tox=1000e-10
.model TFTD nmos level=12 vto=2.5 uo=85.0 tox=1000e-10
```

FIG. 6.7.2. Circuit description of a poly-Si TFT inverter with depletion load.

description with the kink effect included. We apply a DC Analysis where `vin` is swept from 0 to 30 V with increments of 0.1 V.

To remove the kink effect, we set the model parameter V_0 to a large value, for example, 1000 V, and run the simulation again. The results for both cases are shown in Fig. 6.7.3. From the simulation results, we observe that the kink effect increases the width of the transition region and degrades the characteristics. One way to reduce this effect is to scale back the power supply to prevent the transistors from entering the kink regime.

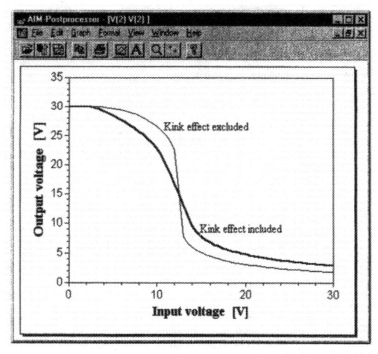

FIG. 6.7.3. Simulated inverter transfer characteristics with and without the kink effect included.

6.8. BSIM3 MOSFET MODEL

BSIM3 (Berkeley short-channel IGFET model) is the third generation of short-channel MOSFETs from the University of California at Berkeley. The latest version of the model (version 3 at the moment of writing) is a physics-based deep-submicrometer model for digital and analog circuit design. The model is based on a single I–V expression to describe the current and the output conductance from subthreshold to strong inversion as well as from the linear to the saturation regimes of operation.

The BSIM3 model includes short- and narrow-channel effects on the threshold voltage, the effect of nonuniform doping in both lateral and vertical directions, mobility reduction due to the vertical field, bulk charge effect, carrier velocity saturation, drain-induced barrier lowering, channel length modulation, substrate current-induced body effect, subthreshold conduction, source/drain parasitic resistances, and scaling of model parameters with device dimensions. Here we review some of the features of the BSIM3 model. For additional details, we refer to the BSIM3 Version 3.0 Manual (1995), Huang et al. (1993), and Cheng et al. (1995).

6.8.1. Threshold Voltage

Accurate modeling of the threshold voltage is one of the most important issues in compact MOSFET modeling. The threshold voltage model in BSIM3 is based on the standard model of a long and wide MOSFET but is extended to describe the effects of nonuniform doping and small gate dimensions. Using BSIM3 notation, the standard model for the threshold voltage, including the effect of vertical nonuniform doping, can be expressed as (see Section 6.3.3)

$$V_{T\text{Standard}} = V_{To} + K_1(\sqrt{\Phi_s - V_{bs}} - \sqrt{\Phi_s}) - K_2 V_{bs} \qquad (6.8.1)$$

where V_{To}, K_1, and K_2 are model parameters and $\Phi_s = 2\varphi_b$. To model the effects of small gate dimensions, additional terms are added to Eq. (6.8.1), and the following expression is obtained for the threshold voltage:

$$V_T = V_{T\text{Standard}} + \Delta V_{T\text{NUL}} + \Delta V_{T_L} + \Delta V_{T_W} + \Delta V_{T\text{DIBL}} \qquad (6.8.2)$$

Here $\Delta V_{T\text{NUL}}$, ΔV_{T_L}, ΔV_{T_W}, and $\Delta V_{T\text{DIBL}}$ correspond to the change in threshold voltage due to non-uniform channel doping, short gate lengths, narrow channel widths, and DIBL, respectively.

Nonuniform doping in the lateral direction is also observed for some technologies, where the doping concentration near drain and source may be higher than in the middle of the channel. Hence, the voltage required to turn the device on will be higher. This effect becomes more pronounced as the channel length is scaled down since the average doping density increases. In BSIM3, the effect of lateral nonuniform doping is described by the following expression:

$$\Delta V_{T\text{NUL}} = K_1 \left(\sqrt{1 + \frac{\text{NLX}}{L_{\text{eff}}}} - 1 \right) \sqrt{\Phi_s} \qquad (6.8.3)$$

where NLX is an empirical parameter and L_{eff} is the effective gate length. We note from Eq. (6.8.3) that a reduction in channel length causes an increase of the threshold voltage.

Ideally, the threshold voltage should be independent of the channel length. However, as the channel length becomes shorter, the threshold voltage exhibits a greater dependence on the channel length. From experimental data (see, e.g., Fig. 6.2.5), V_T is observed to scale exponentially with the channel length and is modeled as follows:

$$\Delta V_{T_L} = -D_{VT0} \left[\exp\left(-D_{VT1}\frac{L}{2l_t}\right) + 2 \exp\left(-D_{VT1}\frac{L}{l_t}\right) \right] (V_{\text{bi}} - \Phi_s) \qquad (6.8.4)$$

Here D_{VT0} and D_{VT1} are empirical parameters, V_{bi} is the built-in voltage, and l_t is a characteristic length given by

$$l_t = \sqrt{\frac{\epsilon_s d_i d_{\text{dep}}}{\epsilon_i}} (1 + D_{VT2} V_{bs}) \qquad (6.8.5)$$

where

$$d_{\text{dep}} = \sqrt{\frac{2\epsilon_s(\Phi_s - V_{bs})}{qN_{\text{ch}}}} \qquad (6.8.6)$$

is the channel depletion width, D_{VT2} is an additional empirical parameter, ϵ_s and ϵ_i are the dielectric permeabilities of silicon and the gate oxide, respectively, d_i is the oxide thickness, and N_{ch} is the channel doping density. Since the doping is usually nonuniform in the vertical direction, D_{VT2} is introduced to model how the doping density at the distance d_{dep} from the oxide depends on V_{bs}.

As the channel becomes narrower, the effects of fringing fields increase and, consequently, a higher gate bias is required to turn the device on. In BSIM3, this increase in threshold voltage is modeled as

$$\Delta V_{TW} = (K_3 + K_{3b}V_{bs})\frac{d_i}{W_{\text{eff}} + W_o}\Phi_s \qquad (6.8.7)$$

An expression similar to Eq. (6.8.4) is used to model DIBL and charge-sharing effects:

$$\Delta V_{T\text{DIBL}} = -\left[\exp\left(-D_{\text{sub}}\frac{L_{\text{eff}}}{2l_{to}}\right) + 2 \exp\left(-D_{\text{sub}}\frac{L_{\text{eff}}}{l_{to}}\right) \right] (E_{tao} + E_{tab}V_{bs})V_{ds} \qquad (6.8.8)$$

where D_{sub}, E_{tao}, and E_{tab} are empirical fitting parameters and $l_{to} = l_t$ ($V_{bs} = 0$).

Finally, we should mention that in the SPICE implementation of BSIM3, an effective body bias expression is used to avoid unreasonable values of V_{bs} during a SPICE simulation. The expression assures that the body bias never exceeds a user-specified value V_{BM}, which by default is -10 V.

6.8.2. Effective Carrier Mobility

In general, the mobility of the carriers in a MOSFET channel depends on process parameters and bias conditions such as, for example, the oxide thickness, the channel-doping density, the threshold voltage, the gate voltage, and the substrate voltage. In BSIM3, a Taylor expansion of the formulation by Sabnis and Clemens (1979) is used. The resulting expression can be written as

$$\mu_{eff} = \frac{\mu_o}{1 + (U_a + U_c V_{bs}) \dfrac{V_{gs} \pm V_T}{d_i} + U_b \left(\dfrac{V_{gs} \pm V_T}{d_i} \right)^2} \qquad (6.8.9)$$

where U_a, U_b, and U_c are empirical fitting parameters. The upper and lower signs in the denominator are for enhancement and depletion mode devices, respectively.

6.8.3. Unified Drain Current

BSIM3 uses the concept of a unified drain current expression, where one single expression covers all regimes of operation—below and above threshold and below and above the saturation voltage. To achieve unification, the model uses two effective voltages, V_{gsteff} and V_{dseff}. The term V_{gsteff} is an effective gate voltage overdrive that tends to $V_{gs} - V_T$ well above threshold and decreases exponentially toward zero below threshold, not unlike V_{gte} of Eq. (6.3.5). The expression for the effective drain–source voltage V_{dseff} is similar to that used for V_{DSe} in Section 6.2.2. It approaches V_{ds} below saturation and the saturation voltage well above the saturation point.

The following expression for the unified drain current is used by BSIM3:

$$I_d = \frac{I_{do}}{1 + R_t I_{do}/V_{dseff}} \left(1 + \frac{V_{ds} - V_{dseff}}{V_A} \right) \left(1 + \frac{V_{ds} - V_{dseff}}{V_{ASCBE}} \right) \qquad (6.8.10)$$

where I_{do} is the ideal drain current described by

$$I_{do} = \frac{W_{eff} \mu_{eff} c_i V_{gseff} \{ 1 - A_{bulk}[V_{dseff}/2(V_{gseff} + 2V_{th})] \} V_{dseff}}{L_{eff}(1 + V_{dseff}/F_s L_{eff})} \qquad (6.8.11)$$

c_i is the gate oxide capacitance per unit area, A_{bulk} is the bulk charge parameter, and F_s is the magnitude of the electric field where the carrier velocity saturates. In Eq. (6.8.10), the term $1/(1 + R_t I_{do}/V_{dseff})$ accounts for the effects of the parasitic source and drain resistances ($R_t = R_s + R_d$). The voltages V_A and V_{ASCBE} are parameters used for modeling the output conductance in saturation caused by channel length modulation, DIBL, and substrate current-induced body effect.

6.8.4. Intrinsic Capacitance Model

In previous versions of the BSIM model (BSIM1 and BSIM2), the intrinsic capacitance model had several shortcomings. For example, long-channel charge models were utilized that did not scale properly with the effective gate length. The resulting model overestimated the capacitance for short-channel devices. Furthermore, the old models exhibited discontinuities in the gate capacitance at the threshold voltage. The new BSIM3 formulation of the capacitances addresses these issues by providing a completely new charge-based model. The major features of the new model are as follows:

- Separate, effective channel length and width are used for the I–V and the C–V models, thereby assuring proper scaling with gate length.
- Accurate short-channel capacitance estimates down to a gate length of 0.2 μm.
- Simple enough to be suitable for implementation in circuit simulators.
- Utilizes one single equation for each nodal charge that covers all regimes of operation. This ensures continuity of all derivatives and enhances convergence properties of the model.

Similar to the charge-based capacitance model described in Sections 6.2 and 6.3, BSIM3 model uses terminal charges instead of terminal voltages as state variables to guarantee charge conservation. The four terminal charges Q_g, Q_d, Q_s, and Q_b are described analytically in terms of the terminal voltages, and the capacitances are given as derivatives of each terminal charge with respect to each of the terminal voltages.

Assuming zero charge at the drain side of the channel in the saturation region is a good approximation for long-channel devices. However, when the gate length is scaled below approximately 2 μm, the effect of velocity saturation reduces the saturation voltage below the pinch-off voltage and the above assumption of zero charge breaks down. To properly take this effect into account, BSIM3 defines a new saturation voltage $V_{dsat,cv}$ that lies between the pinch-off voltage and the saturation voltage used in the I–V model. For V_{ds} greater than $V_{dsat,cv}$ the channel charge near drain becomes constant (but not zero). An empirical expression is used for $V_{dsat,cv}$ that scales properly with gate length.

6.8.5. Non-Quasi-Static Model

Most of the MOSFET models used in SPICE are based on the quasi-static assumption (QSA), where an instantaneous charging of the inversion layer is assumed. However, QSA breaks down in devices operating close to the cut-off frequency.

Hence, circuit simulations will fail to accurately predict the performance of high-speed circuits, especially switched-capacitor-type circuits. Also, unrealistically large drain current spikes frequently occur independent of circuit type.

The channel of a MOSFET is analogous to a bias-dependent distributed RC network, as shown in Fig. 6.8.1. In QSA, the gate channel capacitance is assumed to be connected directly to the intrinsic source and drain nodes, ignoring the finite charging time arising from the RC product associated with the channel resistance and the gate channel capacitance.

A first-order approach toward a non-quasi-static (NQS) model is to use the so-called Elmore equivalent circuit shown in Fig. 6.8.2. With this equivalent circuit, the channel charge build-up is modeled with reasonable accuracy because the lowest frequency pole of the original RC network is retained. The Elmore resistance R_{Elmore} is calculated from the channel resistance in strong inversion as

$$R_{Elmore} \approx \frac{L_{eff}^2}{\epsilon \mu_{eff} Q_{ch}} \tag{6.8.12}$$

where ϵ is the Elmore constant with a theoretical value close to 5 and Q_{ch} is the total charge in the channel.

This formulation is only valid above threshold, where the drift current dominates. To obtain a unified expression, including the subthreshold diffusion current, a relaxation-time-based approach is adapted. The overall relaxation time for channel charging and discharging can be written as a combination of the contributions due to drift and diffusion as follows:

$$\frac{1}{\tau} = \frac{1}{\tau_{drift}} + \frac{1}{\tau_{diff}} \tag{6.8.13}$$

where

$$\tau_{drift} = R_{Elmore} C_i \tag{6.8.14}$$

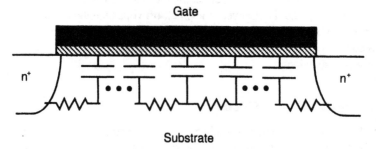

Gate

Substrate

FIG. 6.8.1. Equivalent RC network representing the MOSFET channel.

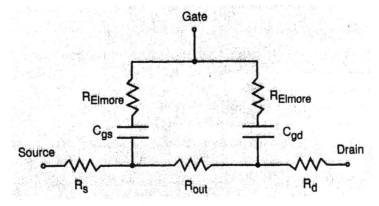

FIG. 6.8.2. Elmore NQS equivalent circuit.

Below threshold, the conduction is mainly due to diffusion, and the relaxation time constant can be approximated by

$$\tau_{\text{diff}} = \frac{q(L_{\text{eff}}/4)^2}{\mu_{\text{eff}} k_B T} \tag{6.8.15}$$

Based on this relaxation time concept, the NQS effect in BSIM3 is implemented with the subcircuit shown in Fig. 6.8.3. The variable Q_{def} is an additional node created to keep track of the amount of deficit or surplus channel charge necessary to achieve equilibrium [the quasi-static equilibrium channel charge is Q_{ch} in Eq. (6.8.12)]. Here, Q_{def} will decay exponentially into the channel with a bias-dependent NQS relaxation time τ, and the terminal currents can be written as

$$I_d = I_d(\text{dc}) + X_d \frac{Q_{\text{def}}}{\tau} \tag{6.8.16}$$

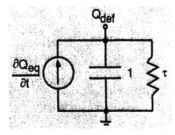

FIG. 6.8.3. Non-quasi-static subcircuit implementation in BSIM3. Note that with the resistance and capacitance values chosen, the RC time constant is τ.

$$I_s = -I_d(\text{dc}) + X_s \frac{Q_{\text{def}}}{\tau} \qquad (6.8.17)$$

$$I_g = -\frac{Q_{\text{def}}}{\tau} \qquad (6.8.18)$$

Here $X_d = 1 - F_p$ and $X_s = F_p$, where F_p is the charge-partitioning factor introduced in Section 6.2.2.

REFERENCES

N. Arora, MOSFET *Models for VLSI Circuit Simulation. Theory and Practices*, Springer-Verlag, Wien (1993).

M. Bohr, "MOS Transistors: Scaling and Performance Trends," *Semiconductor International*, vol. 18, no. 6, pp. 75–80 (1995).

J. R. Brews, W. Fichtner, E. H. Nicollian, and S. M. Sze, "Generalized Guide for MOSFET Miniaturization," *IEEE Electron Device Letters*, vol. EDL-1, p. 2 (1980).

BSIM3 Version 3.0 Manual, University of California at Berkeley (1995).

Y. Byun, K. Lee, and M. Shur, "Unified Charge Control Model and Subthreshold Current in Heterostructure Field Effect Transistors," *IEEE Electron Device Letters*, vol. EDL-11, no. 1, pp. 50–53 1990. Erratum IEEE Electron Device Letters, vol. EDL-11, no. 6, p. 273 (1990).

Y. Byun, M. Shur, M. Hack, and K. Lee, "New Analytical Poly-Silicon Thin-Film Transistor Model for CAD and Parameter Characterization," *Solid State Electronics*, vol. 35, No. 5, pp. 655–663 (1992).

P. C. Chao, M. Shur, R. C. Tiberio, K. H. G. Duh, P. M. Smith, J. M. Ballingall, P. Ho, and A. A. Jabra, "DC and Microwave Characteristics of Sub-0.1 μm Gate-Length Planar-Doped Pseudomorphic HEMTs," *IEEE Transactions on Electron Devices*, vol. ED-36, no. 3, pp. 461–473 (1989).

Y. Cheng and T. A. Fjeldly, "Unified Physical *I–V* Model Including Self-Heating Effect for Fully Depleted SOI/MOSFET's," *IEEE Transactions on Electron Devices*, vol. ED-43, pp. 1291–1296 (1996).

Y. Cheng, C. Hu, K. Chen, M. Chan, M. Jeng, Z. Liu, J. Huang, and P. K. Ko, "A Unified BSIM *I–V* Model for Circuit Simulation," *Proceedings of the 1995 International Semiconductor Device Research Symposium, ISDRS'95*, Charlottesville, VA, pp. 603–606 (1993).

B. Faughan, "Subthreshold Model of a Polycrystalline Silicon Thin-Film Field-Effect Transistor," *Applied Physics Letters*, vol. 50, no. 5, pp. 290–292 (1987).

T. A. Fjeldly and M. Shur, "Unified CAD Models for HFETs and MESFETs," in *Proceedings of the 11th European Microwave Conference (Workshop Volume)*, Stuttgart, pp. 198–205, Sept. (1991).

T. A. Fjeldly and M. Shur, "Threshold Voltage Modeling and the Subthreshold Regime of Operation of Short-Channel MOSFETs," *IEEE Transactions on Electron Devices*, vol. 40, no. 1, pp. 137–145 (1993).

T. A. Fjeldly, B. Moon, and M. Shur, "Analytical Solution of Generalized Diode Equation," *IEEE Transactions on Electron Devices*, vol. ED-38, no. 8, pp. 1976–1977 (1991).

T. A. Fjeldly, M. Shur, and T. Ytterdal, "Field Effect Transistor Modeling Issues," *Physica Scripta*, vol. T69, pp. 30–39 (1997).

L. Geppert, "Semiconductor Lithography for the Next Millennium," *IEEE Spectrum*, pp. 33–38, (April 1996).

C. Hu, "MOSFET Scaling in the Next Decade and Beyond," *Semiconductor International*, vol. 17, no. 6, pp. 105–114 (1994).

J. H. Huang, Z. H. Liu, M. C. Jeng, P. K. Ko, and C. Hu, "A Robust Physical and Predictive Model for Deep-Submicrometer MOS Circuit Simulation," in *Proceedings of the IEEE Custom Integrated Circuits Conference*, pp. 14.2.1–14.2.4 (1993).

B. Iñiguez and T. A. Fjeldly, "Unified Substrate Current Model for MOSFETs," *Solid-State Electronics*, vol. 41, no. 1, pp. 87–94 (1997).

B. Iñiguez and E. G. Moreno, "An Improved C ∞-Continuous Small-Geometry MOSFET Modeling for Analog Applications," *Analog Integrated Circuits and Signal Processing*, vol. 13, 1–13 (1997).

O. G. Johannessen, T. A. Fjeldly, and T. Ytterdal, "Unified Capacitance Modeling of MOSFETs," *Physica Scripta*, vol. T54, pp. 128–130 (1994).

K. Lee, M. Shur, T. A. Fjeldly, and T. Ytterdal, *Semiconductor Device Modeling for VLSI*, Prentice-Hall, Englewood Cliffs, NJ (1993).

J. E. Meyer, "MOS Models and Circuit Simulation," *RCA Review*, vol. 32, pp. 42–63 (1971).

Y. Mii, S. Rishton, Y. Taur, D. Kern, T. Lii, K. Lee, K. A. Jenkins, D. Quinlan, T. Brown Jr., D. Danner, F. Sewell, and M. Polcari, "Experimental High Performance Sub-0.1 μm Channel nMOSFET's," *IEEE Electron Devices Letters*, vol. EDL-15, no. 1, pp. 28–30 (1994).

M. Nawaz and T. A. Fjeldly, "A Charge Conserving Capacitance Model for GaAs MESFETs for CAD Applications," *Physica Scripta*, vol. T69, 242–246 (1997).

A. A. Owusu, M. D. Jacunski, M. S. Shur, and T. Ytterdal, "SPICE Model for the Kink Effect in Polysilicon TFTs," 1996 Electrochemical Society Fall Meeting, San Antonio, TX, Oct. (1996).

C. K. Park, C. Y. Lee, K. R. Lee, B. J. Moon, Y. Byun, and M. Shur, "A Unified Charge Control Model for Long Channel *n*-MOSFETs," *IEEE Transactions on Electron Devices*, vol. ED-38, pp. 399–406 (1991).

K. M. Rho, K. Lee, M. S. Shur and T. A. Fjeldly, "Unified Quasi-Static MOSFET Capacitance Model," *IEEE Transactions on Electron Devices*, ED-40, pp. 131–136 (1993).

A. G. Sabnis and J. T. Clemens, "Characterization of Electron Velocity in the Inverted <100> Si Surface," *Tech. Dig. Int. Electron Device Meeting*, pp. 18–21 (1979).

M. Shur, *GaAs Devices and Circuits*, Plenum, New York (1987).

M. Shur, M. Hack, and J. G. Shaw, "New Analytic Model for Amorphous Silicon Thin Film Transistors," *Journal of Applied Physics*, vol. 66, no. 7, pp. 3371–3380 (1989).

M. Shur, T. A. Fjeldly, T. Ytterdal, and K. Lee, "Unified GaAs MESFET Model for Circuit Simulations," *International Journal of High Speed Electronics*, vol. 3, pp. 201–233 (1992a).

M. Shur, T. A. Fjeldly, T. Ytterdal, and K. Lee, "Unified MOSFET Model," in *Solid-State Electronics*, 35, No. 12, pp. 1795–1802 (1992b).

M. Shur, M. Hack, and Y. H. Byun, "Circuit Model and Parameter Extraction Technique for

Polysilicon Thin Film Transistors", in *Proceedings of the 1993 International Semiconductor Device Research Symposium, ISDRS'93,* Charlottesville, VA, pp. 165–168 (1993).

M. Shur, M. Jacunski, H. Slade, M. Hack, "Analytical Models for Amorphous and Polysilicon Thin Film Transistors for High Definition Display Technology," *Journal of the Society for Information Display,* vol. 3, no. 4, p. 223 (1995).

H. Slade, "Device and Material Characterization and Analytical Modeling of Amorphous Silicon Thin Film Transistors," Ph.D. Thesis, University of Virginia (1997).

H. Statz, P. Newman, I. W. Smith, R. A. Pucel, and H. A. Haus, *IEEE Transactions on Electron Devices,* vol. ED-34, p. 160 (1987).

F. Stern, "Self-Consistent Results for *n*-type Si Inversion Layers," *Physical Review,* vol. B-5, no. 12, pp. 4891–4899 (1972).

D. E. Ward, "Charge Based Modeling of Capacitance in MOS Transistors," Ph.D. Thesis, Stanford University (1981).

D. E. Ward and R. W. Dutton, "A Charge-Oriented Model for MOS Transistor Capacitances," *IEEE Journal of Solid-State Circuits,* SC-13, 703–708 (1978).

T. Ytterdal, M. Shur, and T. A. Fjeldly, "Sub-0.1 μm MOSFET Modeling and Circuit Simulation," *Electronics Letters,* vol. 30, pp. 1545–1546 (1994).

T. Ytterdal, S.-H. Kim, K. Lee, and T. A. Fjeldly, "A New Approach for Modeling of Current Degradation in Hot-Electron Damaged LDD NMOSFETs," *IEEE Transactions on Electron Devices,* vol. ED-42, pp. 362–364 (1995a).

T. Ytterdal, B-J. Moon, T. A. Fjeldly, and M. S. Shur, "Enhanced GaAs MESFET CAD Model for a Wide Range of Temperatures," *IEEE Transactions on Electron Devices,* vol. ED-42, pp. 1724–1734 (1995b).

S. M. Sze, *Physics of Semiconductor Devices,* 2nd ed., Wiley, New York (1981).

PROBLEMS

6.1.1. List possible effects that may limit the ultimate minimum CMOS feature size.

6.1.2. What is the order of magnitude of the barrier height formed in the channel in the subthreshold regime (see Fig. 6.1.2).

6.2.1. Comment on how the constant parasitic gate–drain and gate–source capacitances will modify the C–V characteristics shown in Fig. 6.2.3.

6.2.2. (a) Use Eq. (6.2.17) and the simple charge control I–V model for MOSFETs in Eq. (5.3.12) to derive an expression for the output conductance in the saturation regime near the onset of saturation.

(b) Repeat the calculation for the velocity saturation I–V model in Eq. (5.3.27).

Hint: Express ΔL in terms of V_{DS}, replace L by $L - \Delta L$ in the expression for the saturation current, and solve for I_{sat} versus V_{DS}. Note that ΔL and $V_{DS} - V_p$ approach zero when V_{DS} approaches V_{SAT}. Use this fact to expand the final result to lowest order near the onset of saturation.

6.3.1. The FET *I–V* characteristic shown in Fig. 6.2.1 is assumed to come from a MOSFET with $L = 0.25$ μm, $d_i = 6.5$ nm, $V_T = 0.5$ V, and $V_{gt} = 2.5$ V. Use AIM-Spice with MOSFET Level 7 (universal MOSFET model) to extract from this characteristic the device width, the source and drain series resistances, and the parameter λ that determines the finite-output conductance in the saturation regime.

Hint: Start with default parameters, except for those specified above, and simulate the characteristic. Then optimize the fit progressively by concentrating on the parts of the characteristic most influenced by each of the parameters to be determined; that is, adjust the device width to fit the saturation current, then adjust the series resistances to fit the channel conductance in the linear region, and finally adjust parameter λ to fit the output conductance in the saturation regime. Repeat the procedure (iterate) if necessary.

6.3.2. Use the universal MOSFET model (Level 7) to calculate the maximum *p*-channel MOSFET transconductance per millimeter of gate width as a function of the gate length. Start from a gate length of 10 μm and a gate width of 50 μm. Scale the gate length down to 0.2 μm. Scale the oxide thickness and the gate width proportionally to the gate length. Start from the following circuit description:

```
Circuit for problem 6.3.2
m1 2 3 0 0 mp w=50u 1=10u
vds 1 0 dc -3
vids 1 2 dc 0
vgs 3 0 dc 0
.model mp pmos level=7 vto=-0.6 gammas0=0.514 phi=0.635 pb=0.635
+ js=3.2e-5 tox=1000e-10 nsub=3e15 xj=0.4e-6 ld=0.3e-6 vmax=2.5e4
+ uo=195 eta=1.745 lambda=0.0429 m=3 cj=190E-6 mj=0.3 cjsw=260E-12
+ cgso=1.27E-10 cgdo=1.27E-10 cgbo=7.66E-11 mjsw=0.33 rsh=76.45
+ vsigma=0.3 sigma0=0.005
```

Hint: To calculate the transconductance, sweep the gate–source voltage from 0 to –8 V and use AIM-Postprocessor to plot the derivative of the drain current.

6.3.3. Repeat Problem 6.3.2 for an *n*-channel MOSFET using the following circuit description:

```
Circuit for problem 6.3.3
m1 2 3 0 0 mn w=50u 1=10u
vds 1 0 dc 3
vids 1 2 dc 0
vgs 3 0 dc 0
.model mn nmos level=7 vto=0.6 gammas0=0.825 phi=0.686 pb=0.8
+ js=3.2e-5 tox=1000e-10 nsub=8e15 xj=0.25e-6 ld=0.24e-6 vmax=5e4
+ uo=418 eta=1.745 lambda=0.00969 m=2.9 cj=570E-6 mj=0.3
+ cjsw=260E-12 cgso=1.27E-10 cgdo=1.27E-10 cgbo=7.66E-11
+ mjsw=0.33 rsh=12.865 vsigma=0.3 sigma0=0.005
```

6.3.4. Use the circuits in Problems 6.3.2 and 6.3.3 to calculate the propagation de-lay of a CMOS inverter as a function of gate length. Start from a gate length of 10 μm and a gate width of 50 μm. Scale the gate length down to 0.2 μm. Scale the oxide thickness and gate width proportionally to the gate length.

There are several ways to simulate propagation delays. Probably the most flexible is to simulate a ring oscillator and calculate the propagation delay based on the oscillation frequency. A ring oscillator consists of a chain of an odd number of identical inverters. The output of one inverter is connected to the input of the next, and the output of the last inverter is connected to the in-put of the first one, thereby forming a ring. When a power supply is connect-ed to each inverter, the circuit starts to oscillate at a frequency determined by the propagation delay of each inverter. From a transient simulation of the ring oscillator, we determine the oscillation frequency, from which the prop-agation delay t_{pd} of each inverter is given by

$$t_{pd} = \frac{1}{2Nf}$$

Here, N is the number of inverters and f is the oscillation frequency.

6.3.5. (a) Review the example in Section 6.3.4.

(b) The differential small-signal gain depends on the bias current Ibias. In-vestigate how the magnitude of Ibias influences the gain.

Hint: To easily calculate the gain, choose a value for Ibias and run an AC Analysis at low frequencies (1 Hz).

6.3.6. (a) Review the example in Section 6.3.4.

(b) Calculate the small-signal frequency response of the op-amp by running an AC Analysis from 1 Hz to 1 GHz. What is the cut-off frequency?

6.4.1. Use the universal AIM-Spice MESFET model (Level 2) and parameters specified in the caption of Fig. 6.4.2 to simulate the dependence of the drain saturation current at $V_{gs} = 0$ V and $V_{ds} = 2$ V on the gate length for 0.09 μm $< L < 1$ μm.

Hint: Scale the channel thickness d proportionally with L and the doping N_d proportionally with L^2.

6.4.2. (a) Review the example in Section 6.4.4.

(b) Calculate the propagation delay of a single inverter in the ring oscillator for different load widths. The procedure for calculation of propagation delays is given in the text of Problem 6.3.4.

6.4.3. (a) Review the example in Section 6.4.4.

(b) In modern high-speed digital integrated circuits, interconnects play an important role in determining the highest frequency of operation. In SPICE, interconnects can be modeled to a first-order approximation as a single capacitance inserted between the output node of a logic gate and ground. A reasonable value is 0.2 fF per micrometer of interconnect length L_i. As can be seen from the circuit description in Fig. 6.4.7, an interconnect capacitance of 20 fF, which corresponds to $L_i \approx 10$ μm, is used for the 11-stage ring oscillator. Calculate and plot the propagation delay of a single logic gate (one inverter) in the ring oscillator versus L_i. Start with $L_i = 1$ μm.

6.4.4. (a) Review the example in Section 6.4.4.
(b) Calculate the propagation delay of a single logic gate (one inverter) in the ring oscillator for fan-outs of 1, 2, and 3. In the circuit description of Fig. 6.4.7, we have a fan-out of 1. To increase the fan-out, add extra inverters at the output of each inverter in the original chain of inverters. Let the output nodes of the extra inverters be floating.

6.5.1. Use the universal HFET model (see Model Parameter Specifications A in Appendix A2) and calculate the transfer curve (V_{out} versus V_{in}) of a direct-coupled field-effect transistor logic (DCFL) inverter (see Fig. P6.5.1) as a function of the gate length. Start with a gate length of 10 μm and a gate width of 200 μm for both transistors and scale the gate length down to 0.2 μm. Use the threshold voltages $V_T = 0.1$ V and $V_T = -1$ V for the switching (enhancement mode) and load (depletion mode) transistors, respectively. Scale the AlGaAs thickness and the gate width proportionally with the gate length. Repeat the calculation for power supply voltages V_{dd} of 3 and 1.5 V. Use the following model definitions:

```
.model hfetdrv nhfet vto=0.1 di=0.04e-6 rgs=90 rgd=90
+ js1d=1 m1d=1.32 js2d=1.15e6 m2s=6.9 js1s=1 m1s=1.32
+ js2s=1.15e6 m2d=6.9

.model hfetload nhfet vto=-1 di=0.04e-6 rgs=90 rgd=90
+ js1d=1 m1d=1.32 js2d=1.15e6 m2s=6.9 js1s=1 m1s=1.32
+ js2s=1.15e6 m2d=6.9
```

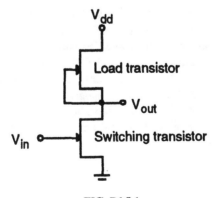

FIG. P6.5.1

6.5.2. Use the universal HFET model to calculate the delay time of an HFET DCFL inverter for different values of interconnect capacitance. Use the inverter from Problem 6.5.1.

   Hint: See Problem 6.3.4 and 6.4.3 for help on how to calculate delay times and add interconnect capacitances.

6.6.1. (a) Review the example in Section 6.6.2.
   (b) Simulate the *I–V* curves of the switching transistor in the example for operating temperatures of 25 and 50°C. Explain why the current level increases with increasing temperature while for a crystalline MOSFET the opposite is true.

6.6.2. Use AIM-Spice to simulate the room temperature transfer characteristic of the a-Si inverter shown in Fig. P6.6.2 for $V_{ds}1 = 15$ V and for $V_{ds2} = V_{ds1}$, V_{ds1} + 5 V, and V_{ds1} + 10 V. The width of the switching and load transistors are 40 and 80 μm, respectively. Use default values for other parameters.

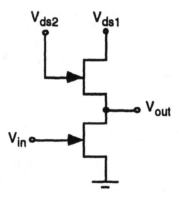

FIG. P6.6.2

6.6.3. The equivalent circuit of a liquid crystal display (LCD) pixel based on a-Si TFTs is shown in Fig. P6.6.3. Assume that the capacitor has been charged to 30 V during the previous clock cycle. The current state of the circuit is shown in the figure. How small should the TFT off current be in order to keep C_{LCD} charged at more than 90% its initial value of 30 V for longer than 10 ms? Assume a gate width of 150 μm, a gate length of 50 μm, and $C_{LCD} = 1$ pF. Use the model parameters of the example in Section 6.6.3 and simulate the pixel circuit in AIM-Spice. What is the highest operating temperature that satisfies the above requirement.

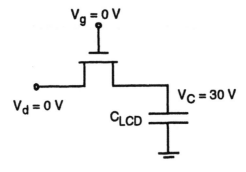

$V_g = 0$ V

$V_d = 0$ V

C_{LCD}

$V_C = 30$ V

FIG. P6.6.3

6.7.1. Use AIM-Spice to simulate and compare the output $(I_d–V_{ds})$ and transfer $(I_d–V_{gs})$ characteristics of the following three devices: Si n-channel MOS-FET, a-Si TFT, and poly-Si TFT, all with $L = 1$ μm, $W = 20$ μm, and $V_T = 0.2$ V. Other parameters of these devices are default AIM-Spice parameters for the respective universal models.

6.7.2. Solve Problem 6.6.2 for poly-Si TFTs using the default AIM-Spice parameters.

6.8.1. (a) Use the BSIM3 model in SPICE to calculate the delay time of a single CMOS logical gate for different values of the interconnect capacitance. Assume $L = 0.7$ μm for both NMOS and PMOS; $W = 10$ μm for NMOS and $W = 30$ μm for PMOS. Otherwise, use default model parameter values. Why should we select a wider gate length for the PMOS transistor?
(b) Repeat the calculation for a fan-out of 2 and 3.
Hint: See Problems 6.3.3 and 6.4.3 for details on the calculations.

6.8.2. To test the automatic parameter scaling feature of BSIM3, use this model in AIM-Spice (MOSFET Level 14) to calculate the delay time of a single CMOS logic gate for different values of the device gate length in the range 2 μm $> L > 0.2$ μm. Use default model parameter values.

APPENDIX A1

AIM-SPICE USERS MANUAL

A1.1. INTRODUCTION

The purpose of this Appendix is to provide a manual to the circuit simulator Automatic Integrated Circuit Modeling Spice (AIM-Spice). This circuit simulator is based on version 3e.1 of the popular circuit simulator SPICE. The original version of SPICE was developed at Berkeley in the 1970s (see Nagel, 1975). SPICE is a general-purpose analog simulator that contains models for most circuit elements and can handle complex nonlinear circuits. The simulator can calculate dc operating points; perform transient analyses; locate poles and zeros for different kinds of transfer functions; find the small-signal frequency response, small-signal transfer functions, and small-signal sensitivities; and perform Fourier, noise, and distortion analyses. There are many versions and modifications of SPICE. In Appendix A2, we provide a reference with basic information about the features of version 3e.1 that are fully retained in AIM-Spice. Appendices A1 and A2 contain material from the on-line help of AIM-Spice [Ytterdal et al. (1993), see also the book *Semiconductor Device Modeling for VLSI* by Lee et al. (1993).]

AIM-Spice also incorporates advanced and intermediate device models, some of which are presented in this book, and some more fully described in Lee et al., (1993). As demonstrated in that book, these models can be used not only for device and circuit simulation but also for straightforward parameter extraction, which makes AIM-Spice very convenient for practical applications, including such challenging tasks as yield and statistical analysis.

AIM-Spice has a simple and user-friendly interface and can be used for interactive circuit simulations. AIM-Spice displays the results of the simulation in progress by plotting the output during the run. It also has extensive capabilities for postprocessing data manipulation. In short, we tried to make this program as versatile and simple to use as possible.

AIM-Spice is a Windows application that can be run on one of the Microsoft Windows operating systems. A student version of AIM-Spice can be downloaded from the AIM-Spice home page on the Web (http://www.aimspice.com). (The student version was used for most of the device and circuit simulation examples included in this book.) Information on the complete, professional version of AIM-Spice can be obtained by contacting any of the authors.

In Section A1.2, we give a detailed introduction on how to use AIM-Spice. Section A1.3 is a users' guide for the AIM-Postprocessor that comes with AIM-Spice.

A1.2. AIM-SPICE

As already mentioned, AIM-Spice is based on version 3.e1 of the popular circuit simulator SPICE developed at the University of California at Berkeley.

The AIM-Spice simulation package consists of two applications running under the Microsoft Windows family of operating systems: AIM-Spice itself and a graphic postprocessor called AIM-Postprocessor. An overview of the simulator package is shown in Fig. A1.2.1.

AIM-Spice features are as follows:

- Runs under the Microsoft Windows Graphical Environment, which gives you a simple and user-friendly interface.
- Allows for Interactive Simulation Control. AIM-Spice displays simulation results in progress by plotting the output during the run with the option to cancel a simulation at any time.

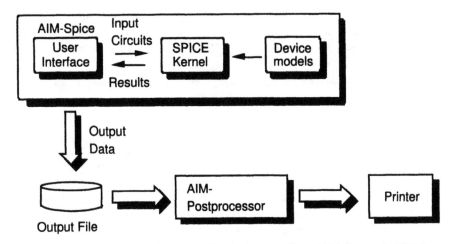

FIG. A1.2.1. Overview of the AIM-Spice simulator package. (After Lee et al., 1993.)

- Incorporates advanced and intermediate-level device models described elsewhere in this book and in *Semiconductor Device Modeling for VLSI* (Lee et al., 1993). These models can be used not only for device and circuit simulation but also for straightforward parameter extraction.

Some updated features of Berkeley SPICE version 3.e1 are as follows:

- Analyses supported are DC, AC, Transient, Transfer Function, Pole-Zero, and Noise Analysis.
- New models from Berkeley are BSIM3, lossy transmission lines, and MOS level 6.
- Improved DC Operating Point analysis.

AIM-Spice is a full-blown Microsoft Windows application and takes full advantage of all Microsoft Windows features, for example, a user-friendly interface and multitasking that allows the user to switch to other applications to do meaningful work while a lengthy simulation runs in the background.

This manual gives a detailed step-by-step description of how to use AIM-Spice. We refer to Appendix A2 for a full documentation on input syntax, devices, and device models supported.

The student version of AIM-Spice can be downloaded from the AIM-Spice home page on the WEB. The address is

`http://www.aimspice.com`

We have used the student version for most of the device and circuit simulation examples included in this book. Information on the complete, professional version of AIM-Spice can be obtained from any of the authors.

A1.2.1. Getting Started

Software Requirement. AIM-Spice is a Windows application and can only be run on one of the Microsoft Windows operating systems. This means that either Windows 3.1, Windows 95, or Windows NT must be installed on your computer.

Hardware Requirement. Minimum—PC with 386 or compatible processor with 4 MB RAM. Recommended—PC with 486 or Pentium processor with 8 MB RAM.

AIM-Spice Installation. AIM-Spice installation depends on which operating system you are using. Please refer to the AIM-Spice home page for details.

Running AIM-Spice. Select the AIM-Spice program group. To run AIM-Spice, double click the program icon. (If you do not have a mouse, use the arrow keys to move the highlight to the program icon and press Enter.)

The main AIM-Spice window appears together with an untitled document window; see Fig. A1.2.2. This document window is a text editor where you can enter your circuit description.

When AIM-Spice is finished loading, the text editor will be active. This is made clear by the blinking insertion point in the upper left corner of the window. You are now ready to type a new circuit description.

The AIM-Spice Toolbar. As shown in Fig. A1.2.2, AIM-Spice displays a toolbar right below the menu bar that gives you instant access to the most frequently used commands. If you are running the Windows 95/NT version of AIM-Spice, tool tips are shown whenever you move the cursor over one of the buttons in the toolbar. The tool tip explains which command each button performs if it is clicked. Unfortunately, the Windows 3.1 version of AIM-Postprocessor does not have this nice tool tip feature. Therefore, we have listed the toolbar button commands in Fig. A1.2.3.

A1.2.2. Editing the Circuit Description (Netlist)

The circuit topology is defined in terms of a list of circuit elements called the netlist. This description is made in a simple text editor integrated into AIM-Spice.

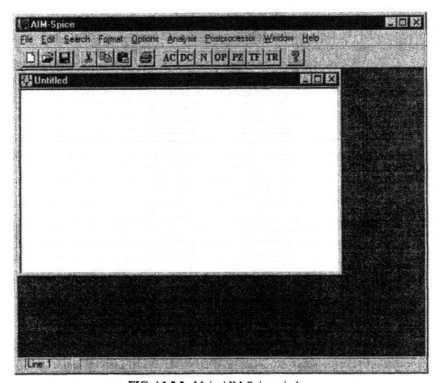

FIG. A1.2.2. Main AIM-Spice window.

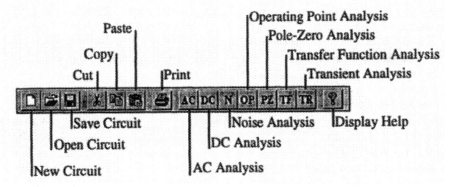

FIG. A1.2.3. Toolbar button commands for edit mode.

This editor uses standard editing rules known from popular Windows applications such as Notepad and Word for Windows and therefore does not require a detailed presentation here.

A1.2.3. Working with AIM-Spice Circuit Files

We now turn to a discussion of AIM-Spice circuit files. In AIM-Spice, a circuit is defined in terms of three main parts: the circuit description (netlist), which is placed in the text editor in a document window (also called a circuit window); analysis parameters and options; and information about the last run on the circuit. These pieces of information are stored in a text file with a special format, the so-called circuit file. To work with circuit files, use commands in the File menu. These commands are used to create, open, and save circuit files in the AIM-Spice format.

Opening Circuit Files. You can open new or existing circuit files in AIM-Spice. Theoretically, there is no limit to how many circuit files you can have open at any time. A new circuit window is created for every circuit file you open, and the circuit description is placed in that window.

When you want to close a circuit window after having made changes to the circuit description or to other circuit information since opening the file, AIM-Spice will ask if you want to save the changes. Use the following table to determine your response:

To	Choose
Save changes	Yes
Discard changes	No
Continue working in the current file	Cancel

1. *Creating a New Circuit File*

- Choose the File New command.

A new circuit window is opened.

2. *Opening an Existing Circuit File.* Only files saved in the AIM-Spice format can be opened. The default extension for the names of these files are CIR. You open an existing AIM-Spice file with the following procedure:

- Choose the File Open command. The dialog box shown in Fig. A1.2.4 is displayed.
- Select the name of the file you want to open from the list box named Files.
- Choose OK.

When you use the mouse, you can open the file in one operation.

- Double-click the filename of the file you want to open.

3. *Viewing a Circuit File in Another Directory.* The listbox with files only lists files with the extension CIR in the current drive and directory. It is possible to list other files as well. To list other files:

- Select the drive, directory, or the group of files you want in the dropdown listbox named Look in or type this information in the text box at the top of the dialog box. For example, you can type *.CKT to see a list of all files with that extension.
- Choose Open.

The listbox named Files lists the files in the drive, directory, or group of files you specified.

4. *Opening Existing Standard Text Files.* It is possible to open standard text files. When you specify such files in the File Open dialog box and choose the Open command, a message box is displayed that warns you that the file you want to open

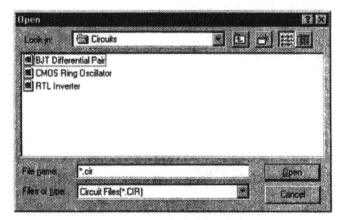

FIG. A1.2.4. File Open dialog box.

is not an AIM-Spice circuit file. You can choose to cancel the operation or continue opening the file.

A standard text file is displayed in the same way as circuit files, but you have lost the benefits of the integrated environment. When you now look at the menus, you see that most of the commands are dimmed. All the menu items in the Options menu are dimmed, and only one command from the Analysis menu is available— the Run Standard Spice File command. This command runs the first analysis control line listed in the standard text file.

5. *Importing Files.* Another way of loading standard text files is to use the Import command in the File menu. This command converts a standard text file to the AIM-Spice circuit file format. .OPTIONS control lines and analysis control lines are converted to the format used by AIM-Spice.

After converting the file, trace through the circuit description and check if any lines need manual converting. If you try to simulate a circuit with illegal lines in the circuit description, AIM-Spice will remove these lines.

6. *Saving a Circuit File.* When you create a file or you want to take a break, you can save the file and return to it later. Two commands are available for saving a file: Save As and Save.

7. *Saving a New File.* Use Save As when you want to give a name to the new file. You can also use Save As when you want to save an old file under a new name. To save a new file:

- Select the circuit window you want to save.
- Choose the File Save As command. AIM-Spice displays the dialog box shown in Fig. A1.2.5.
- Type the name that you want to give to the file in the text box. If you do not give an extension, AIM-Spice will give it the extension CIR.
- Choose Save. AIM-Spice saves the file on your disk.

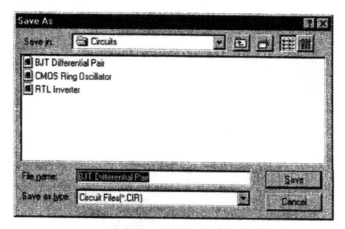

FIG. A1.2.5. Save As dialog box.

The circuit window will remain on the screen, allowing you to continue working with it. The name that you gave the circuit file now appears in the title bar of that window.

8. *Saving Changes.* You can use the Save command when you want to save changes to a circuit you currently are working with and you want to retain the present filename. To save changes:

- Choose the File Save command.

The file on your disk is replaced with the current version of the circuit.

A1.2.4. Circuit Description in AIM-Spice

Above, we discussed how to work with the editor in AIM-Spice and how to manage circuit files. We are now ready to learn how to define circuits and how to prepare a circuit description. When necessary, we will refer to the example of the differential pair circuit shown in Figs. 1.3.1 and 1.3.2 in Chapter 1.

Regarding general syntax rules for circuit descriptions, we note the following :

- The first line (only) is the title line and can contain any text.
- Comment lines are marked by an asterisk in the first column and can contain any text.
- Except for the title line and subcircuit definitions, the ordering of the lines is arbitrary.
- AIM-Spice does not distinguish between upper- and lowercase letters.
- The number of blanks is not significant except in the title line. Commas, parentheses, and tabs are equivalent to blanks.

In the remainder of this section, we discuss the different elements making up a complete circuit, and in the next section, we focus on the set of commands available for simulating the circuit.

Names. The definition of line 7 starting with RC1 defines the resistor RC1 in the circuit example of Fig. 1.3.2. The first field in the line (RC1) is the name of the resistor. Names must start with a letter that signifies the type of circuit element being considered. The rest of the name can be any string or alphanumeric characters.

Nodes. In line 7, the two items following the name (4 and 8) are the nodes to which the resistor RC1 is connected. Node names are not limited to integers, but can be any alphanumeric text string. There is one exception: the ground node must have the name 0. Nodes are not treated as integers but as text strings. Therefore, 000 and 0 are different names.

Values. The last item on line 7 (10K) is the resistor value. Numerical values are written in standard floating-point notation, with optional scale suffixes. Here are some examples of legal values:

$$1.0 \quad 1. \quad 1 \quad 0.5 \quad .5 \quad -1.0 \quad 1E6 \quad 1.6e-9$$

The scale suffixes follows the normal scientific notation, that is,

$$
\begin{aligned}
F &= 10^{-15} \\
P &= 10^{-12} \\
N &= 10^{-9} \\
U &= 10^{-6} \\
MIL &= 25.4 \times 10^{-6} \\
M &= 10^{-3} \\
K &= 10^{+3} \\
MEG &= 10^{+6} \\
G &= 10^{+9} \\
T &= 10^{+12}
\end{aligned}
$$

Units are allowed but are ignored by AIM-Spice. All characters that are not scale suffixes can be used as units

Circuit Elements or Devices. Each circuit element or device in the circuit is represented by a line not beginning with a period. All such lines have the same format:

The name of the device, followed by
two or more nodes, followed by
a model name (not all devices have this), followed by
one or more parameters.

All lines that do not start with a period, except for the title line, represent circuit elements or devices. The first letter in a device name specifies the device type. Names of resistors must start with an R, capacitors with a C, diodes with a D, bipolar transistors with a Q, and so on. The device type specification determines the meaning of the information in rest of the line: how many nodes, if a model name is required, and which parameters are to be specified at the end of the line.

Some of the devices allow or require a model name. A model gives you the option to define model parameters once and then use that set of parameters for as many devices as you want. For example, all the transistors in our tutorial circuit example (see Section 1.3) have the same parameter beta ($\beta = 80$). All refer to the same model, QNL, which defines β in terms of BF=80.

The ordering of the device lines is not significant. How they are connected is determined by the nodes. All device terminals with the same node name are connected to each other.

The rest of this section presents an overview of the device types available in AIM-Spice. A description of the devices is found elsewhere in this book and in the book by LEE et al., (1993).

Passive Devices. The passive devices available in AIM-Spice are resistors, inductors, capacitors, transformers, and transmission lines. They are all linear. Resistors and capacitors can have model names, but this is not required.

Semiconductor Devices. The semiconductor devices available in AIM-Spice are semiconductor resistors, semiconductor capacitors, *RC* transmission lines, *p–n* diodes, Schottky diodes, heterostructure diodes, silicon bipolar transistors, Junction Field Effect Transistors (JFETs), Metal Oxide Semiconductor Field Effect Transistors (MOSFETs), compound semiconductor Metal Semiconductor Field Effect Transistors (MESFETs), Heterostructure Field Effect Transistors (HFETs), amorphous silicon Thin Film Transistors (a-Si TFTs), poly-silicon Thin Film Transistors (poly-Si TFTs), and Heterostructure Bipolar Transistors (HBTs). All these devices require models, many of which are discussed in this book. Additional information, such as device geometry, can be specified.

Voltage and Current Sources. These devices are the only sources generating power. There are two types of sources: controlled and independent.

1. *Controlled Sources.* All combinations of controlled sources are available in AIM-Spice: current-controlled voltage source, current-controlled current source, voltage-controlled voltage source, and voltage-controlled current source. They perform the following functions:

$$v = e \times v \quad i = g \times v \quad v = h \times i \quad i = f \times i$$

where the constants e, g, h, and f represent voltage gain, transconductance, current gain, and transresistance, respectively.

2. *Independent Sources.* Independent sources can have different values for different types of analyses. One value can be specified for a Transient Analysis, another for an AC Analysis, and so on. A value for a DC Analysis must be prefixed by the keyword DC; the keyword for an AC Analysis is AC. For a Transient Analysis, use one of the following keywords: EXP, PULSE, PWL, SFFM, or SIN.

The voltage sources VIN, VCC, and VEE are used in the circuit example from Chapter 1. From the netlist in Fig. 1.3.2, we infer that VCC and VEE have only dc values. In our case, VIN does not have a specified dc value: It will be swept over a voltage range during a DC Analysis. On the other hand, this source is specified for both an AC Analysis (amplitude 1 V and phase 0°) and a Transient Analysis (pulsed with initial voltage –0.5 V, pulsed value 0.5 V, delay time 0.1 μs, rise time 1 ns, fall time 1 ns, pulse 0.1 μs, and period 2 μs). VCC and VEE will be assigned a value of 0 V during an AC Analysis and with their specified dc values during a Transient Analysis or DC Analysis.

The following rules apply when specifying independent sources:

- Power supplies, such as VCC and VEE, can be specified without using the keyword DC.
- The inputs to the circuit, such as VIN, may contain, for example, input waveforms and clocks.
- For a voltage source without specified values for a given analysis, the values will be set to zero and the circuit response will not be affected. However, the current through such a source can be monitored, allowing the source to be used as a current meter.

Switches. Switches allow us to change circuit connections during an analysis. They can be either voltage or current controlled. Switches require model names: SW for voltage-controlled switches and CSW for current-controlled switches.

Models. Many of the device types use model statements for specifying the parameters used in describing the device. The .MODEL statement has the following form:

```
.MODEL NAME TYPE (PARAMETER=VALUE PARAMETER=VALUE ....)
```

The model statement in our example circuit is common for all the transistors in the circuit.

Appendix A2 provides a complete list of all models used by AIM-Spice. Each model has its own set of parameters. Since default values are assigned to all parameters, specification of one or more parameter values can be omitted. For example, if only default values are used for a BJT, the model description becomes

```
.MODEL QNL NPN()
```

Subcircuits. When a circuit contains many identical blocks or subcircuits, it is convenient to be able to write a block once and then reference it when needed. You can define a subcircuit as a "super" device and reference it many times without retyping the block. Logical elements, for example, are prime candidates for subcircuit specifications.

A subcircuit is defined in terms of a block of lines that start with the line .SUBCKT and ends with the line .ENDS. Between these lines, there are one or more devices, models, calls to other subcircuits, and even new subcircuit definitions. When a subcircuit has been defined, it can be referenced as a device with a name that starts with the letter X.

Nodes can be defined as terminals for a subcircuit, making it possible to connect the subcircuit to the rest of the circuit. Node names used in subcircuit definitions are local names, and they will not come in conflict with global node names in the main circuit. The following is an example of a CMOS inverter subcircuit:

```
.SUBCKT INV 1 2 3
M1 3 2 1 1 MOSP W=24U L=1.4U
M2 3 2 0 0 MOSN W=12U L=1.0U
.ENDS
```

A1.2.5. Circuit Analysis with AIM-Spice

AIM-Spice supports seven different types of circuit analyses:

1. Operating Point Analysis (dc analysis for given source values)
2. DC Transfer Analysis (dc analysis for voltage or current sweeps)
3. AC Small Signal Analysis (calculates frequency responses)
4. Transient Analysis (calculates the time-domain response)
5. Pole-Zero Analysis (locates poles and zeros in the small-signal transfer function)
6. Transfer Function Analysis (calculates the dc small-signal transfer function, input resistance, and output resistance)
7. Noise Analysis (analysis of the device-generated noise in the circuit)

All analyses are available as commands from the Analysis menu. All analysis commands, except for the DC Operating Point Analysis, require additional control parameters to be specified. However, before choosing one of the commands in the Analysis menu, you should decide which circuit to analyze and make the corresponding circuit window the active window.

When choosing one of the commands from the Analysis menu, a dialog box appears. You specify the control parameters in the dialog box before starting the simulation. All dialog boxes have one command button labeled Run. You choose this button to initiate a simulation. The dialog box for the Transient Analysis is shown in Fig. A1.2.6.

The dialog boxes contain three command buttons: Cancel, Save, and Run. After completing the parameter fields, you can choose among these command buttons. If

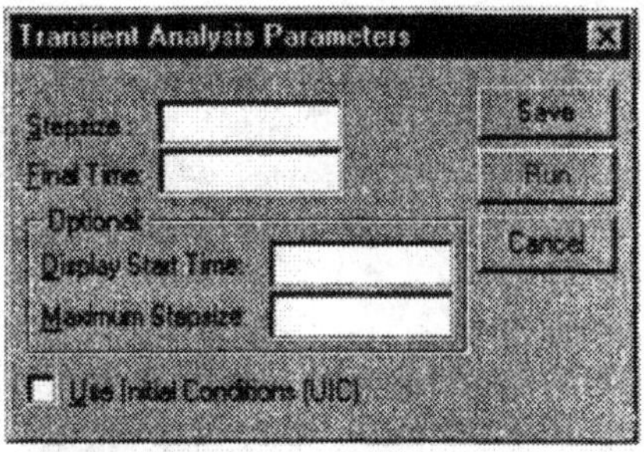

FIG. A1.2.6. Dialog box for the Transient Analysis.

you choose Cancel, the parameters entered will be discarded. If you choose Save, the parameters will be stored together with the rest of the circuit information. Run performs the same operation as Save and, in addition, initiates the analysis.

We will now discuss the control parameters for the different types of analyses available in AIM-Spice.

Operating Point. This analysis calculates the dc operating point. No control parameters are required.

DC Transfer Analysis. In this analysis, one or two sources (voltage or current sources) are swept over a user-defined interval. The dc operating point of the circuit is calculated for each value of the source(s). The DC Transfer Analysis is useful for finding the logic swing of logic gates and the *I–V* characteristics of a transistor, for example. The dialog box for the DC Transfer Analysis is shown in Fig. A1.2.7.

The first parameter in a dc analysis is the sweep variable. To specify a sweep source, open the combo box (drop-down list box) next to the source name field to see a list of all sources in the circuit and select one of them. If you specify two sources, the first one will be in the inner loop; that is, it varies faster than the other. The other parameters are the start, stop, and increment values for the source(s).

For example, if we want to find the *I–V* characteristics of a MOSFET, we can use the parameter values shown in the dialog box in Fig. A1.2.7. The drain–source voltage source *vds* is in the inner loop and sweeps from 0 to 5 V every time the gate–source voltage *vgs* changes value (the start, stop, and increment values of *vgs* are specified by selecting the 2. Source button in the dialog box).

AC Small Signal Analysis. This analysis calculates the frequency response of the circuit by linearizing the circuit equations around the operating point.

The dialog box for the AC Small Signal Analysis is shown in Fig. A1.2.8.

The first parameter determines the number of frequencies at which the analysis is performed. The option buttons in the dialog box determine the distribution of the frequencies. If you choose LIN, the number you specify will be the total number of frequencies. If you choose OCT, the value corresponds to the number of frequencies

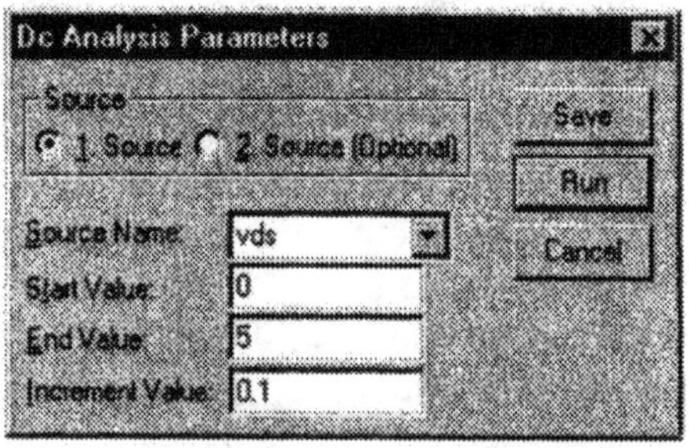

FIG. A1.2.7. Dialog box for the DC Transfer Analysis.

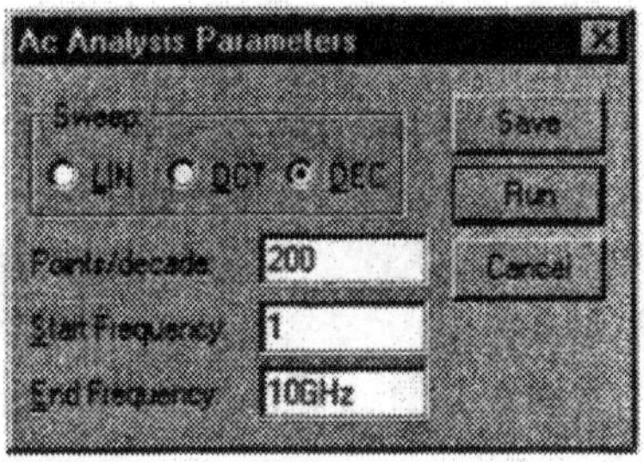

FIG. A1.2.8. Dialog box for the AC Small Signal Analysis.

per octave, and if you choose DEC, the value gives the number of frequencies per decade.

The selection shown in the dialog box specifies that the analysis starts at 1 Hz and ends at 10 GHz and that the response is calculated at 200 frequencies per decade.

Transient Analysis. The time-domain response of the circuit is calculated from time $t = 0$ to a user-defined upper time limit. The dialog box in Fig. A1.2.9 specifies a Transient Analysis that ends at 200 ns with a suggested step size of 2 ns. This analysis also has two optional parameters: the display start time and the maximum step size. The first optional parameter specifies that the generation of output starts at a time value different from zero, and the second sets an upper limit on the timesteps used by AIM-Spice.

Just as for the AC Small Signal Analysis, source values are taken from the device lines in the circuit description. In our example circuit in Chapter 1, the voltage source VIN is specified with a pulsed value during a Transient Analysis.

Pole-Zero Analysis. AIM-Spice is able to locate poles and zeros in the small-signal ac transfer function. First the dc operating point is calculated, and then the circuit is linearized around the bias point. The resulting circuit is used for locating poles and zeros.

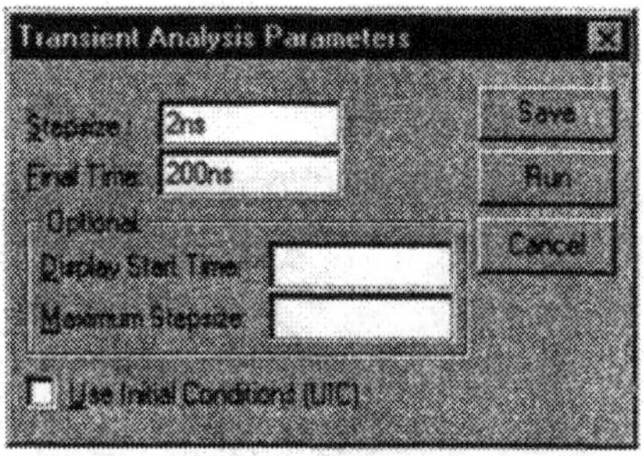

FIG. A1.2.9. Dialog box for the Transient Analysis.

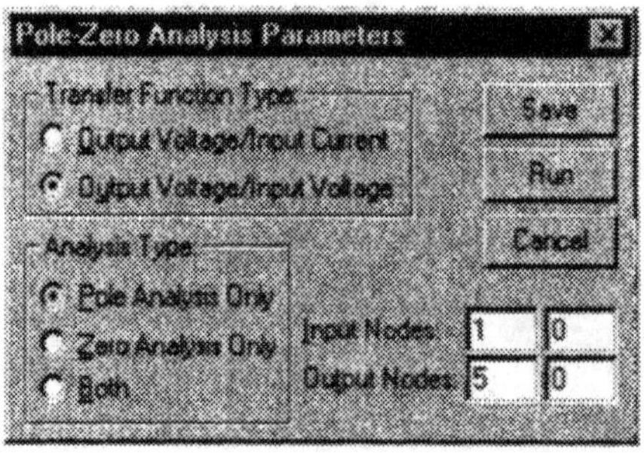

FIG. A1.2.10. Dialog box for the Pole-Zero Analysis.

The dialog box for the Pole-Zero Analysis is shown in Fig. A1.2.10. In this dialog box, you specify which type of transfer function you want, if you want to locate both poles and zeros or only one type, and the nodes defining the input and output of the circuit.

The Pole-Zero Analysis can be performed on circuits containing all types of devices except transmission lines.

Transfer Function Analysis. The dialog box for the Transfer Function Analysis; is shown in Fig. A1.2.11. This analysis computes the dc small-signal value of the transfer function, the input resistance, and the output resistance. In the example in the dialog box, AIM-Spice would compute the ratio v(8)/vin—the small signal input resistance for vin—and the small-signal output resistance measured across nodes 8 and 0.

Noise Analysis. With this selection, AIM-Spice does a Noise Analysis of the circuit. The dialog box for this analysis is shown in Fig. A1.2.12.

The format of the parameter Output Noise Variable is V(OUTPUT<REF>), where OUTPUT is the node at which the total output noise is sought. The parameter Input Source is an independent source to which input noise is referred. The next three parameters contain information relating to frequency, identical to those of the AC Analysis. The last parameter is an optional integer. If specified, the noise contribution of each noise generator is produced at every Points per Summary frequency point.

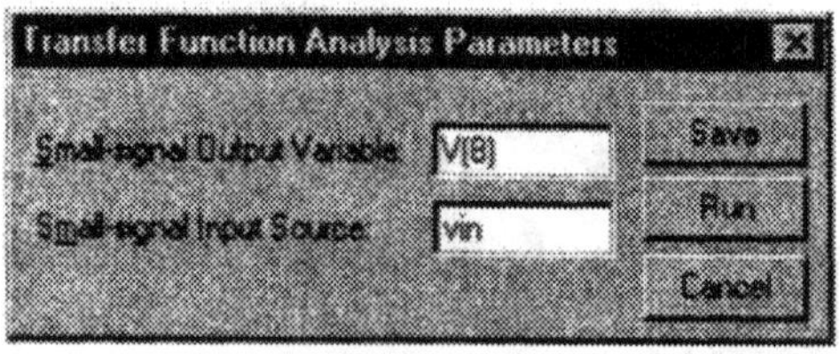

FIG. A1.2.11. Dialog box for the Transfer Function Analysis.

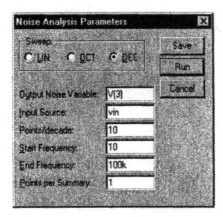

FIG. A1.2.12. Dialog box for the Noise Analysis.

Setting Initial Conditions. Initial conditions are currents and/or voltages specified to help locate the bias point of a circuit or to force the bias point to satisfy one or more conditions. One reason for giving AIM-Spice initial conditions is to select one out of two or more stable states, for example, in bistable circuits.

There are three different ways of specifying initial conditions: the .IC statement, the .NODESET statement, and the IC= specification on individual device lines. All these statements are specified in the editor together with the netlist.

.IC. This control statement is used to specify initial values for a Transient Analysis. This statement can be interpreted in two ways, depending on whether the UIC is selected (see the dialog box for Transient Analysis in Fig. A1.2.9). If UIC is selected, the node voltages in the .IC statement will be used to compute initial voltages for capacitors, diodes, and transistors. This is equivalent to specifying IC= for each element but is much more convenient. IC= can still be specified and will override the .IC values. However, AIM-Spice will not perform any operating point analysis when this statement is used. Hence, the statement should be used with care.

AIM-Spice will perform an operating point analysis before the transient analysis if UIC is not specified. Then, the .IC statement has no effect.

.NODESET. This control statement helps AIM-Spice locate the dc operating point. Specified node voltages are used as a first guess for the dc operating point. This statement can be useful with bistable circuits. However, normally it is not needed.

.NODESET is active during all bias point calculations, not only in transient analyses. .IC has higher priority than .NODESET for a Transient Analysis.

IC=. Using this statement, capacitors, inductors, transmission lines, diodes, and transistors can be given initial voltage (e.g., capacitors) or current (e.g., inductors) values on the device line. The UIC option must be active in the Transient Analysis dialog box (see Fig. A1.2.9) in order to use these initial values. AIM-Spice will skip the calculation of the bias point and go directly to the Transient Analysis when the UIC option is active. All devices without assigned IC= value are assumed to have a zero initial value.

Options. A set of options that control different aspects of a simulation is available from the Options menu. These are divided among the following four logical groups:

- General
- Analysis specific
- Device specific
- Numeric specific

Each circuit has its own set of options, and before you reset any of the them, decide which circuit you want to work with and make the corresponding circuit window the active window.

To reset one of the options, choose one of the commands in the Options menu. All options in a group are listed in a dialog box together with their default values.

The different options are explained in detail in Appendix A2.

Interactive Simulation Control. During a simulation, different commands can be executed depending on the type of analysis being performed. DC Operating Point, Pole-Zero, and Transfer Function Analysis are so-called one-vector plots: that is, they produce only one data point. Therefore, these analyses are executed immediately after you choose the Run command in the dialog box, and the results are presented after the simulation is completed. The results are presented in a table for DC Operating Point and Transfer Function Analysis and in a graph for Pole-Zero Analysis. The other analysis types produce output during the simulation, and you have to select which variables to monitor before you start the simulation. The following paragraphs explain the procedures for the different analyses.

DC Operating Point. As mentioned, the DC Operating Point Analysis produces a so-called one-vector plot. The results are presented in a table as soon as the simulation is over. An example of such a presentation is shown in Fig. A1.2.13.

Pole-Zero Analysis. Like the Operating Point Analysis, the Pole-Zero Analysis produces a one-vector plot. But unlike the Operating Point Analysis, this analysis presents its results in a graph. An example is shown in Fig. A1.2.14.

Transfer Function Analysis. This analysis also produces a one-vector plot, and the results are presented in a table, as shown in Fig. A1.2.15.

Noise Analysis. The Noise Analysis produces both one-vector plots and multivector plots. We have chosen to use the same interface as for the analysis types listed above. The results are presented after the simulation is completed and only one-vector plots are displayed. To display the other plots, use AIM-Postprocessor.

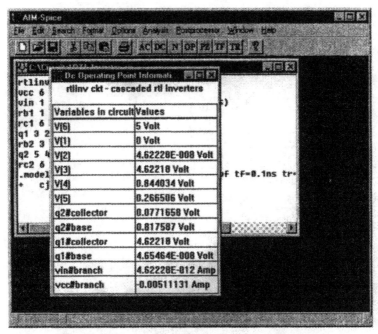

FIG. A1.2.13. Presentation of DC Operating Point results.

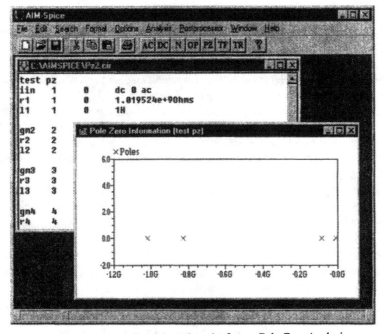

FIG. A1.2.14. Presentation of results from a Pole-Zero Analysis.

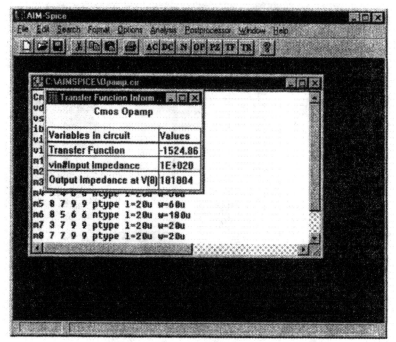

FIG. A1.2.15. Presentation of transfer function results.

AC, DC Sweep, and Transient Analysis. AIM-Spice changes mode when you choose the Run command from one of the analysis dialog boxes. The menus listed in the menu bar will change (see Fig. A1.2.16), the toolbar buttons will change, and the status bar will give you information about what happens at any time. The toolbar button commands in the analysis mode are shown in Fig. A1.2.17.

When the application changes to the analysis mode, the circuit description, together with options and analysis parameters, are loaded into the Spice kernel. While reading the circuit into the Spice kernel, the status bar will show the text "Parsing circuit, Please wait. . . ." After the input and error-checking operation is done, the status bar shows which analysis is selected.

The simulation can only begin after you have specified which circuit variables to plot. All commands needed for the preliminary work before starting a simulation are located in the Control menu. This menu is shown in Fig. A1.2.18.

Selection of Variables to Plot and Specifying Plot Limits. In AIM-Spice you can open as many plot windows as you like. A plot window contains one or more circuit variables that will be plotted graphically during a simulation. To open a new plot window, choose the command Select Variables to Plot in the Control menu. This command displays the dialog box shown in Fig. A1.2.19. This dialog box is divided into several functions. To the left is a list of all the variables of the

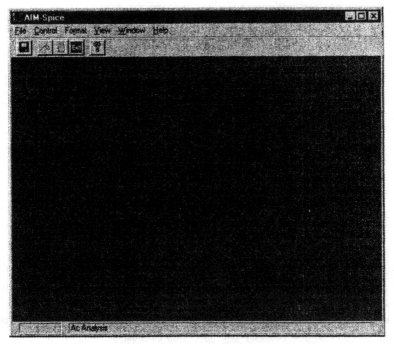

FIG. A1.2.16. AIM-Spice window in the analysis mode.

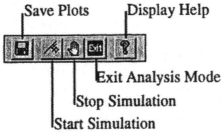

FIG. A1.2.17. Toolbar button commands in the analysis mode.

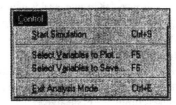

FIG. A1.2.18. Control menu for plot variables and limits.

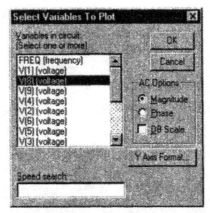

FIG. A1.2.19. Command menu dialog box for selecting plot variables.

circuit. The first element in this list is always the independent variable. The dialog box in Fig. A1.2.19 corresponds to an AC Analysis of our sample circuit (see Section 1.2.4). In an AC Analysis, the frequency is the independent variable. The node voltages in the circuit are listed after the independent variable. You can plot several variables in the same window by selecting more than one variable from the list.

If it is too cumbersome to scroll through the list of variables, you can use the speed search feature. Start entering the name of the variable you want to select in the speed search field and the variable is automatically scrolled into view.

The area to the right is active only for AC Analysis, in which case you can choose to plot either the amplitude value or the phase value. If you choose to plot the amplitude, you can select a decibel scale. This area is dimmed if other types of analyses are selected.

After having specified which variables to plot in a given plot window, you have to choose y-axis limits for the windows. AIM-Spice has no way of guessing limits for the plots. The plot limits can be specified by selecting the button called Y Axis Format, in which case the dialog box shown in Fig. A1.2.20 appears. You have to specify lower and upper limits and increments for the vertical axis (y axis) of the plots. It is wise to specify large intervals as a first guess. As the simulation progresses, the plot traces will appear on the screen, and you can reset the limits manually or by the Auto Scale command in the Format menu (Alt+U).

Complete the Y Axis Format dialog box and choose OK to return to the Select Variables to Plot dialog box. Choose OK again and a new plot window is opened. The title bar of the new window indicates which variables will be plotted in that window. To open more plot windows, choose Select Variables to Plot again. In Fig. A1.2.21, we have opened two plot windows.

It is not possible to use the command Select Variables to Plot to add more variables when the simulation is running or after the simulation is completed. As soon as you start the simulation, this command is disabled.

To study your simulation results more carefully after the simulation is completed,

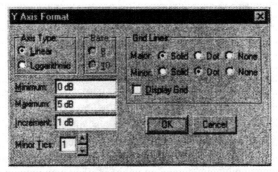

FIG. A1.2.20. Dialog box for setting plot limits.

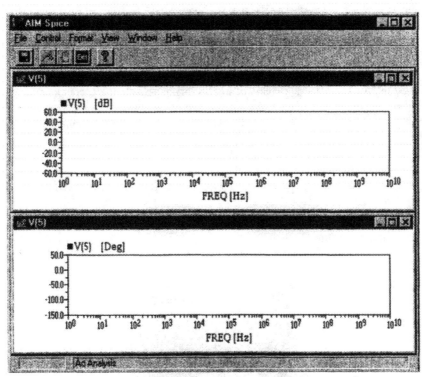

FIG. A1.2.21. Main window with two plot windows opened.

save the results to a file and use AIM-Postprocessor. We have chosen this approach because we wanted the graphical operations in the simulator to be as fast as possible and to relax the functionality requirements.

Selection of Variables to Save. The command Select Variables to Save is handy when you are simulating large circuits containing hundreds of nodes. The default is to save every circuit variable for the entire simulation. In some cases, this produces large amounts of data. However, this command lets you specify which variables you are interested in and want to save. Specifying a small number of variables to save reduces both the simulation time and the size of the output file. The command displays the dialog box in Fig. A1.2.22. It contains a list of all circuit variables together with the two useful command buttons labeled Select All and Clear All. Here you have the option of going through the list and selecting the variables you wish to save.

To select all variables in the list, choose the command button labeled Select All. To deselect all variables in the list, choose the command button labeled Clear All. Note that the independent variable (always the first variable in the list) is always saved, regardless of whether you select it or not.

Formatting Axes and Labels. When AIM-Spice creates new graph windows, it uses default axis and label formats. To change the format, use commands from the Format menu. You can use these commands as long as you are in the analysis mode. To format, for example, the *x* axis of a graph, first activate the corresponding plot window and then choose the X-Axis command. If you have a mouse, you may also double click the *x* axis to change its format. You can change the following properties of an axis:

- Axis type (linear or logarithmic)
- Base (when you select logarithmic axis, you have to choose which base number to use, 8 or 10).
- Minimum value
- Maximum value

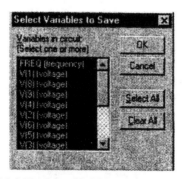

FIG. A1.2.22. Select Variables to Save dialog box.

- Increment value (distance between axis labels)
- Minor tics (number of tic marks between labels)
- Grid lines (you specify line styles for grid lines drawn on major and minor tic marks and you may turn grid on or off)

To format a label, use commands from the Format menu or double click a label. You can change the following properties of labels:

- Notation (select among three numeric formats in the label: AIM-Spice scale factors, decimal, or scientific)
- Number of decimal digits used in the numeric format

Arranging Plot Windows. Commands for arranging the plot windows inside the main window are located in the Windows menu (see the menu bar in the main window). When you choose Cascade Windows, the plot windows will be arranged in a stack. To place a given plot window at the top of the stack, choose the title of that window from the Window menu. Choose Tile Windows Vertical to arrange all the windows side by side or Tile Windows Horizontal to arrange all the windows vertically.

The plot windows can also be moved and resized with the mouse or with keyboard commands from the system menu.

Starting a Simulation. To start a simulation, choose the command Start Simulation from the Control menu of the main window. This command is dimmed until you have opened at least one plot window. Once the simulation is launched, AIM-Spice plots the selected variables in the plot windows as soon as they are available from the simulator. Figure A1.2.23 shows a snapshot of typical simulation plots for a simulation in progress.

Stopping a Simulation. The instantaneous simulation results in the plot windows may tell you that something needs to be altered in your circuit description or in the parameter settings. Then you may open the Control menu and stop the simulation at any time. With a simulation in progress, the Control menu is slightly altered from that shown in Fig. A1.2.18: The command Start Simulation is replaced by Stop Simulation, and all other commands are dimmed and not available.

Resetting the Plot Limits During a Simulation. If during a simulation you realize that the axis limits specified are unsuitable, they can be reset at any time without having to stop the simulation. The simplest procedure is to select Autoscale from the Format menu (Alt+U). Alternatively, use either the other commands available in the Format menu, the Zoom command from the View menu, or double click on the appropriate axis and complete the dialog box.

To make the Zoom command active, choose the command once. A check mark appears next to the command name. Now you can use the mouse to reset limits.

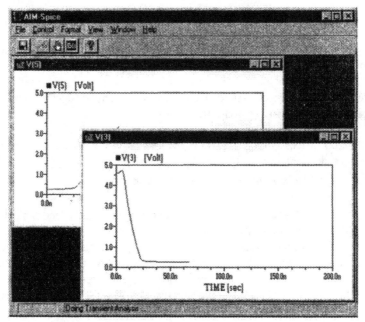

FIG. A1.2.23. Plots from a Transient Analysis in progress.

Place the mouse cursor in the upper left corner of the new viewing rectangle. Then press and hold the right mouse button while you drag the mouse cursor to the lower right-hand corner of the new viewing rectangle. Release the mouse button and AIM-Spice will redraw your plot with the new viewing rectangle. When you choose the Zoom command once more, you deactivate the command, and the graph is redrawn with the original axis limits.

Saving Results after Completing a Simulation. AIM-Spice indicates that a simulation is completed by displaying "Simulation Done ..." in the status bar. When a simulation is over, you can save the results in a file by choosing the Save Plots command from the File menu of the main window. This command displays the dialog box shown in Fig. A1.2.24.

This dialog box displays a list with plots accumulated since the last time you saved plots. If you want to save all the plots in the list, choose the command Save All Plots. If you want to save only a selection of the listed plots, select the plots you want to save, and then choose the command Save Selected Plots. If you are not interested in any of the plots, choose the command Destroy All Plots. When one of the commands Save All Plots or Save Selected Plots is selected, a Save As dialog box is displayed. Complete the entries in the dialog box, and choose OK.

Hint: If you want to compare results from different simulations in AIM-Post-

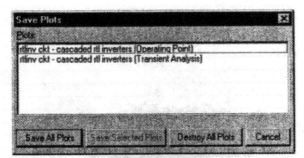

FIG. A1.2.24. Dialog box for saving of plots.

processor, run all the simulations before you save the results. That way you can collect the results from all simulations in a single output file. When you load this output file into the Postprocessor, you are able to plot variables from different simulations in the same graph.

Exiting after Completing a Simulation. To leave the analysis mode, choose the command Exit Analysis Mode. The menu bar changes back to the original menu and the circuit description reappears in the main window. To exit from AIM-Spice, choose the command Exit from the File menu.

Error Reporting. Error messages are written to an error file as soon as they are detected. When an error occurs, AIM-Spice interrupts the simulation and displays the message box shown in Fig. A1.2.25. If you choose the button labeled Yes, AIM-Spice displays the contents in the error file in a popup window, as shown in Fig. A1.2.26. This popup window is closed the same way as the main application window. The error window stays on top of all windows belonging to AIM-Spice.

To correct an error, you quit the analysis mode, return to the circuit window, and make the necessary changes to the circuit with the error window visible. The error window automatically closes when you start another Run.

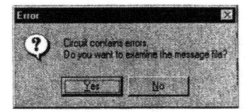

FIG. A1.2.25. Message box for error reporting.

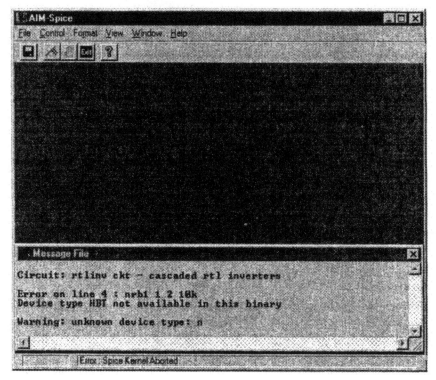

FIG. A1.2.26. Popup window with error information.

A1.3. AIM-POSTPROCESSOR

AIM-Postprocessor is an application containing routines for further processing of data obtained from the various analyses in AIM-Spice and for graphical presentation. It works independently of the analysis part of AIM-Spice but still within the Windows environment.

Although AIM-Spice has facilities to plot circuit variables graphically, AIM-Postprocessor has a much more powerful plotting engine that includes the following features:

- Plotting of sums and differences, derivatives, integrals, and mathematical functions of circuit variables
- Cursors to select numerical values and to calculate differences between variables
- Import of experimental data
- Hard copies

This section is divided into a tutorial on postprocessing with AIM-Postprocessor and a complete command reference.

A1.3.1. AIM-Postprocessor: Tutorial

In this tutorial, you will learn step by step how to use the basics of the post-processor to produce plots for presentation. A full documentation of AIM-Postprocessor is deferred to Section A1.3.2, the Command Reference section.

To get the most out of this tutorial, read it in front of your computer with Windows running and follow the steps outlined below. Text lines in italic are your actions and we strongly encourage you to try these actions.

Loading AIM-Postprocessor. Double click on the AIM-Postprocessor icon and the main window is displayed, as shown in Fig. A1.3.1.

The first task after loading the postprocessor is to open a data file. The data file format is a special binary format used by AIM-Spice and AIM-Postprocessor. As shown in Fig. A1.3.1, AIM-Postprocessor displays a toolbar right below the menu that gives you instant access to the most frequently used commands. If you are running the Windows 95/NT version of AIM-Postprocessor, tool tips are shown whenever you move the cursor over one of the buttons in the toolbar. The tool tips explain which command each button performs if it is clicked. Unfortunately, the Windows 3.1 version of AIM-Postprocessor does not have this feature. The toolbar button commands are shown in Fig. A1.3.2.

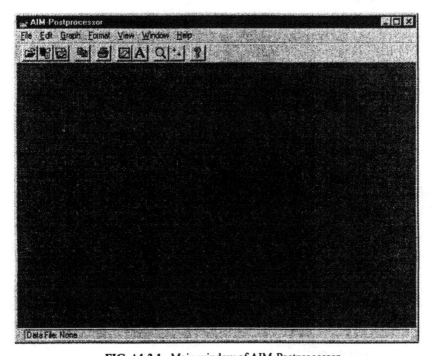

FIG. A1.3.1. Main window of AIM-Postprocessor.

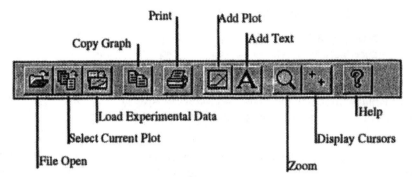

FIG. A1.3.2. AIM-Postprocessor toolbar button commands.

Opening an AIM-Spice Data File. An AIM-Spice data file contains one or more plots. A plot consists of three main parts: plot information, a list of output variables, and a list of data vectors.

To open an AIM-Spice data file, follow these steps:

- Choose the File Open command.
- A standard File Open dialog box is displayed. The default extension for data files is OUT. The Files list box contains a list of files in the current directory with extension OUT.
- Select the file you want to open and choose OK. AIM-Postprocessor opens and reads the contents of the file.

Every plot that contains only one data vector is displayed in a table or a graph immediately after the file is loaded. Plots from the Operating Point Analysis, the Transfer Function Analysis, the Pole-Zero Analysis and the Noise Analysis, are such one-vector plots.

We now turn to a tutorial example included in AIM-Spice.

Open the file "tutorial.out."

Select Current Plot. After the file is read, a dialog box appears containing a list of all plots saved in the file (this dialog box is not displayed if the file contains only one plot). This dialog box is shown in Fig. A1.3.3.

You are asked to select one plot as the current plot. To select the current plot, select the corresponding list box item and then choose OK. You can change the current plot at any time by choosing the File Change Current Plot command.

After a plot has been selected, and provided this is not a one-vector plot, you are able to create line graphs of the variables contained in the plot. If you select a one-vector plot as the current plot, the graph or table window with the plot is made the active window. If the window does not exist, AIM-Postprocessor creates a new window and displays the plot.

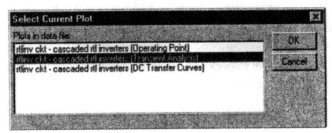

FIG. A1.3.3. Select Current Plot dialog box.

Select the second plot.

The information part of a plot can be viewed at any time by choosing the View Plot Info command. This command displays information about the current plot in two dialog boxes, one of which is the simulation statistics window discussed earlier.

Creating Graphical Plots. A graphical plot is a document window that contains a graph with one or more traces.

You create graphical plots by choosing the Graph Add Plot command. When you choose this command, the Add Plot dialog box is displayed as shown in Fig. A1.3.4. The Add Plot dialog box is divided into four main fields:

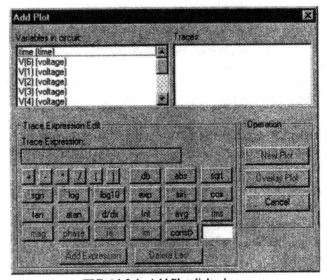

FIG. A1.3.4. Add Plot dialog box.

- A list box with the variables in the current circuit.
- A list box with the traces you want to add to the graphical plot. This list box is initially empty.
- A Trace Expression Editor (TEE) that lets you edit the traces selected.
- Three command buttons.

The following gives a description of the trace expression editing facility.

Trace Expression Editing (TEE). Trace expression editing gives you the ability to create complex expressions, including mathematical functions, derivatives, and integrals of the variables in a plot. These expressions are plotted graphically when you choose the New Plot or Overlay Plot command.

TEE works much like a calculator. With a calculator, you add operands, operators, and functions of operands, and then you press "=" to calculate your expression. In the TEE the "=" symbol is replaced by the command Add Expression, which saves the expression you have edited in the list of traces. The trace editor, like a calculator, also has a display window that shows the current expression. With a backspace key you can delete the last operation in the trace editor. The editor is intelligent in the sense that it ignores operations that are not allowed. For example, every expression starts with an operand or a function, and accordingly, the editor does not allow you to start with an operator. Operator buttons are inactive when operators are not allowed, the list of variables is grayed when operands are not allowed, and so on.

Here are three example expressions based on our example circuit and the recipe to create and save them in the trace editor:

(i)   V(3)
(ii) V(3) + V(5)
(iii) avg (V(3))

Follow these steps to create the three traces shown in Fig. A1.3.5:

Open the Add Plot dialog box.

Select the variable V(3) from the list of circuit variables. The variable V(3) appears in the TEE display.

Choose the command button Add Expression. The trace is added in the trace list and the TEE display is cleared and is ready to receive a new expression.

Select the variable V(3) from the list of circuit variables. The variable V(3) appears in the TEE display.

Choose the operator "+" from the operator buttons. The plus operator appears in the TEE display.

Select the variable V(5) from the list of circuit variables.

Choose the command button Add Expression.

Select the function "avg" from the function buttons.

Select the variable V(3).

Select the closing bracket.

Choose the command button Add Expression.

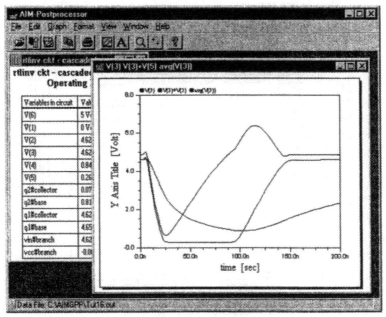

FIG. A1.3.5. Main window of AIM-Postprocessor with open graph and table windows.

The trace list now contains three entries.

Choose the command button New Plot.

A new graph window is created and the graph with the three traces shown in Fig. A1.3.5 is displayed. Note that the Overlay Plot command was grayed in the dialog box. This command is available only when you have already created one or more graphs. You use it when you want to add new traces to an already existing graph.

On the screen, the traces in Fig. A1.3.5 can be distinguished by different colors, symbols, or line styles. It is also possible to drag the legends close to the corresponding traces, which is especially useful when creating hard copies.

So far, we have seen how to create a graph. Next, we will take a closer look at graph editing, that is, the operations used for preparing the graph for presentation.

Graph Editing. A graph consists of two axes, two axis labels, two axis titles, the trace area, text, and legends to identify the traces. Figure A1.3.6 illustrates the different parts of a graph. It is possible to change the appearance of all these parts with the graph-editing facilities in AIM-Postprocessor.

Note that legends are linked to the graph and are moved whenever the trace area is moved or resized. This link is broken when you move or resize one of the legends (the link is broken only for the legend moved or resized).

The graph edit facilities are as follows:

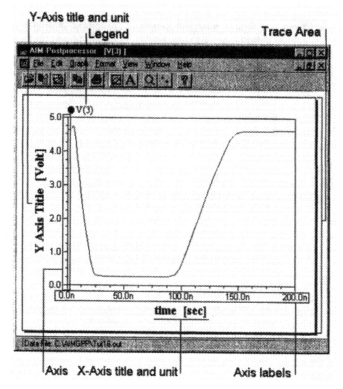

FIG. A1.3.6. Different parts of a graph.

- Formatting axes
- Formatting axes titles
- Formatting labels
- Formatting the trace area
- Adding and formatting text
- Formatting legends
- Changing x-axis expression

Formatting Axes. To format an axis, first select a graph window as the active document window. Then choose the appropriate command from the Format menu. Note that it is also possible to format an axis by double clicking the axis. If you choose to format the *x* axis of a graph, a dialog box similar to the one shown in Fig. A1.3.7 appears. This dialog box has the following fields:

Axis Type	Choose between linear or logarithmic axis.
Base	If you choose a logarithmic axis, you can select the base number to be 8 or 10.

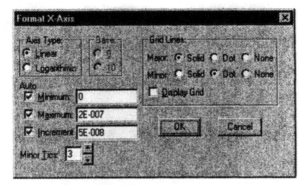

FIG. A1.3.7. The Format Axis dialog box.

Minimum	Specifies the minimum value for the axis.
Maximum	Specifies the maximum value for the axis.
Increment	Specifies the distance between axis labels.
Minor Tics	Specifies number of tic marks drawn between labels.
Grid Lines	You can specify line styles for grid lines drawn on major tic marks and on grid lines drawn on minor tic marks. The check box Display Grid turns the grid on or off.

A faster way of changing the axis limits is to zoom with the mouse using the following steps:

- Turn zooming on by choosing the View Zoom command.
- Position the mouse cursor where you want the upper left corner of your view area to be. Press and hold the right mouse button.
- Drag the mouse cursor down to the lower right corner of your view area and release the button.
- The graph is then redrawn with the new axis limits.

When you turn the zooming off by choosing the View Zoom command, you reset the axis limits to their values before the zoom operation.

Formatting Axis Titles. To format an axis title, first select a graph window as the active document window. Then choose the appropriate command from the Format menu. Note that it is also possible to format an axis title by double clicking the title. If you choose to format the x-axis title of a graph, a dialog box like the one shown in Fig. A1.3.8 appears.

This dialog box has the following fields:

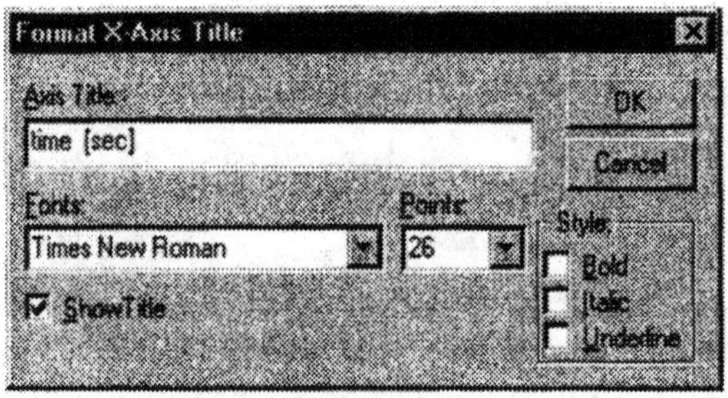

FIG. A1.3.8. Format Axis Title dialog box.

Axis Title Contains the current axis title.

Fonts Contains a list off all fonts available in the printer currently selected. The font currently used for the label is displayed in the edit box. To select another font, type the name of the font in the edit box or open the list of fonts by clicking on the down arrow in the drop-down list box, and then select a font from the list.

Points Contains a list of point sizes. The point size currently used for label text is displayed in the edit box. To use another point size for label text, type a new point size in the edit field or open the list of point sizes by clicking on the down arrow in the drop-down list box; then select a point size from the list.

Style Specifies the text style used for the label. You can specify bold, italic, and/or underlined text.

Show Specifies if the title is displayed or not.

Formatting Labels. To format a label, first select a graph window as the active document window. Then choose the appropriate command from the Format menu. Note that it is also possible to format a label by double clicking the label. If you choose to format the *x*-axis label of a graph, a dialog box such as the one shown in Fig. A1.3.9. appears.

This dialog box has the following field in addition to the Fonts, Points, and Styles fields discussed in conjunction with Fig. A1.3.8:

Notation Specifies the numeric format used in the label. The number 1 million is displayed as 1M if you specify AIM-Spice scale factors, 1000000 if you specify Decimal, and 1E+006 if you specify Scientific. The Digits text box lets you specify how many digits to use in the numeric format.

Formatting the Trace Area. It is possible to move and resize the borders of the rectangular trace area. To move the rectangle, click and hold the left mouse button anywhere inside it. Then drag the rectangle to the new position and release the

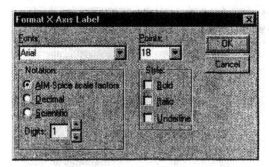

FIG. A1.3.9. The Format Axis Label dialog box.

mouse button. To resize the rectangle, click the left mouse button anywhere inside it. This operation selects the rectangle. Then place the mouse cursor over one of the handles and click and hold the left mouse button. Then drag the rectangle borders to their new positions and release the mouse button. The handle selected determines which rectangle border to move.

Double clicking anywhere inside the trace area restores the trace area to the default position and size. Legends are also restored to their default positions and links to the trace area are made active again.

Adding and Formatting Text. To add text to your graph, choose the Graph Add Text command. When you place the cursor over a graph window, the cursor will change to reflect that you have selected to add text. Position the cursor on the location where you want the upper left corner of your text to be located and then click the left mouse button. The dialog box in Fig. A1.3.10. appears.

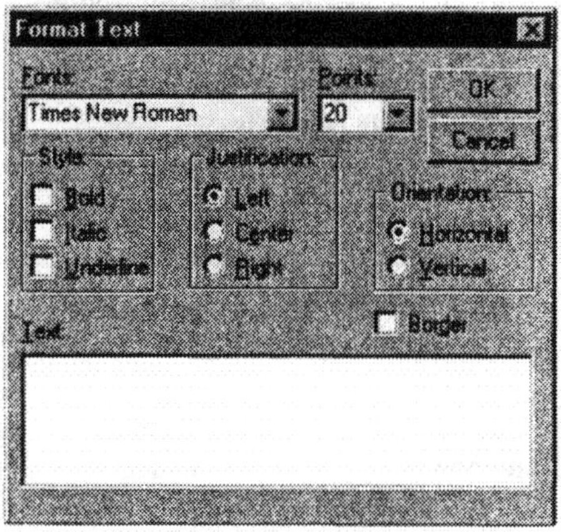

FIG. A1.3.10. Format Text dialog box.

This dialog box has the following fields in addition to the Fonts, Points, and Styles fields discussed in conjunction with Fig. A1.3.8:

Justification	Specifies the horizontal justification of the text within the text rectangle.
Orientation	Specifies the orientation of the text.
Border	If selected, a rectangle is drawn around the text.
Text	Here, you type the text you want to add to the graph.

Complete the dialog box and choose OK to add the text with the format you have selected. It is possible to cancel the Add Text command by choosing the command once more before you click the mouse button.

You can format a text at any time by double clicking within the text area with the left mouse button to select the text. In response, the Text Format dialog box is displayed. You are now able to make changes in the selected text. To make your changes visible, choose OK.

It is also possible to move and resize text. To move a text, click and hold the left mouse button anywhere in the text area. Then drag the text to the new position and release the mouse button. To resize a text, click the left mouse button anywhere in the text area. Then place the mouse cursor over one of the handles and click and hold the left mouse button. Then drag the rectangle borders to their new positions and release the mouse button.

Formatting Legends. Legends are used to identify the different traces contained in a graph. The legends are positioned at the upper left corner of the trace area with a symbol identifying the trace by color. You can move and resize the legend and change its format. To move or resize the legend, follow the procedure described above for moving or resizing text.

To format a legend, point the cursor to it and double click. The dialog box in Fig. A1.3.11 appears. This dialog box has the following fields in addition to the Fonts, Points, Styles and Border fields discussed in conjunction with Figs. A1.3.8 and A1.3.10:

Legend Text	Specifies the legend text.
Symbols	Contains a list of predefined symbols used to identify traces and data points generated by AIM-Spice.
Display Symbol with Text	If selected, the currently selected symbol is drawn to the left of the legend text.
Display Symbols in Graph	If selected, the currently selected symbol is drawn on data points.
Symbol Period	Active only if Display Symbols in Graph is selected. The number you type in this field specifies the symbol frequency. If you specify 1, a symbol is drawn for every data point. If you specify 10, a symbol is drawn for every tenth data point.

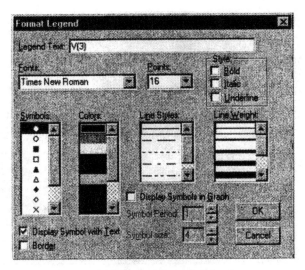

FIG. A1.3.11. Format Legend dialog box.

Symbol Size   Determines the size of the symbol. The lowest value you can specify is 1.

Colors   Contains a list of predefined colors used to identify the traces. The color you select is used when drawing the trace and the currently selected symbol.

Line Styles   Contains a list of different line styles. AIM-Postprocessor uses the line style you have selected when it draws lines between data points to create a line graph.

When you use the screen as the output device, the use of different colors is the best way to identify the different traces in a graph. However, when you want to print a graph, you will normally not use colors. One way of identifying the different traces without using colors is to drag the legend into the trace area and position it close to the trace. When you do so and you do not want to show the symbol next to the legend text, you turn the option Display Symbol with Legend Text off.

Changing the x-*Axis Expression.* You do not have to use the x-axis variable from the simulation. You are free to change the x axis to a general expression like the ones we described in the section Creating Graphical Plots. To change the expression for the x axis, choose the Graph X-Axis Expression command and the dialog box in Fig. A1.3.12 appears.

This dialog box is almost identical to the Add Plot dialog box. The only difference is that now you edit only one expression.

This completes the discussion of graph editing.

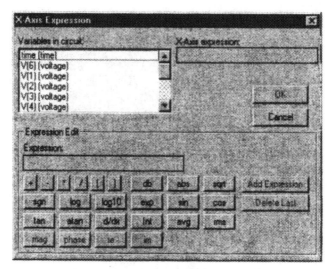

FIG. A1.3.12. X-Axis Expression dialog box.

Cursors. Cursors are tools for extracting numerical values from your graphs. With two cursors, you are given the opportunity to calculate differences between traces. To turn cursors on, choose the View Show Cursors command. The cursor information window appears in the upper right corner of the main window, as shown in Fig. A1.3.13. To remove the cursors, select the View Hide Cursors command.

Printing. To print a graph or a table, follow these steps:

- Format the graph or table you want to print.
- Choose File Print.

If you want to change the printer settings before you print, choose the File Printer Setup command. The command lets you, for example, choose a different printer, change the orientation of the page, or the page size, as shown in the dialog box in Fig. A1.3.14.

1.3.2. Command Reference

This section gives you a description of the commands and dialog boxes in AIM-Postprocessor. Each menu has its own set of commands. We start by reviewing the File menu shown in Fig. A1.3.15.

The File Menu

- *Open:* Displays the dialog box in Fig. A1.3.16. This command lets you specify which AIM-Spice data file to open. This is a standard Windows File Open dialog box, and Section A1.2 gives a full description of how to open files.

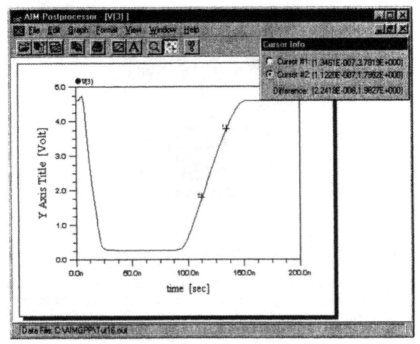

FIG. A1.3.13. Main window of AIM-Postprocessor with cursors active.

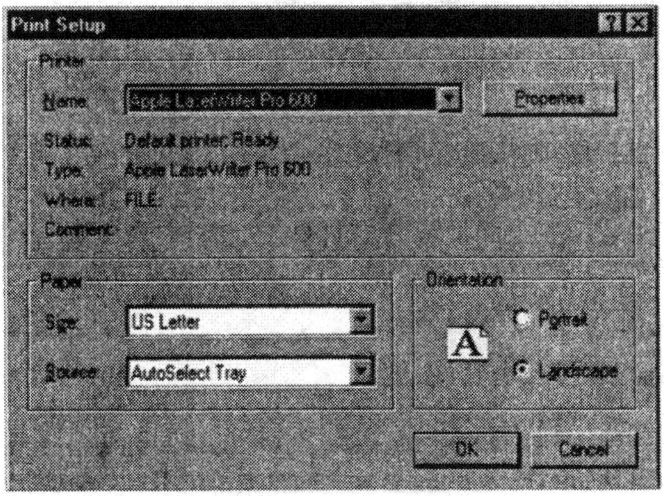

FIG. A1.3.14. Print Setup dialog box.

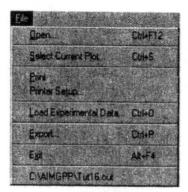

FIG. A1.3.15. File menu.

- *Select Current Plot:* In general, an AIM-Spice data file contains more than one plot. This command lets you select the current plot. When you create graphs, the circuit variables and data vectors are taken from the current plot. This command displays the dialog box shown in Fig. A1.3.3. The list box in the dialog box contains a list of all plots in the data file. You select the current plot by selecting the corresponding list box item and then choose OK.
- *Print:* Prints the active graph or table window.
- *Printer Setup:* Lets you specify printer and printer parameters.
- *Load Experimental Data:* Loads a text file with experimental data. Displays a standard File Open dialog box to let you specify the file that contains the experimental data. The data are presented in the active graph window as centered symbols. The text file has the following format:

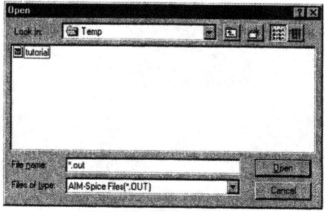

FIG. A1.3.16. File Open dialog box.

```
Number of data rows (nrows)
Number of data columns (ncol)
Legend text for column #1
Legend text for column #2
  .
  .
  .
Legend text for column #ncol
[Experimental data formatted in nrows and ncol with one
or more spaces between columns]
```

An example file is shown here:

```
21
3
V(3)
V(5)
V(1)
0.000E+00 4.622E+00 2.665E-01 0.000E+00
1.000E-08 3.463E+00 2.674E-01 5.000E+00
2.000E-08 8.559E-01 2.879E-01 5.000E+00
3.000E-08 2.847E-01 5.395E-01 5.000E+00
4.000E-08 2.715E-01 1.277E+00 5.000E+00
5.000E-08 2.679E-01 2.085E+00 5.000E+00
6.000E-08 2.667E-01 2.866E+00 5.000E+00
7.000E-08 2.662E-01 3.576E+00 5.000E+00
8.000E-08 2.660E-01 4.185E+00 5.000E+00
9.000E-08 2.852E-01 4.671E+00 0.000E+00
1.000E-07 6.477E-01 4.922E+00 0.000E+00
1.100E-07 1.558E+00 4.729E+00 0.000E+00
1.200E-07 2.522E+00 3.773E+00 0.000E+00
1.300E-07 3.419E+00 2.318E+00 0.000E+00
1.400E-07 4.172E+00 8.201E-01 0.000E+00
1.500E-07 4.572E+00 2.887E-01 0.000E+00
1.600E-07 4.606E+00 2.728E-01 0.000E+00
1.700E-07 4.612E+00 2.686E-01 0.000E+00
1.800E-07 4.616E+00 2.673E-01 0.000E+00
1.900E-07 4.618E+00 2.668E-01 0.000E+00
2.000E-07 4.620E+00 2.666E-01 0.000E+00
```

After the file is read, the data are displayed in the active graph window.

- *Export:* Lets you export data from an AIM-Spice data file or from the active window to a text file. If the active window is a graph window, the dialog box in Fig. A1.3.17 is shown. If you want to export from the currently loaded AIM-Spice data file, select the button labeled From raw data. The other op-

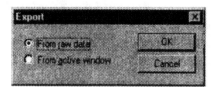

FIG. A1.3.17. Export dialog box.

tion, From active window, lets you pick data from the currently active graph window to save to a text file. If you select to export from the currently loaded AIM-Spice data file, the dialog in Fig. A1.3.18 appears. The list contains all the variables in the current plot. Select the variables you want to export and choose OK. A Save As dialog box is displayed that lets you specify a filename to export to.

If you select to export from the active graph window, the dialog box shown in Fig. A1.3.19 appears. The list contains all the traces in the active window. Select the traces you want to export and choose OK. A Save As dialog box is displayed that lets you specify a filename to export to.

- *Exit:* Terminates AIM-Postprocessor.

The Edit Menu. The Edit menu is displayed in Fig. A1.3.20.

- *Copy Graph:* This command copies the graph in the active windows to the clipboard.
- *Delete Text:* Deletes the selected text. This command can also be invoked with the accelerator key Del.

The Graph Menu. The Graph menu is displayed in Fig. A1.3.21.

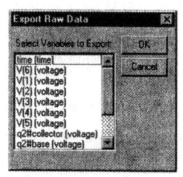

FIG. A1.3.18. Export Raw Data dialog box.

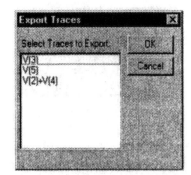

FIG. A1.3.19. Export Traces dialog box.

- *Add Plot:* Lets you create graphs. The command is described in detail in Section A1.3.1 on Creating Graphical Plots.
- *Add FFT Plot:* This command lets you create FFT (fast Fourier transform) graphs from the graph that was active when you chose the command. The command is useful for examining the spectrum of the output on nonlinear circuits. The dialog box in Fig. A1.3.22 is displayed in response of this command.

The drop-down list box contains a list of all traces contained in the active graph. One of the traces is selected [V(3) is selected in Fig. A1.3.22]. The other two fields in the dialog box display the number of data points in V(3) and the suggested number of samples to use when transforming V(3). AIM-Postprocessor transforms each trace individually and creates two new graphs for each trace. One contains the amplitude of the original trace as a function of frequency and the other contains the phase.

When AIM-Postprocessor transforms a trace, it creates a new set of data points where the number of points is equal to the number of samples specified in the dialog box. The data points are equally spaced in the time interval. This is done using linear interpolation on the original set of data points. The new set of data points is then transformed.

The resolution in frequency is the reciprocal of extent in time, and extent in frequency is proportional to the number of samples used. So, if you want higher resolution in the frequency domain, you have to run the transient analysis for a longer time in AIM-Spice. Run the circuit for many cycles if necessary.

FIG. A1.3.20. Edit menu.

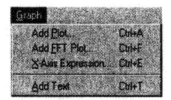

FIG. A1.3.21. Graph menu.

Note that AIM-Postprocessor can transform all kinds of traces, not only those that are functions of time. It is your responsibility to create meaningful transforms.

- *X-Axis Expression:* This command lets you change the x-axis expression. A dialog box similar to the Add Plot dialog box is displayed. The command is described in detail in Section A1.3.1 on Changing the X-Axis Expression.
- *Add Text:* This command lets you add text to your graphs. The command is described in detail in Section A1.3.1 on Adding and Formatting Text.

The Format Menu. The Format menu is displayed in Fig. A1.3.23.

- *X-Axis:* Lets you change the x-axis format of the graph in the active graph window. This command is described in detail in Section A1.3.1 on Formatting Axes.
- *Y-Axis:* Lets you change the y-axis format of the graph in the active graph window. This command is described in detail in Section A1.3.1 on Formatting Axes.
- *X-Label:* Lets you change the label format for the x axis of the graph in the active graph window. This command is described in detail in Section A1.3.1 on Formatting Labels.

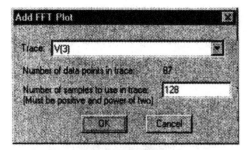

FIG. A1.3.22. Add FFT Plot dialog box.

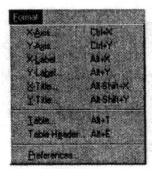

FIG. A1.3.23. Format menu.

- *Y-Label:* Lets you change the label format for the *y* axis of the graph in the active graph window. This command is described in Section A1.3.1 on Formatting Labels.
- *Table:* Lets you change the format of the table in the active table window. This command displays the dialog box shown in Fig. A1.3.24. The dialog box has the fields Fonts, Points, Style, Justification, and Borders discussed previously in conjunction with Figs. A1.3.9 and A1.3.10.
- *Table Header:* Lets you change the table header format of the table in the active table window. This command displays the dialog box shown in Fig. A1.3.25.

A table has three headers, one for each type of justification. AIM-Postprocessor uses the Center Header as default when creating tables. You can also choose to use

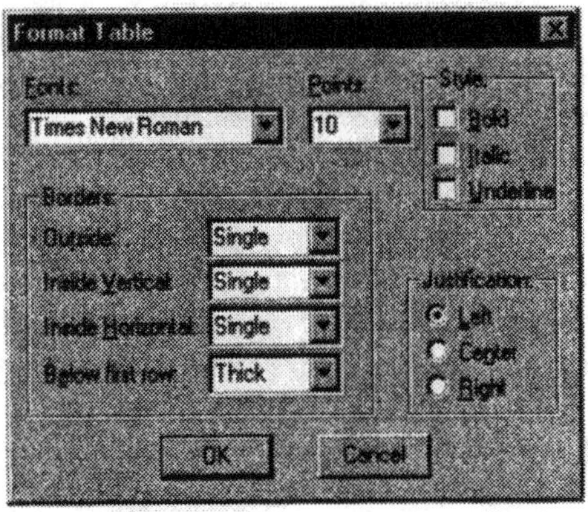

FIG. A1.3.24. Format Table dialog box.

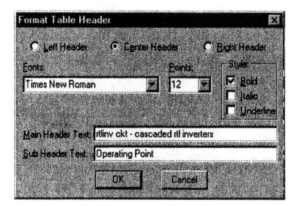

FIG. A1.3.25. Format Table Header dialog box.

headers with left or right justification. With one header selected, the dialog box has the following fields, in addition to Fonts, Points, and Style fields discussed in conjunction with Fig. A1.3.8:

Main Header Text   Contains the text displayed as the first line of the header.

Sub Header Text   Contains the text displayed as the second line of the header.

- *Preferences:* Lets you change preference values. This command displays the dialog box in Fig. A1.3.26.

  Your preferences will be saved on exit if "Save settings on exit" is active. The values are restored the next time you load AIM-Postprocessor.

The View Menu. The View menu is displayed in Fig. A1.3.27.

- *Plot Info:* Displays information on the current plot. This information is presented in two dialog boxes, the second of which is displayed when you select the Statistics command from the first dialog box.
- *Zoom:* This command lets you change axis limits with the mouse by drawing a rectangle that defines the limits of the axes. To perform the zooming, follow these steps: (1)Turn zooming on by selecting the View Zoom command. (2) Position the mouse cursor where you want the upper left corner of your view area to be. Press and hold the right mouse button. (3) Drag the mouse cursor down to the lower right corner of your view area and release the button.

  The graph is then redrawn with the new axis limits.

  When you turn zooming off by selecting the View Zoom command once more, you reset the axis limits to their previous values.
- *Show Cursors:* For a complete description of this command, see Section A1.3.1 on Cursors.

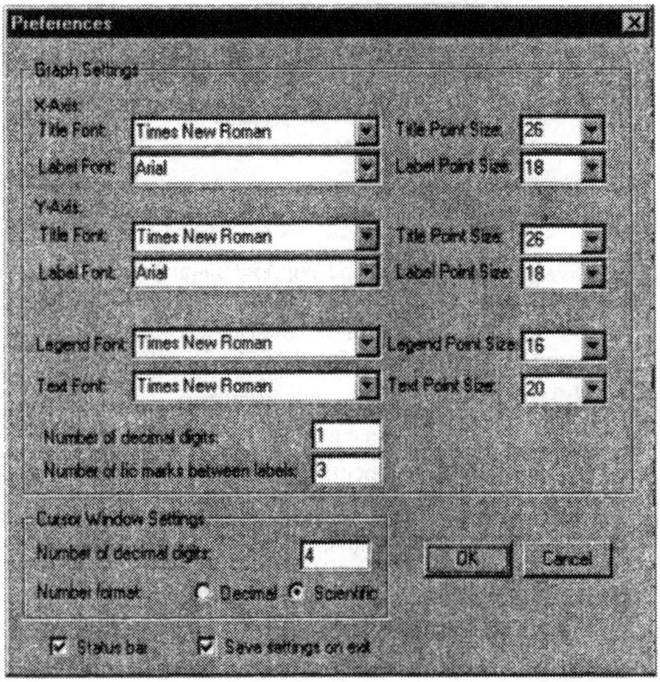

Preferences
Graph Settings
X-Axis:
Title Font: Times New Roman
Title Point Size: 26
Label Font: Arial
Label Point Size: 18
Y-Axis:
Title Font: Times New Roman
Title Point Size: 26
Label Font: Arial
Label Point Size: 18
Legend Font: Times New Roman
Legend Point Size: 16
Text Font: Times New Roman
Text Point Size: 20
Number of decimal digits: 1
Number of tic marks between labels: 3
Cursor Window Settings
Number of decimal digits: 4
Number format: Decimal Scientific
OK
Cancel
Status bar
Save settings on exit

FIG. A1.3.26. Preferences dialog box.

View
Plot Info... Ctrl+I
Zoom Ctrl+Z
Show Cursors Ctrl+U
Hide Legends Ctrl+L

FIG. A1.3.27. View menu.

Window
Tile Horizontal Alt+H
Tile Vertical Alt+V
Cascade Alt+C
Arrange Icons Alt+I
Close All Alt+A
1 V[3]
2 V[3]+V[5]
3 avg(V[3])

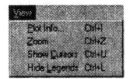

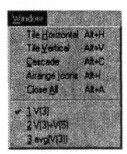

FIG. A1.3.28. Window menu.

FIG. A1.3.29. Help menu.

The Window Menu. Commands in the Window menu (Fig. A1.3.28) act on document windows and their icons only.

- *Tile Horizontal:* Tiles windows side by side horizontally.
- *Tile Vertical:* Tiles windows side by side vertically.
- *Cascade:* Stacks the windows in a cascade.
- *Arrange Icons:* Arranges the document icons at the bottom row of the main window.
- *Close All:* Closes all graph and table windows.

The menu items below the separator are the window titles of all the graph and table windows. When you select one of these titles, you make the corresponding graph or table window the active window. This window is placed in front of all the other windows.

The Help Menu. The Help menu is displayed in Fig. A1.3.29.
- *Contents:* Displays the on-line help manual.
- *About AIM-Postprocessor:* Displays the About dialog box.

REFERENCES

K. Lee, M. Shur, T. A. Fjeldly, and T. Ytterdal, *Semiconductor Device Modeling for VLSIs*, Prentice-Hall, Englewood Cliffs, NJ (1993).

L. W. Nagel, "SPICE 2: A Computer Program to Simulate Semiconductor Circuits," Memorandum No. ERL-M520, Electronic Research Laboratory, College of Engineering, University of California, Berkeley (1975).

T. Ytterdal et al., *AIM-Spice* (1983).

AIM-SPICE REFERENCE

A2.1. INTRODUCTION

This reference manual contains a complete listing of all analyses, options, device and control statements in AIM-Spice. Analyses and options are specified by choosing menu commands, and devices and control statements are specified in the circuit description. The reference first lists all analyses and options, then device and control statements.

Notation Used in this Reference

Item	Example	Description
Name	M12	A name field is an alphanumeric string. It must begin with a letter and cannot contain any delimiters.
Node	5000	A node field may be arbitrary character strings. The ground node must be named 0. Node names are treated as character strings, thus 0 and 000 are different names.
Scale scuffix		$T = 10^{12}$, $G = 10^{9}$, $MEG = 10^{6}$, $K = 10^{3}$, $MIL = 25.4 \times 10^{-6}$, $M = 10^{-3}$, $U = 10^{-6}$, $N = 10^{-9}$, $P = 10^{-12}$, $F = 10^{-15}$
Units suffix	V	Any letter that is not a scale factor or any letters that follows a scale suffix
Value	1KHz	Floating-point number with optional scale and/or units suffixes

Item	Example	Description
(text)	(option)	Comment
<Item>	<OFF>	Optional item
{Item}	{model}	Required item

A2.2. TYPES OF ANALYSES

AC ANALYSIS

AC Analysis is used to calculate the frequency response of a circuit over a range of frequencies.

DEC, OCT, and LIN stand for decade, octave variation, and linear variation, respectively. The specification of number of points, changes with the selection mode. If DEC is specified, the number of points is per decade. If OCT is specified, the number of points is per octave. If LIN is specified, the number of points is the total number across the whole frequency range. The frequency range is specified with the Start Frequency and End Frequency parameters. Note that in order for this analysis to be meaningful, at least one independent source must be specified with an ac value.

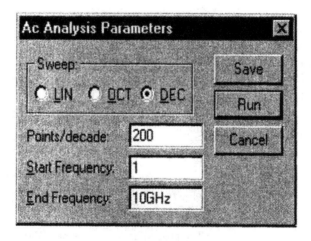

If the circuit has only one ac input, it is convenient to set that input to unity and zero phase. Then the output variable will be the transfer function of the output variable with respect to the input.

DC OPERATING POINT

This analysis calculates the dc operating point of a circuit. It has no parameters.

DC TRANSFER CURVE ANALYSIS

In a DC Transfer Curve Analysis, one or two sources (voltage or current sources) are swept over a user-defined interval. The dc operating point of the circuit is calculated for every value of the source(s).

Source Name is the name of an independent voltage or current source, Start Value, End Value, and Increment Value are the starting, final, and increment values, respectively. In the example below, the voltage source vds is swept from 0 to 5 V in increments of 0.1 V.

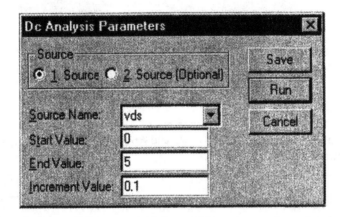

A second source may optionally be specified with associated sweep parameters. In such a case, the first source is swept over its range for each value of the second source. This option is useful for obtaining semiconductor device output characteristics.

NOISE ANALYSIS

Noise Analysis computes device-generated noise in a circuit. The Output Noise Variable parameter has the form V(OUTPUT<,REF>), where OUTPUT is the node at which the total output noise is desired. If REF is specified, the noise voltage V(OUTPUT) - V(REF) is calculated. By default, REF is assumed to be ground. In Input Source the name of an independent source parameter, to which input noise is referred, is specified. The next three parameters are the same as for AC Analysis. The last parameter is an optional integer; if specified, the noise contribution of each noise generator is produced every "Points per Summary" frequency point.

This analysis produces two plots. One for the noise spectral density curves and one for the total integrated noise over the specified frequency range. All noise voltages/currents are in squared units (V^2/Hz and A^2/Hz for spectral density, V^2 and A^2 for integrated noise).

POLE-ZERO ANALYSIS

Pole-Zero Analysis computes poles and/or zeros in the small-signal ac transfer function. You may instruct AIM-Spice to locate only poles or only zeros. This feature may allow one of the sets to be determined if there is a convergence problem with finding both.

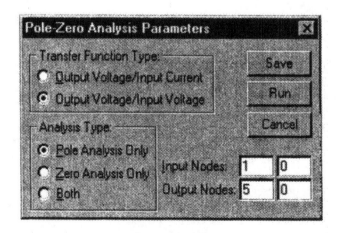

TRANSFER FUNCTION ANALYSIS

When you select this analysis, AIM-Spice computes the dc small-signal value of the transfer function, the input resistance, and the output resistance. In the example shown in the dialog box below, AIM-Spice will compute the ratio v(8)/vin, the small-signal input resistance at vin, and the small-signal output resistance measured across nodes 8 and 0.

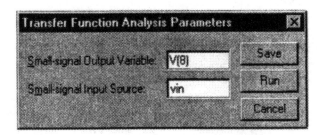

TRANSIENT ANALYSIS

Transient Analysis in AIM-Spice computes the time-domain response of a circuit. The parameter Stepsize is the suggested computing increment. Final Time is the last time point computed. Transient Analysis always starts at time zero. If you are not interested in the results until at a later time, you specify this time with the parameter Display Start Time. The parameter Maximum Stepsize is useful when you want to limit the internal stepsize used by AIM-Spice.

When selecting the Use Initial Conditions (UIC) option, you indicate that you do not want AIM-Spice to solve for the quiescent operating point before starting the Transient Analysis. Initial transient conditions can then be specified in the circuit description using an IC= control command (see Appendix A1). Alternatively, an .IC control line can be entered, specifying node voltages used to compute the initial conditions for the devices. (When UIC is not selected, the .IC and the IC= statements have no effect.)

OPTIONS

A set of options that controls different aspects of a simulation is available through the Options menu. The options are divided between the following four logical groups:

- General
- Analysis specific
- Device specific
- Numeric specific

Each option group has a separate dialog box. The options are listed in the table below. The numbers in parentheses indicate the group to which they belong.

Option	Description	Default
GMIN	Minimum allowed conductance (1)	1.0×10^{-12}
RELTOL	Relative error tolerance (1)	0.001
ABSTOL	Absolute current error tolerance (1)	1 nA
VNTOL	Absolute voltage error tolerance (1)	1 mV
CHGTOL	Charge tolerance (1)	1.0×10^{-14}
TNOM	Nominal temperature. The value can be overridden by a temperature specification on any temperature-dependent device model. (1)	27
TEMP	Operating temperature of the circuit. The value can be overridden by a temperature specification on any temperature-dependent instance. (1)	27
TRYTOCOMPACT	Applicable only to the LTRA model. When specified, the simulator tries to condense LTRA transmission lines past history of input voltages and currents. (1)	Not set
TRTOL	Transient analysis error tolerance (2)	7.0
ITL1	Maximum number of iterations in computing the dc operating point (2)	100
ITL2	Maximum number of iterations in dc transfer curve analysis (2)	50
ITL4	Transient analysis time point iteration limit (2)	10
DEFL	Default channel length for a MOS transistor (3)	100 μm
DEFW	Default channel width for a MOS transistor (3)	100 μm
DEFAD	Default drain diffusion area for a MOS transistor (3)	0.0
DEFAS	Default source diffusion area for a MOS transistor (3)	0.0
PIVTOL	Minimum value for an element to be accepted as a pivot element. (4)	1.0×10^{-13}

Option	Description	Default
PIVREL	The minimum relative ratio between the largest element in the column and an accepted pivot element (4)	1.0×10^{-13}
METHOD	Sets the numerical integration method used by AIM-Spice. Possible methods are Gear or Trapezoidal. (4)	Trap

---

TITLE LINE

General Form

Any text.

Example

```
SIMPLE DIFFERENTIAL PAIR
MOS OPERATIONAL AMPLIFIER
```

The title line must be the first line in the circuit description.

COMMENT LINE

General Form

* (Any text.)

Example

```
* MAIN CIRCUIT STARTS HERE
```

An asterisk in the first column indicates that this line is a comment line. Comment lines may be placed anywhere in the circuit description.

A2.3. MODEL PARAMETER SPECIFICATIONS

A   HETEROSTRUCTURE FIELD-EFFECT TRANSISTORS (HFETS)

General Form

```
AXXXXXXX ND NG NS MNAME <L=VALUE> <W=VALUE> <OFF>
+ <IC=VDS,VGS>
```

Example

```
al 7 2 3 hfeta l=1u w=10u
```

ND, NG, and NS are the drain, gate, and source nodes, respectively. MNAME is the model name, L is the channel length, W is the channel width, and OFF indicates an optional initial value for the element in a dc analysis. The optional initial value IC=VDS, VGS is meant to be used together with UIC in a transient analysis. See the description of the .IC control statement for a better way to set transient initial conditions. If length and/or width is not specified, AIM-Spice will use default values, $L = 1$ μm and $W = 20$ μm.

HFET Model

```
.MODEL {model name} NHFET <model parameters>
.MODEL {model name} PHFET <model parameters>
```

The HFET model is a unified extrinsic model as described in Section 6.5 and in the book by Lee et al. (1993). The model parameters are listed below. Note that the default values used correspond to the n-channel device considered in Lee et al. (1993).

Name	Parameter	Unit	Default
VTO	Pinch-off voltage	V	0.15
RD	Drain ohmic resistance	Ω	0
RS	Source ohmic resistance	Ω	0
RDI	Internal drain ohmic resistance	Ω	0
RSI	Internal source ohmic resistance	Ω	0
DI	Thickness of interface layer	m	0.04×10^{-6}
LAMBDA	Output conductance parameter	V^{-1}	0.15
ETA	Subthreshold ideality factor	—	1.28
M	Knee shape parameter	—	3
MC	Capacitance transition parameter	—	3
GAMMA	Capacitance parameter	—	3
SIGMA0	DIBL parameter	—	0.057
VSIGMAT	DIBL parameter	V	0.3
VSIGMA	DIBL parameter	V	0.1
MU	low-field mobility	$m^2/V\ s$	0.4
DELTA	Transition width parameter	—	3
VS	Saturation velocity	m/s	1.5×10^5
NMAX	Maximum sheet charge density in the channel	m^{-2}	2×10^{16}
DELTAD	Thickness correction	m	4.5×10^{-9}
EPSI	Dielectric constant for interface layer	F/m	1.084×10^{-10}
JS1D	Forward gate drain diode saturation current density	A/m^2	1.0

Name	Parameter	Unit	Default
JS2D	Reverse gate drain diode saturation current density	A/m^2	1.15×10^6
JS1S	Forward gate source diode saturation current density	A/m^2	1.0
JS2S	Reverse gate source diode saturation current density	A/m^2	1.15×10^6
M1D	Forward gate drain diode ideality factor	—	1.32
M2D	Reverse gate drain diode ideality factor	—	6.9
M1S	Forward gate source diode ideality factor	—	1.32
M2S	Reverse gate source diode ideality factor	—	6.9
RGD	Gate–drain ohmic resistance	Ω	90
RGS	Gate–source ohmic resistance	Ω	90
ALPHAG	Drain–source correction current gain	—	0

Either intrinsic or extrinsic models can be selected by proper use of the parameters RD, RS, RDI, RSI. If values for RD and RS are specified, the intrinsic model is selected with parasitic resistances applied externally. The extrinsic model is selected by specifying values for RDI and RSI.

Supported Analyses

Distortion, Noise, and Pole-Zero Analysis not supported.

B NONLINEAR DEPENDENT SOURCES

General Form

```
BXXXXXXX N+ N- <I=EXPR> <V=EXPR>
```

Example

```
b1 0 1 i=cos(v(1))+sin(v(2))
b1 0 1 v=ln(cos(log(v(1,2)^2)))-v(3)^4+v(2)^v(1)
b1 3 4 i=17
b1 3 4 v=exp(pi^i(vdd))
```

N+ and N- are the positive and negative nodes, respectively. The values of the V and I parameters determine the voltages and currents across and through the device, respectively. If I is given, then the device is a current source, and if V is given, the device is a voltage source. One and only one of these parameters must be given.

During an AC Analysis, the source acts as a linear dependent source with a proportionality constant equal to the derivative of the source at the dc operating point.

The expressions given for V and I may be any function of node voltages and/or currents through voltage sources in the system. The following are allowed functions of real variables:

abs	asinh	cosh	sin
acos	atan	exp	sinh
acosh	atanh	ln	sqrt
asin	cos	log	tan

The following operations are defined:

$$+ \quad - \quad * \quad / \quad \wedge \quad . \quad \text{unary} \, -$$

If the argument of log, ln, or sqrt becomes less than zero, the absolute value of the argument is used. If a divisor becomes zero or the argument of log or ln becomes zero, an error will result. Other problems may occur when the argument of a function in a partial derivative enters a region where that function is undefined.

To introduce time into an expression, you can integrate the current from a constant-current source with a capacitor and use the resulting voltage. For a correct result, you have to set the initial voltage across the capacitor.

Nonlinear capacitors, resistors, and inductors may be synthesized using nonlinear dependent sources. Here is an example on how to implement a nonlinear capacitor:

```
* Bx: calculate f(input voltage)
Bx 1 0 v=f(v(pos,neg))
* Cx: linear capacitance
Cx 2 0 1
* Vx: Ammeter to measure current into the capacitor
Vx 2 1 DC 0 Volts
* Drive the current through Cx back into the circuit
Fx pos neg Vx 1
```

Supported Analyses

All.

C CAPACITORS

General Form

```
CXXXXXXX N+ N- VALUE <IC=Initial values>
```

Examples

```
cl 66 0 70pf
CBYP 17 23 10U IC=3V
```

N+ and N− are the positive and negative element nodes, respectively. VALUE is the capacitance in farads.

The optional initial value is the initial time-zero value of the capacitor voltage in volts. Note that the value is used only when the option UIC is specified in a transient analysis.

Semiconductor Capacitors

General Form

```
CXXXXXXX N1 N2 <VALUE> <MNAME> <L=LENGTH> <W=WIDTH>
+ <IC=VALUE>
```

Examples

```
CMOD 3 7 CMODEL L=10U W=1U
```

This is a more general model for the capacitor than the one presented above. It gives you the possibility of modeling temperature effects and calculating capacitance values based on geometric and process information. VALUE, if given, defines the capacitance, and information on geometry and process will be ignored. If MNAME is specified, the capacitance value is calculated based on information on process and geometry. If VALUE is not given, then MNAME and LENGTH must be specified. If WIDTH is not given, then the model default width will be used.

Capacitor Model

```
.MODEL {model name} C <model parameters>
```

The model allows calculation of the capacitance value based on information on geometry and process by the expression

```
C=CJ · (L−NARROW) · (W−NARROW) +2CJSW · (L+W−2 · NARROW)
```

where the parameters are defined below.

Name	Parameter	Unit	Default
CJ	Junction bottom capacitance	F/m^2	
CJSW	Junction side-wall capacitance	F/m	
DEFW	Default width	m	1×10^{-6}
NARROW	Narrowing due to side etching	m	0.0

Supported Analyses

All.

D DIODES

General Form

```
DXXXXXXX N+ N- N MNAME <AREA> <OFF> <IC=VD> <TEMP=T>
```

Examples

```
DBRIDGE 2 10 DIODE1
DCLMP 3 7 DMOD 3.0 IC=0.2
```

N+ and N- are the positive and negative nodes, respectively. MNAME is the model name, AREA is the area factor, and OFF indicates an optional initial value during a DC Analysis. If the area factor is not given, 1 is assumed. The optional initial value IC=VD is meant to be used together with a UIC in a Transient Analysis. The optional TEMP value is the temperature at which this device operates. It overrides the temperature specified as an option.

Diode Model

```
.MODEL {model name} D <model parameters>
```

AIM-Spice has two diode models. Level 1, the default model, is an expanded version of the standard diode model supplied from Berkeley (extended to include high-level injection and generation/recombination current; see Section 3.2). Level 2 is a GaAs/AlGaAs heterostructure diode model described in Section 3.3. To select the heterostructure diode model, specify LEVEL=2 on the model line.

Level 1 model parameters are as follows:

Name	Parameter	Unit	Default
IS	Saturation current (Level 1 only)	A	1.0×10^{-14}
RS	Ohmic resistance	Ω	0
N	Emission coefficient	—	1
TT	Transit time	s	0
CJO	Zero-bias junction capacitance	F	0
VJ	Junction potential	V	1
M	Grading coefficient	—	0.5
EG	Activation energy	eV	1.11
IKF	Corner for high-injection current roll-off	A	Infinite
ISR	Recombination saturation current	A	0
NR	Recombination emission coefficient	—	2
XTI	Saturation current temperature exponent	—	3.0

Name	Parameter	Unit	Default
KF	Flicker noise coefficient	—	0
AF	Flicker noise exponent	—	1
FC	Coefficient for forward-bias depletion capacitance formula	—	0.5
BV	Reverse-breakdown voltage	V	Infinite
IBV	Current at breakdown voltage	A	1.0×10^{-3}
TNOM	Parameter measurement temperature	°C	27

Level 2 model parameters are as follows (in addition to those for Level 1):

Name	Parameter	Unit	Default
DN	Diffusion constant for electrons	m²/s	0.02
DP	Diffusion constant for holes	m²/s	0.000942
LN	Diffusion length for electrons	m	7.21×10^{-7}
LP	Diffusion length for holes	m	8.681×10^{-7}
ND	Donor doping density	m⁻³	7.0×10^{24}
NA	Acceptor doping density	m⁻³	3×10^{22}
DELTAEC	Conduction band discontinuity	eV	0.6
XP	p-Region width	m	1 μm
XN	n-Region width	m	1 μm
EPSP	Dielectric constant on p-side	F/m	1.0593×10^{-10}
EPSN	Dielectric constant on n-side	F/m	1.1594×10^{-10}

Temperature Effects

Temperature appears explicitly in the exponential terms.

Temperature dependence of the saturation current in the junction diode model is determined by

$$IS(T_1) = IS(T_0)\left(\frac{T_1}{T_0}\right)^{XTI/N} \exp\left(\frac{qEG(T_1 - T_0)}{NkT_1T_0}\right)$$

where k is the Boltzmann constant, q is the electronic charge, EG is the energy gap parameter (in electron-volts), XTI is the saturation current temperature exponent, and N is the emission coefficient. The last three quantities are model parameters.

For Schottky barrier diodes, the value for XTI is usually 2.

Supported Analyses

All.

E   LINEAR VOLTAGE-CONTROLLED VOLTAGE SOURCES

General Form

```
EXXXXXXX N+ N- NC+ NC- VALUE
```

Example

```
E1 2 3 14 1 2.0
```

N+ and N- are the positive and negative nodes, respectively. NC+ and NC- are the positive and negative controlling nodes, respectively. VALUE is the voltage gain.

Supported Analyses

All.

F   LINEAR CURRENT-CONTROLLED CURRENT SOURCES

General Form

```
FXXXXXXX N+ N- VNAME VALUE
```

Example

```
F1 14 7 VIN 5
```

N+ and N- are the positive and negative nodes, respectively. Current flows from the positive node through the source to the negative node. VNAME is the name of the voltage source where the controlling current flows. The direction of positive control current is from the positive node through the source to the negative node of VNAME. VALUE is the current gain.

Supported Analyses

All.

G   LINEAR VOLTAGE-CONTROLLED CURRENT SOURCES

General Form

```
GXXXXXXX N+ N- NC+ NC- VALUE
```

Example

```
G1 2 0 5 0 0.1MMHO
```

N+ and N− are the positive and negative nodes, respectively. Current flows from the positive node through the source to the negative node. NC+ and NC− are the positive and negative controlling nodes, respectively. VALUE is the transconductance in mhos.

Supported Analyses

All.

H   LINEAR CURRENT-CONTROLLED VOLTAGE SOURCES

General Form

```
HXXXXXXX N+ N− VNAME VALUE
```

Example

```
HX1 6 2 Vz 0.5K
```

N+ and N− are the positive and negative nodes, respectively. Current flows from the positive node through the source to the negative node. VNAME is the name of the voltage source where the controlling current flows. The direction of positive control current is from the positive node through the source to the negative node of VNAME. VALUE is the transresistance in ohms.

Supported Analyses

All.

I   INDEPENDENT CURRENT SOURCES

General Form

```
IYYYYYYY N+ N− <<DC>DC/TRAN VALUE> <AC<ACMAG <ACPHASE>>>
+ <DISTOF1 <F1MAG <F1PHASE>> <DISTOF2 <F2MAG <F2PHASE>>>
```

Examples

```
isrc 23 21 ac 0.333 45.0 sffm(0 1 10K 5 1k)
```

N+ and N- are the positive and negative nodes, respectively. Positive current flows from the positive node through the source to the negative node.

DC/TRAN is the source value during a DC or a Transient Analysis. The value can be omitted if it is zero for both analyses. If the source is time invariant, its value can be prefixed with DC.

ACMAG is the amplitude and ACPHASE is the phase of the source during an AC Analysis. If ACMAG is omitted after the keyword AC, 1 is assumed. If ACPHASE is omitted, 0 is assumed.

All independent sources can be assigned time-varying values during a Transient Analysis. If a source is assigned a time-varying value, its value at $t = 0$ is used during a DC Analysis. There are five predefined functions for time-varying sources: pulse, exponent, sine, piecewise linear, and single-frequency FM. If the parameters are omitted, the default values shown below will be assumed. DT and T2 are increment time and final time in a Transient Analysis, respectively.

Pulse

General Form

PULSE(I1 I2 TD TR TF PW PER)

Parameter	Default	Unit
I1 (initial value)	None	A
I2 (pulsed value)	None	A
TD (delay time)	0.0	s
TR (rise time)	DT	s
TF (fall time)	DT	s
PW (pulse width)	T2	s
PER (period)	T2	s

Example

IB 3 0 PULSE(1 5 1S 0.1S 0.4S 0.5S 2S)

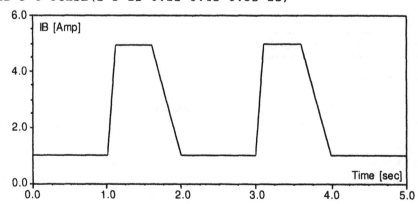

Sine

General Form

```
SIN(IO IA FREQ TD THETA)
```

Parameter	Default	Unit
IO (offset)	None	A
IA (amplitude)	None	A
FREQ (frequency)	1/T2	Hz
TD (delay)	0.0	s
THETA (attenuation factor)	0.0	s^{-1}

The shape of the waveform is as follows:

$0 < time < T_D$

$I = I_0$

$T_D < time < T_2$

$I = I_0 + I_A \sin[2\pi \cdot \text{FREQ} \cdot (time + T_D)]\exp[-(time - T_D) \cdot \text{THETA}]$

Example

```
IB 3 0 SIN(2 2 5 1S 1)
```

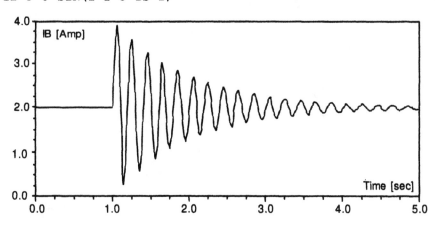

Exponent

General Form

```
EXP(I1 I2 TD1 TAU1 TD2 TAU2)
```

Parameter	Default	Unit
I1 (initial phase)	None	A
IA (pulsed value)	None	A
TD1 (rise delay time)	0.0	s
TAU1 (rise time constant)	DT	s
TD2 (delay fall time)	TD1 + DT	s
TAU2 (fall time constant)	DT	s

The shape of the waveform is as follows:

$0 < time < T_{D1}$

$$I = I_1$$

$T_{D1} < time < T_{D2}$

$$I = I_1 + (I_2 - I_1)\{1 - \exp[-(time - T_{D1}) / \text{TAU}_1]\}$$

$T_{D2} < time < T_2$

$$I = I_1 + (I_2 - I_1)\{1 - \exp[-(time - T_{D1}) / \text{TAU}_1]\}$$
$$+ (I_1 - I_2)\{1 - \exp[-(time - T_{D2}) / \text{TAU}_2]\}$$

Example

```
IB 3 0 EXP(1 5 1S 0.2S 2S 0.5S)
```

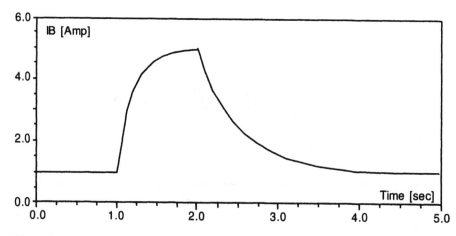

Piecewise Linear

General Form

```
PWL(T1 I1 <T2 I2 T3 I3 T4 I4 T5 I5 ....>)
```

Parameters and Default Values

Pairs of values (T_i, I_i) specify the value of the source, I_i, at T_i. The value of the source between these values is calculated using a linear interpolation.

Example

```
IB 7 5 PWL(0 0 1 0 1.2 4 1.6 2.0 2.0 5.0 3.0 1.0)
```

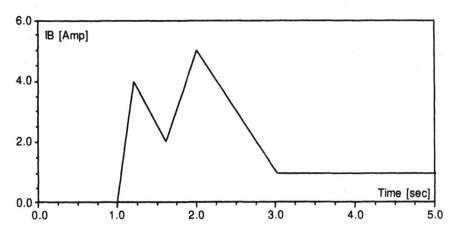

Single-Frequency FM

General Form

```
SFFM (IO IA FC MDI FS)
```

Parameter	Default	Unit
IO (offset)	None	A
IA (amplitude)	None	A
FC (carrier frequency)	1/T2	Hz
MDI (modulation index)	None	—
FS (signal frequency)	1/T2	Hz

The shape of the waveform is

$$I = I_0 + I_A \sin[(2\pi F_C \cdot \text{time}) + \text{MDI} \cdot \sin(2\pi F_S \cdot \text{time})]$$

Example

```
IB 12 0 SFFM(0 1M 20K 5 1K)
```

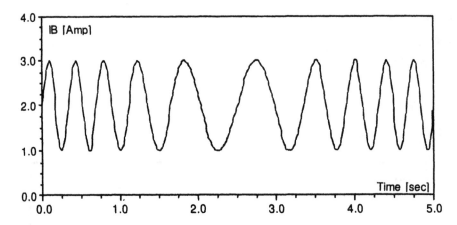

Supported Analyses

All.

J   JUNCTION FIELD-EFFECT TRANSISTORS (JFETS)

General Form

```
JXXXXXXX ND NG NS MNAME <AREA> <OFF> <IC=VDS,VGS> <TEMP=T>
```

Example

```
J1 7 2 3 JM1 OFF
```

ND, NG, and NS are the drain, gate, and source nodes, respectively. MNAME is the model name, AREA is the area factor, and OFF indicates an optional initial value for the element in a DC Analysis. If the area factor is omitted, 1.0 is assumed. The optional initial value IC=VDS,VGS is meant to be used together with UIC in a transient analysis. See the description of the .IC control statement for a better way to set transient initial conditions. The optional TEMP value is the temperature at which this device operates. It overrides the temperature specified in the option value.

JFET Model

```
.MODEL {model name} NJF <model parameters>
.MODEL {model name} PJF <model parameters>
```

Name	Parameter	Unit	Default
VTO	Threshold voltage	V	−2.0
BETA	Transconductance parameter	A/V^2	1.0×10^{-4}
LAMBDA	Channel length modulation parameter	1/V	0
RD	Drain resistance	Ω	0
RS	Source resistance	Ω	0
CGS	Zero-bias gate–source junction capacitance	F	0
CGD	Zero-bias gate–drain junction capacitance	F	0
PB	Gate junction potential	V	1
IS	Gate junction saturation current	A	1.0×10^{-14}
FC	Coefficient for forward-bias depletion	—	0.5
TNOM	Parameter measurement temperature	°C	27

Temperature Effects

The temperature appears explicitly in the exponential terms.

The temperature dependence of the saturation current in the two gate junctions of the model is determined by

$$I_S(T_1) = I_S(T_0) \exp\left(\frac{1.11(T_1/T_0 - 1)}{V_{\text{th}}} \right)$$

where V_{th} is the thermal voltage.

Supported Analyses

All.

K COUPLED INDUCTORS (TRANSFORMERS)

General Form

```
KXXXXXXX LYYYYYYY LZZZZZZZ VALUE
```

Examples

```
k43 laa lbb 0.9999
kxfrmr 11 12 0.82
```

LYYYYYYY and LZZZZZZZ are the names of the two coupled inductors, and VALUE is the coupling coefficient K, which must be greater than 0 and less than or equal to 1. Using the dot convention, place a dot on the first node of each inductor. The relation between the coupling coefficient and the mutual inductance is given by $K = M_{ij} / \sqrt{L_i L_j}$, where L_i and L_j are the coupled pair of inductors and M_{ij} is the mutual inductance between L_i and L_j.

Supported Analyses

All.

L   INDUCTORS

General Form

```
LYYYYYYY N+ N- VALUE <IC=Initial values>
```

Examples

```
llink 42 69 1uh
lshunt 23 51 10u ic=15.7ma
```

N+ and N− are the positive and negative element nodes, respectively. VALUE is the inductance in henrys. The optional initial value is the $t = 0$ value of the inductor current in amperes that flows from N+ through the inductor to N−. Notice that the value is used only when the option UIC is specified in a Transient Analysis.

Supported Analyses

All.

M   MOSFETs

General Form

```
MXXXXXXX ND NG NS NB MNAME <L=VALUE> <W=VALUE> <AD=VALUE>
+ <AS=VALUE> <PD=VALUE> <PS=VALUE> <NRD=VALUE>
+ <NRS=VALUE> <OFF> <IC=VDS,VGS,VBS> <TEMP=T>
```

Example

```
M1 24 2 0 20 TYPE1
m15 15 15 12 32 m w=12.7u 1=207.8u
M1 2 9 3 0 MOD1 L=10U W=5U AD=100P AS=100P PD=40U PS=40U
```

ND, NG, NS, and NB are the drain, gate, source, and bulk (substrate) nodes, respectively. MNAME is the model name, L and W are the channel length and width in meters, respectively. AD and AS are the drain and source diffusion areas in square meters. If any of L, W, AD, or AS are not specified, default values are used, PD and PS are the perimeters of the drain and source diffusion areas. NRD and NRS are the relative resistivities of drain and source in number of squares, respectively. Default values of PD and PS are 0.0, while default values of NRD and NRS are 1.0. OFF indicates an optional initial value for the element in a DC Analysis. The optional initial value IC=VDS, VGS, VBS is meant to be used together with UIC in a Transient Analysis. See the description of the .IC control statement for a better way to set transient initial conditions. The optional TEMP value is the temperature at which this device operates. It overrides the temperature specified in the option value.

Note: The substrate node is ignored in Levels 11 and 12, and so is the VBS initial voltage.

MOSFET Model

```
.MODEL {model name} NMOS <model parameters>
.MODEL {model name} PMOS <model parameters>
```

AIM-Spice supports 12 MOSFET models. The parameter LEVEL selects which model to use. The default is LEVEL=1.

LEVEL=1   Schichman–Hodges (described in Section 5.3)

LEVEL=2   Geometric based analytical model (described in Section 5.3)

LEVEL=3   Semiempirical short-channel model (described in Section 5.3)

LEVEL=4   BSIM (Berkeley Short Channel IGFET Model)

LEVEL=5   BSIM2 (described in Jeng, 1990)

LEVEL=6   MOS6 (described in Sakurai and Newton, 1990)

LEVEL=7   Universal extrinsic short-channel MOS model (described in Section 6.3)

LEVEL=8   Unified long-channel MOS model (described in Lee et al. 1993, Sections 3.10 and 3.11)

LEVEL=9   Short-channel MOS model (described in Lee et al., 1993, Sections 3.10 and 3.11)

LEVEL=10   Unified intrinsic short-channel model (described in Lee et al., 1993, Sections 3.10 and 3.11)

LEVEL11   Unified extrinsic amorphous silicon thin-film transistor model 1 (described in Lee et al., 1993, Section 5.2)

LEVEL=12   Polysilicon thin-film transistors model (described in Section 6.7)

LEVEL=14   BSIM3 (described in Section 6.8)

LEVEL=15   Unified extrinsic amorphous silicon thin-film transistor model 2 (described in Section 6.6)

LEVEL=16   Polysilicon thin-film transistors model 2 (described in Shur et al., 1995)

Effects of charge storage based on the model by Meyer is implemented in Levels 1–3, 6–12, 15, and 16. In the universal MOSFET model (Level 7), a second, unified charge storage model based on the charge-conserving Meyer-like approach proposed by Turchetti et al. (1986) is implemented (see Chapter 6). The BSIM models (Levels 4, 5, and 14) use charge based models owing to Ward and Dutton (1978).

Effects of a thin-oxide capacitance are treated slightly differently in Level 1. Voltage-dependent capacitances are included only if TOX is specified.

A redundancy exists in specifying junction parameters. For example, the reverse current can be specified either with the IS parameter (in amperes) or with JS (in amperes per square meter). The first choice is an absolute value while the second choice is multiplied by AD and AS to give the reverse current at the drain and source junctions, respectively. The latter approach is preferred. The same is also true for the parameters CBD, CBS, and CJ. Parasitic resistances can be specified in terms of RD and RS (in ohms) or RSH (in ohms/square). RSH is multiplied by number of squares NRD and NRS.

The following are parameters for Levels 1, 2, 3 and 6:

Name	Parameter	Unit	Default
VTO	Zero-bias threshold voltage	V	0.0
KP	Transconductance parameter	A/V^2	2.0×10^{-5}
GAMMA	Bulk threshold parameter	$\sqrt{V}$	0.0
PHI	Surface potential	V	0.6
LAMBDA	Channel length modulation (only MOS1 and MOS2)	1/V	0.0
RD	Drain resistance	Ω	0.0
RS	Source resistance	Ω	0.0
CBD	Zero-bias B-D junction capacitance	F	0.0
CBS	Zero-bias B-S junction capacitance	F	0.0
IS	Bulk junction saturation current	A	1.0×10^{-14}
PB	Bulk junction potential	V	0.8
CGSO	Gate–source overlap capacitance per meter channel width	F/m	0.0
CGDO	Gate–drain overlap capacitance per meter channel width	F/m	0.0
CGBO	Gate–bulk overlap capacitance per meter channel width	F/m	0.0
RSH	Drain and source diffusion sheet resistance	Ω/sq.	0.0
CJ	Zero-bias bulk junction bottom capacitance per square meter of junction area	F/m^2	0.0
MJ	Bulk junction bottom grading coefficient	—	0.5
CJSW	Zero-bias bulk junction sidewall capacitance per meter of junction perimeter	F/m	0.0
MJSW	Bulk junction sidewall grading coefficient	—	0.50 (Level 1) 0.33 (Level 2)

Name	Parameter	Unit	Default
JS	Bulk junction saturation current per square meter of junction area	A/m^2	0
TOX	Gate oxide thickness	m	1.0×10^{-7}
NSUB	Substrate doping	cm^{-3}	0.0
NSS	Surface state density	cm^{-2}	0.0
NFS	Fast surface state density	cm^{-2}	0.0
TPG	Type of gate material (+1, opposite of substrate; −1, same as substrate; 0, Al gate)	—	1.0
XJ	Metallurgical junction depth	m	0.0
LD	Lateral diffusion	m	0.0
U0	Surface mobility	cm^2/V s	600
UCRIT	Critical field for mobility degradation (only MOS2)	V/cm	1.0×10^4
UEXP	Critical field exponent in mobility degradation (only MOS2)	—	0.0
UTRA	Transverse field coefficient (deleted for MOS2)	—	0.0
VMAX	Maximum drift velocity for carriers	m/s	0.0
NEFF	Total channel charge (fixed and mobile) coefficient (only MOS2)	—	1.0
KF	Flicker noise coefficient	—	0.0
AF	Flicker noise exponent	—	1.0
FC	Coefficient for forward-bias depletion capacitance formula	—	0.5
DELTA	Width effect on threshold voltage (only MOS2 and MOS3)	—	0.0
THETA	Mobility modulation (MOS3 only)	V^{-1}	0.0
ETA	Static feedback (only MOS3)	—	0.0
KAPPA	Saturation field factor (MOS3 only)	—	0.2
TNOM	Parameter measurement temperature	°C	27

MOSFET Levels 4, 5, and 14 are BSIM models (Berkeley Short Channel IGFET Models). The parameters for these models should be obtained from process characterization.

Parameters marked with an asterisk in the l/w column in the tables below have length and width dependency (only Levels 4 and 5). For example, for the flat-band voltage VFB the dependence on the gate electrode geometry can be expressed in terms of the additional flat band parameters, LVFB and WVFB, measured in volt micrometers:

$$VFB = VFB0 + \frac{LVFB}{L_{\text{effective}}} + \frac{WVFB}{W_{\text{effective}}}$$

where

$$L_{\text{effective}} = L_{\text{input}} - \text{DL}$$

$$W_{\text{effective}} = W_{\text{input}} - \text{DW}$$

Note that BSIM 1 and 2 models are meant to be used together with a process characterization system. None of the parameters in these models have default values, and leaving one out is registered as an error.

The following are model parameters for Levels 4 and 5:

Name	Parameter	Unit	l/w
TOX	Gate oxide thickness (only Levels 4 and 5)	μm	
VFB	Flat-band voltage	V	
PHI	Surface inversion potential	V	*
K1	Body effect coefficient	$V^{\frac{1}{2}}$	*
K2	Drain–source depletion charge-sharing coefficient	—	*
DL	Shortening of channel (only Levels 4 and 5)	μm	
DW	Narrowing of channel (only Levels 4 and 5)	μm	
N0	Zero-bias subthreshold slope coefficient (only Levels 4 and 5)	—	*
NB	Sensitivity of subthreshold slope to substrate bias	—	*
ND	Sensitivity of subthreshold slope to drain bias (only Levels 4 and 5)	—	*
TEMP	Temperature at which parameters were measured (TNOM for Level 14)	C	
VDD	Measurement bias range (only Levels 4 and 5)	V	
CGDO	Gate–drain overlap capacitance per meter channel width	F/m	
CGSO	Gate–source overlap capacitance per meter channel width	F/m	
CGBO	Gate–bulk overlap capacitance per meter channel width	F/m	
XPART	Gate–oxide capacitance charge model flag	—	
RSH	Drain and source diffusion sheet resistance	Ω/sq.	
JS	Source–drain junction current density	A/m^2	
PB	Built-in potential of source–drain junction	V	
MJ	Grading coefficient of source–drain junction	—	
PBSW	Built-in potential of source–drain junction sidewall	V	
MJSW	Grading coefficient of source–drain junction sidewall	—	
CJ	Source–drain junction capacitance per unit area	F/m^2	
CJSW	Source–drain junction sidewall capacitance per unit length	F/m	
WDF	Source–drain junction default width (only Levels 4 and 5)	m	
DELL	Source–drain junction length reduction (only Levels 4 and 5)	m	

XPART=0 selects a 40/60 drain–source partitioning of the gate charge in saturation, while XPART=1 selects a 0/100 drain–source charge partitioning (see Chapter 6).

The following are specific BSIM1 (Level 4) parameters:

Name	Parameter	Unit	l/w
ETA	Zero-bias drain-induced barrier-lowering coefficient	—	*
MUZ	Zero-bias mobility	cm^2/V s	
U0	Zero-bias transverse-field mobility degradation coefficient		*
U1	Zero-bias velocity saturation coefficient	μm/V	*
X2MZ	Sensitivity of mobility to substrate bias at V_{ds}=0	cm^2/V^2 s	*
X2E	Sensitivity of drain-induced barrier-lowering effect to substrate bias		*
X3E	Sensitivity of drain-induced barrier-lowering effect to drain bias at V_{ds}=V_{dd}		*
X2U0	Sensitivity of transverse-field mobility degradation effect to substrate bias	V^{-2}	*
X2U1	Sensitivity of velocity saturation effect to substrate bias	μm/V^2	*
MUS	Mobility at zero substrate bias and at V_{ds}=V_{dd}	cm^2/V^2 s	
X2MS	Sensitivity of mobility to substrate bias at V_{ds}=V_{dd}	cm^2/V^2 s	*
X3MS	Sensitivity of mobility to drain bias at V_{ds}=V_{dd}	μm/V^2	*
X3U1	Sensitivity of velocity saturation effect on drain bias at V_{ds}=V_{dd}	μm/V^2	*

The following are specific BSIM2 (Level 5) parameters:

Name	Parameter	Unit	l/w
ETA0	V_{ds} dependence of threshold voltage at V_{ds}=0 V	—	*
ETAB	V_{bs} dependence of ETA	V^{-1}	*
MU0	Low-field mobility at V_{ds}=0 V, V_{gs}=V_{th}	cm^2/V s	
MU0B	V_{bs} dependence of low-field mobility	cm^2/V^2 s	*
MUS0	Mobility at V_{ds}=V_{ds}, V_{gs}=V_{th}	cm^2/V s	*
MUSB	V_{bs} dependence of MUS0	cm^2/V^2 s	*
MU20	V_{ds} dependence of mobility in tanh term	—	*
MU2B	V_{bs} dependence of MU2	V^{-1}	*
MU2G	V_{gs} dependence of MU2	V^{-1}	*
MU30	V_{ds} dependence of mobility in linear term	cm^2/V^2 s	*
MU3B	V_{bs} dependence of MU3	cm^2/V^3 s	*
MU3G	V_{gs} dependence of MU3	cm^2/V^3 s	*
MU40	V_{ds} dependence of mobility in linear term	cm^2/V^3 s	*
MU4B	V_{bs} dependence of MU4	cm^2/V^4 s	*
MU4G	V_{gs} dependence of MU4	cm^2/V^4 s	*

Name	Parameter	Unit	l/w
UA0	Linear V_{gs} dependence of mobility	V^{-1}	*
UAB	V_{bs} dependence of UA	V^{-2}	*
UB0	Quadratic V_{gs} dependence on mobility	V^{-2}	*
UBB	V_{bs} dependence on UB	V^{-3}	*
U10	V_{ds} dependence on mobility	V^{-1}	*
U1B	V_{bs} dependence on U1	V^{-2}	*
U1D	V_{ds} dependence on U1	V^{-2}	*
VOF0	Threshold voltage offset at $V_{ds}=0$, $V_{bs}=0$	—	*
VOFD	V_{ds} dependence on VOF	V^{-1}	*
AI0	Prefactor of hot-electron effect	—	*
AIB	V_{bs} dependence on AI	V^{-1}	*
BI0	Exponential factor of hot-electron effect	V	*
BIB	V_{bs} dependence on BI	—	*
VGHIGH	Upper bound of the cubic-spline function	V	*
VGLOW	Lower bound of the cubic-spline function	V	*
VGG	Maximum V_{gs}	V	
VBB	Maximum V_{bs}	V	

The following are BSIM3 (Level 14) parameters):

Name	Parameter	Unit	Default
	DC Parameters		
TOX	Gate oxide thickness	m	150 Å
XJ	Metallurgical junction depth	m	0.15 μm
NCH	Channel doping concentration	cm^{-3}	1.7×10^{17}
NSUB	Doping concentration	cm^{-3}	6×10^{16}
VTH0	Threshold voltage at zero substrate bias	V	0.7 (NMOS) −0.7 (PMOS)
K1	First-order body-effect coefficient	$V^{1/2}$	0.5
K2	Second-order body-effect coefficient	—	−0.0186
K3	Narrow-width effect coefficient	—	80
K3B	Body-effect coefficient of K3	—	0
W0	Narrow-width effect reference width	m	2.5×10^{-6}
NLX	Lateral nonuniform doping coefficient	m	1.74×10^{-7}
DVT0W	First coefficient of narrow-width effect on V_{th} at small L	m^{-1}	0
DVT1W	Second coefficient of narrow-width effect on V_{th} at small L	m^{-1}	5.3×10^6
DVT2W	Body bias coefficient of short-channel effect on V_{th} at small L	V^{-1}	−0.032
DVT0	First coefficient of short-channel effect on V_{th}	—	2.2
DVT1	Second coefficient of short-channel effect on V_{th}	—	0.53

Name	Parameter	Unit	Default
	DC Parameters		
DVT2	Body bias coefficient of short-channel effect on V_{th}	V^{-1}	−0.032
VBM	Maximum substrate bias	V	−5
U0	Low-field mobility at TNOM	cm^2/V s	670 (NMOS) 250 (PMOS)
UA	First-order mobility degradation coefficient	m/V	2.25×10^{-9}
UB	Second-order mobility degradation coefficient	$(m/V)^2$	5.87×10^{-19}
UC	Body effect of mobility degradation coefficient	V^{-1}	*
VSAT	Saturation velocity at TNOM	cm/s	8.0×10^6
A0	Bulk charge effect coefficient for channel length	—	1.0
AGS	Gate bias coefficient of the A_{bulk}	V^{-1}	0
B0	Bulk charge effect coefficient for channel width	m	0
B1	Bulk charge effect width offset	m	0
KETA	Body bias coefficient of the bulk charge effect	V^{-1}	−0.047
A1	First nonsaturation effect coefficient	V^{-1}	0
A2	Second nonsaturation effect coefficient	—	1.0
RDSW	Width coefficient of parasitic resistance	$\Omega \, \mu m^{WR}$	0
PRWG	Gate bias effect coefficient of RDSW	V^{-1}	0
PRWB	Body effect coefficient of RDSW	$V^{-1/2}$	0
WR	Width offset from W_{eff} for R_{ds} calculation	—	1
WINT	Width offset fitting parameter from $I–V$ without bias	m	0
LINT	Length offset fitting parameter from $I–V$ without bias	m	0
DWG	Coefficient of W_{eff} gate dependence	m/V	0
DWB	Coefficient of W_{eff} body bias dependence	$m/V^{1/2}$	0
VOFF	Offset voltage in the subthreshold region at large W and L	V	−0.11
NFACTOR	Subthreshold swing factor	—	1
ETA0	Subthreshold region DIBL coefficient	—	0.08
ETAB	Body bias coefficient for the subthreshold DIBL effect	V^{-1}	−0.07
DSUB	Subthreshold DIBL coefficient exponent	—	DROUT
CIT	Interface trapped charge capacitance	F/m^2	0
CDSC	Drain–source to channel coupling capacitance	F/m^2	2.4×10^{-4}

Name	Parameter	Unit	Default
DC Parameters			
CDSCD	Drain bias sensitivity of CDSC	F/Vm^2	0
PCLM	Channel length modulation parameter	—	1.3
PDIBLC1	First-output resistance DIBL effect correction parameter	—	0.39
PDIBLC2	Second-output resistance DIBL effect correction parameter	—	0.0086
PDIBLCB	Body-effect coefficient of DIBL correction parameters	V^{-1}	0
DROUT	L dependence coefficient of DIBL correction parameters	—	0.56
PSCBE1	First-substrate current body-effect parameter	V/m	4.24×10^8
PSCBE2	Second-substrate current body-effect parameter	m/V	1.0×10^{-5}
PVAG	Gate voltage dependence of R_{out} coefficient	—	0
DELTA	Effective V_{ds} parameter	V^2	0.01
PB	Bottom-junction built-in potential	V	1.0
PBSW	Source–drain side-junction built-in potential	V	1.0
RSH	Source–drain sheet resistance	Ω/sq.	0
JS	Source–drain junction saturation current density	A/m^2	1×10^{-4}
MOBMOD	Mobility model selector	—	1
AC and Capacitance Parameters			
CAPMOD	Flag for the short-channel capacitance model	—	1
XPART	Charge partitioning rate flag	—	0
CGS0	Non-LDD region source–gate overlap capacitance per meter channel length	F/m	Calculated
CGD0	Non-LDD region drain–gate overlap capacitance per meter channel length	F/m	Calculated
CGB0	Gate–bulk overlap capacitance per meter channel length	F/m	0
CJ	Source–drain junction capacitance per unit area	F/m^2	5×10^4
MJ	Grading coefficient of source–drain junction	—	0.5
MJSW	Grading coefficient of source–drain junction sidewall	—	0.33
CJSW	Source–drain junction sidewall capacitance per unit length	F/m	5×10^{-10}
CGS1	Overlap capacitance of lightly doped source–gate region	F/m	0
CGD1	Overlap capacitance of lightly doped drain–gate region	F/m	0
CKAPPA	Coefficient for lightly doped region overlap capacitance fringing-field capacitance	F/m	0.6

Name	Parameter	Unit	Default
CF	Fringing-field capacitance	F/m	Calculated
CLC	Constant term for short-channel model	m	0.1×10^{-6}
CLE	Exponential term for short-channel model	—	0.6
DLC	Length offset fitting parameter from C-V	m	LINT
DWC	Width offset fitting parameter from C-V	m	WINT

Temperature Effect Parameters

TNOM	Temperature at which parameters are extracted	°C	27
PRT	Temperature coefficient for RDSW	$\Omega\ \mu\text{m}/°\text{C}$	0
UTE	Temperature coefficient of mobility	—	-1.5
KT1	Threshold voltage temperature coefficient	V	-0.11
KT1L	Channel length sensitivity of temperature coefficient for threshold voltage	V m	0
KT2	Body bias coefficient of the V_{th} temperature effect	—	0.022
UA1	Temperature coefficient of UA	m/V	4.31×10^{-9}
UB1	Temperature coefficient of UB	$(\text{m/V})^2$	-7.61×10^{-18}
UC1	Temperature coefficient of UC	V^{-1}	-0.056
AT	Temperature coefficient of V_{SAT}	m/s	3.3×10^4

NQS Model Parameters

NQSMOD	NQS model selector	—	0
ELM	Elmore constant of the channel	—	5

W and *L* Parameters

WL	Coefficient of length dependence for width offset	m^{WLN}	0.0
WLN	Power of length dependence of width offset	—	1.0
WW	Coefficient of width dependence for width offset	m^{WWN}	0.0
WWN	Power of width dependence of width offset	—	1.0
WWL	Coefficient of length and width cross term for width offset	$\text{m}^{\text{WWN+WLN}}$	0.0
L1	Coefficient of length dependence for length offset	m^{L1N}	0.0
L1N	Power of length dependence for length offset	—	1.0
LW	Coefficient of width dependence for length offset	m^{LWN}	0.0
LWN	Power of width dependence for length offset	—	1.0
LWL	Coefficient of length and width cross term for length offset	$\text{m}^{\text{LWN+L1N}}$	0.0

Name	Parameter	Unit	Default
	Bound Parameters		
LMIN	Minimum channel length	m	0.0
LMAX	Maximum channel length	m	1.0
WMIN	Minimum channel width	m	0.0
WMAX	Maximum channel width	m	1.0
	Process Parameters		
PHI	Surface potential at strong inversion	V	Calculated
GAMMA1	Body-effect coefficient near the interface	$\sqrt{V}$	Calculated
GAMMA2	Body-effect coefficient in the bulk	$\sqrt{V}$	Calculated
VBX	V_{bs} at which the depletion width equals XT	V	Calculated
XT	Doping depth	m	1.55×10^{-7}
VBI	Drain–source junction built-in potential	V	Calculated
	Noise Model Parameters		
NOISEA	Noise parameter A	—	1×10^{20} (NMOS) 9.9×10^{18} (PMOS)
NOISEB	Noise parameter B	—	5×10^{4} (NMOS) 2.4×10^{3} (PMOS)
NOISEC	Noise parameter C	—	-1.4×10^{-12} (NMOS) 1.4×10^{-12} (PMOS)
EM	Saturated field	V/m	4.1×10^{7}
AF	Frequency exponent	—	1
NOIMOD	Noise model selector	—	1

* If MOBMOD = 1 or 2, UC = -4.65×10^{-11}. If MOBMOD = 3, UC = -0.046

The following are AIM-Spice model parameters for Levels 7–10:

Name	Parameter	Unit	Default
VTO	Zero-bias threshold voltage	V	0.0
PHI	Surface potential	V	0.6
RD	Drain resistance	Ω	0.0
RS	Source resistance	Ω	0.0
CBD	Zero-bias B-D junction capacitance	F	0.0
CBS	Zero-bias B-S junction capacitance	F	0.0
IS	Bulk junction saturation current	A	1.0×10^{-14}
PB	Bulk junction potential	V	0.8

Name	Parameter	Unit	Default
CGSO	Gate–source overlap capacitance per meter channel width	F/m	0.0
CGDO	Gate–drain overlap capacitance per meter channel width	F/m	0.0
CGBO	Gate–bulk overlap capacitance per meter channel width	F/m	0.0
RSH	Drain and source diffusion sheet resistance	Ω/sq.	0.0
CJ	Zero-bias bulk junction bottom capacitance per square meter of junction area	F/m^2	0.0
MJ	Bulk junction bottom grading coefficient	—	0.5
CJSW	Zero-bias bulk junction sidewall capacitance per meter of junction perimeter	F/m	0.0
MJSW	Bulk junction sidewall grading coefficient	—	0.33
JS	Bulk junction saturation current per square meter of junction area	A/m^2	0
TOX	Thin-oxide thickness	m	1.0×10^{-7}
NSUB	Substrate doping	cm^{-3}	0.0
NSS	Surface state density	cm^{-2}	0.0
TPG	Type of gate material (+1, opposite of substrate; −1, same as substrate; 0, Al gate)	—	1.0
XJ	Metallurgical junction depth	m	0.0
LD	Lateral diffusion	m	0.0
U0	Surface mobility	cm^2/V s	280 (level 7) 625 (others, NMOS) 279 (others, PMOS)
VMAX	Maximum carrier drift velocity	m/s	4.0×10^4 (Level 7) 6.0×10^4 (Level 8) 7.8×10^4 (Level 9) 8.6×10^4 (Level 10)
KF	Flicker noise coefficient	—	0.0
AF	Flicker noise exponent	—	1.0
FC	Coefficient for forward-bias depletion capacitance formula	—	0.5
TNOM	Parameter measurement temperature	°C	27

The following are AIM-Spice model parameters for threshold voltage calculations in Levels 8–10:

Name	Parameter	Unit	Default
GAMMA	Bulk threshold parameter	$V^{\frac{1}{2}}$	0.0
DELTA	Width effect on threshold voltage	—	0.0

The following parameters are used for the threshold voltage model of AIM-Spice MOSFET Level 7:

Name	Parameter	Unit	Default
GAMMAS0	Body-effect constant in front of square-root term	$V^{\frac{1}{2}}$	0.0
LGAMMAS	Sensitivity of γ_S on device length	$V^{\frac{1}{2}}$	0.0
WGAMMAS	Sensitivity of γ_S on device width	$V^{\frac{1}{2}}$	0.0
GAMMAL0	Body-effect constant in front of linear term	—	0.0
LGAMMAL	Sensitivity of γ_L on device length	—	0.0
WGAMMAL	Sensitivity of γ_L on device width	—	0.0
L0	Gate length of nominal device	m	2×e-6
W0	Gate width of nominal device	m	2×e-5

The following are the expressions used for the threshold voltage [see Section 3.7 of Lee et al. (1993)]:

$$V_T = V_{T00} + \gamma_S\sqrt{2\varphi_b - V_{bs}} - \gamma_L(2\varphi_b - V_{bs})$$

where

$$\gamma_S = \text{GAMMAS0} + \text{LGAMMAS}\left(1 - \frac{L_0}{L}\right) + \text{WGAMMAS}\left(1 - \frac{W_0}{W}\right)$$

$$\gamma_L = \text{GAMMAL0} + \text{LGAMMAL}\left(1 - \frac{L_0}{L}\right) + \text{WGAMMAL}\left(1 - \frac{W_0}{W}\right)$$

The value of V_{T00} is given by the model parameter VTO as

$$V_{T00} = \text{VTO} - \gamma_S\sqrt{2\varphi_b} + \gamma_L 2\varphi_b$$

These equations are valid for

$$V_{T00} - V_{bs} \leq \left(\frac{\gamma_S}{2\gamma_L}\right)^2$$

Beyond this limit, we assume that the threshold voltage remains constant at

$$V_{TM} = V_{T00} + \gamma_L\left(\frac{\gamma_S}{2\gamma_L}\right)^2$$

Other parameters specified for Level 7 model are as follows:

Name	Parameter	Unit	Default (NMOS)
LAMBDA	Output conductance parameter	V^{-1}	0.048
ETA	Subthreshold ideality factor	—	1.32
M	Knee shape parameter	—	4.0
DELTA	Transition width parameter	—	5.0
THETA	Mobility degradation parameter	$m^3/V^2 s$	0
SIGMA0	DIBL parameter	—	0.048
VSIGMAT	DIBL parameter	V	1.7
VSIGMA	DIBL parameter	V	0.2
RDI	Internal drain resistance	Ω	0
RSI	Internal source resistance	Ω	0
CV	Charge storage model selector	—	1
CVE	Meyer-like capacitor model selector	—	1
FPE	Charge-partitioning scheme selector	—	1
XQC	Charge-partitioning factor	—	0.6
MCV	Transition width parameter used by the charge-partitioning scheme	—	10
ALPHA	Parameter accounting for the threshold dependence on the channel potential	—	1.05
VFB	Flat-band voltage	V	*

* Parameter is calculated if not specified.

Modeling of charge storage for the Level 7 model is selected by specifying a value for the model parameter CV. CV=1 selects the standard Meyer model and CV=2 selects the charge-conserving Meyer-like model.

Allowed values of the Meyer-like capacitor model selector CVE are 1 and 2. CVE=1 selects the standard Meyer capacitors and CVE=2 selects the UCCM capacitors.

The model parameter FPE selects the charge-partitioning scheme used by the Meyer-like charge storage model (CV=2). FPE=1 selects a constant partitioning factor, FPE=2 selects an empirical partitioning scheme, and FPE=3 selects an analytical partitioning scheme.

Either intrinsic or extrinsic models can be selected by proper use of the parameters RD, RS, RDI, and RSI. If values for RD and RS are specified, the intrinsic model is selected with parasitic resistances applied externally. The extrinsic model is selected by specifying values for RDI and RSI.

Parameters specific for the Level 8 model are as follows:

Name	Parameter	Unit	Default
ALPHA	Parameter accounting for the threshold dependence on the channel potential	—	1.164 (NMOS) 1.4 (PMOS)
K1	Mobility parameter	cm^2/V^2 s cm^2/s	28 (NMOS) 1510 (PMOS)
ETA	Subthreshold ideality factor	—	1.3 (NMOS) 1.2 (PMOS)

Parameters specific for the Level 9 model are as follows:

Name	Parameter	Unit	Default
ALPHA	Parameter accounting for the threshold dependence on the channel potential	—	1.2 (NMOS) 1.34 (PMOS)
XI	Saturation voltage parameter (NMOS only)	—	0.79
GAM	Saturation point parameter	—	3.0 (NMOS) 2.35 (PMOS)
K1	Mobility parameter	cm^2/V^2 s cm^2/s	28 (NMOS) 1510 (PMOS)
LAMBDA	Characteristic length of the saturated region of the channel	m	9.94×10^{-8} (NMOS) 1.043×10^{-7} (PMOS)
ZETA	Velocity saturation factor (PMOS only)	—	0.34

Parameters specific for the Level 10 model are as follows:

Name	Parameter	Unit	Default
ALPHA	Bulk charge factor	—	1.05 (NMOS) 1.3 (PMOS)
ETA	Ideality factor in subthreshold	—	1.42 (NMOS) 1.39 (PMOS)
SIGMA	Parameter accounting for DIBL effects	—	0.03 (NMOS) 0.007 (PMOS)
K1	Mobility parameter	cm^2/V^2 s cm^2/s	28 (NMOS) 1510 (PMOS)
LAMBDA	Channel length modulation parameter	m	8.58×10^{-8} (NMOS) 7.25×10^{-8} (PMOS)
XI	Saturation voltage factor (NMOS only)	—	0.76
ZETA	Velocity saturation factor (PMOS only)	—	0.28
GAM	Saturation point parameter (PMOS only)	—	2.35

Parameters common to Levels 11, 12, 15, and 16 are as follows:

Name	Parameter	Unit	Default
VTO	Zero-bias threshold voltage	V	0.0
RD	Drain resistance	Ω	0.0
RS	Source resistance	Ω	0.0
CGSO	Gate–source overlap capacitance per meter channel width	F/m	0.0
CGDO	Gate–drain overlap capacitance per meter channel width	F/m	0.0
TOX	Thin-oxide thickness	m	1.0e-7
U0	Surface mobility (only Levels 11 and 12)	$cm^2/V\ s$	1.0 (Level 11) 100 (Level 12)
ETA	Subthreshold slope	—	6.9
TNOM	Parameter measurement temperature	°C	27

Parameters specific for Level 11 model are as follows [default model parameters correspond to n-channel TFT devices used as examples in Chapter 5 of Lee et al. (1993)]:

Name	Parameter	Unit	Default
EPS	Relative dielectric constant of substrate	—	11.7
EPSI	Relative dielectric constant of gate insulator	—	3.9
M1	Knee shape parameter	—	2.5
M2	Mobility parameter	—	0.5
V2	Characteristic voltage	V	0.086
VSIGMA	DIBL parameter	V	0.2
VSIGMAT	DIBL parameter	V	1.7
SIGMA0	DIBL parameter	—	0.048
LAMBDA	Transconductance parameter	V^{-1}	0.0048
N0	Scaling factor	m^{-2}	1×10^{16}
X0	Fitting parameter	—	V2/5

Parameters specific for Level 12 model are as follows [default model parameters correspond to n-channel TFT devices used as examples in Chapter 5 of Lee et al. (1993)]:

Name	Parameter	Unit	Default
V0	Scaling voltage	V	10.7
GAMMA	Saturation voltage parameter	—	0.03
RSH	Drain and source diffusion sheet resistance	Ω/sq.	0

Parameters specific for Level 15 are as follows:

Name	Parameter	Unit	Default
ALPHASAT	Saturation modulation parameter	—	0.6
DEF0	Dark Fermi-level position	eV	0.6
DELTA	Transition width parameter	—	5
EL	Activation energy of the hole leakage current	eV	0.1
EMU	Field-effect mobility activation energy	eV	0.02
EPS	Relative dielectric constant of substrate	—	11
EPSI	Relative dielectric constant of gate insulator	—	7.4
GAMMA	Power law mobility parameter	—	0.4
GMIN	Minimum density of deep states	$m^{-3}\,eV^{-1}$	1×10^{23}
IOL	Zero-bias leakage current	A	3×10^{-14}
KVT	Threshold voltage temperature coefficient	V/K	-0.01
LAMBDA	Output conductance parameter	V^{-1}	0.0008
M	Knee shape parameter	—	2.5
MUBAND	Conduction band mobility	m^2/V s	0.001
SIGMA0	Minimum-leakage parameter	A	1×10^{-14}
V0	Characteristic voltage for deep states	V	0.12
VAA	Characteristic voltage for field-effect mobility (determined by tail states)	V	7.5×10^3
VDSL	Hole leakage current drain voltage parameter	V	7
VFB	Flat-band voltage	V	-3
VGSL	Hole leakage current gate voltage parameter	V	7
VMIN	Convergence parameter	V	0.3

Parameters specific for Level 16 are as follows:

Name	Parameter	Unit	Default
ASAT	Proportionality constant of V_{sat}	—	1
AT	DIBL parameter 1	m/V	3×10^{-8}
BLK	Leakage barrier-lowering constant	—	0.001
BT	DIBL parameter 2	m V	1.9×10^{-6}
DASAT	Temperature coefficient of ASAT	$°C^{-1}$	0
DD	V_{ds} field constant	m	1400 Å
DELTA	Transition width parameter	—	4.0
DG	V_{gs} field constant	m	2000 Å
DMU1	Temperature coefficient of KMU	cm^2/V s °C	0
DVTO	Temperature coefficient of VTO	V/°C	0
EB	Barrier height of diode	eV	0.68
ETA	Subthreshold ideality factor	—	7
ETAC0	Capacitance subthreshold ideality factor at zero drain bias	—	ETA

Name	Parameter	Unit	Default
ETAC00	Capacitance subthreshold coefficient of drain bias	V^{-1}	0
I0	Leakage current scaling constant	A/m	6.0
I00	Diode reverse saturation current	A/m	150
LASAT	Coefficient for length dependence of A_{SAT}	m	0
LKINK	Kink effect constant	m	19×10^{-6}
MC	Capacitance knee shape parameter	—	3.0
MK	Kink effect exponent	—	1.3
MMU	Low-field mobility exponent	—	3.0
MU0	High-field mobility	$cm^2/V\ s$	100
MU1	Low-field mobility parameter	$cm^2/V\ s$	0.0022
MUS	Subthreshold mobility	$cm^2/V\ s$	1.0
RF	Resistance in series with C_{gd}	Ω	0
RI	Resistance in series with C_{gs}	Ω	0
VFB	Flat-band voltage	V	-0.1
VKINK	Kink effect voltage	V	9.1

Temperature Effects

Temperature appears explicitly in the exponential terms in the equations describing current across the bulk junctions.

Temperature appears explicitly in the value of the junction potential ϕ (in AIM-Spice PHI). The temperature dependence is given by

$$\phi(T) = \frac{kT}{q} \log_e \left(\frac{N_a N_d}{N_i(T)^2} \right)$$

where k is Boltzmann's constant, q is the electronic charge, N_a is the acceptor impurity density, N_d is the donor impurity density, and N_i is the intrinsic carrier concentration.

Temperature appears explicitly in the value of the surface mobility μ_0 (or U0). The temperature dependence is given by

$$\mu_0(T) = \frac{\mu_0(T_0)}{(T/T_0)^{1.5}}$$

Supported Analyses

Levels 1–3	All
Level 4	Distortion and Noise Analysis not supported
Level 5	Noise Analysis not supported
Level 6	AC, Distortion, Noise, and Pole-Zero Analysis not supported
Levels 7–13	Distortion and Noise Analysis not supported
Level 14	Distortion Analysis not supported

N   HETEROJUNCTION BIPOLAR TRANSISTORS (HBTs)

General Form

```
NXXXXXXX NC NB NE <NS> MNAME <AREA> <OFF> <IC=VBE,VCE>
+ <TEMP=T>
```

Example

```
N23 10 24 13 NMOD IC=0.6,5.0
n2 5 4 0 nnd
```

NC, NB, and NE are the collector, base, and emitter nodes, respectively. NS is the substrate node. If this is given, ground is assumed. MNAME is the model name, AREA is the area factor, and OFF indicates an optional initial value for the element in a dc analysis. If the area factor is omitted, 1.0 is assumed. The optional initial value IC=VBE,VCE is meant to be used together with UIC in a transient analysis. See the description of the .IC control statement for a better way to set transient initial conditions. The optional TEMP value is the temperature at which the device operates. It overrides the temperature specified in the option value.

HBT Model

```
.MODEL {model name} HNPN <model parameters>
.MODEL {model name} HPNP <model parameters>
```

The heterojunction bipolar transistor model in AIM-Spice is a modification of the Gummel–Poon bipolar transistor model.

Name	Parameter	Unit	Default
IS	Transport saturation current	A	1×10^{-16}
BF	Ideal maximum forward beta	—	100
NF	Forward-current emission coefficient	—	1.0
ISE	B-E leakage saturation current	A	0
NE	B-E leakage emission coefficient	—	1.2
BR	Ideal maximum reverse beta	—	1
NR	Reverse-current emission coefficient	—	1
ISC	B-C leakage saturation current	A	0
NC	B-C leakage emission coefficient	—	2
RB	Base resistance	Ω	0
RE	Emitter resistance	Ω	0
RC	Collector resistance	Ω	0
CJE	B-E zero-bias depletion capacitance	F	0
VJE	B-E built-in potential	V	0.75
MJE	B-E junction exponential factor	—	0.33

Name	Parameter	Unit	Default
TF	Ideal forward transit time	s	0
XTF	Coefficient for bias dependence of TF	—	0
VTF	Voltage describing VBC dependence of TF	V	Infinite
ITF	High-current parameter for effect on TF	A	0
PTF	Excess phase at $f=1.0/(TF \cdot 2\pi)$ hertz	deg	
CJC	B-C zero-bias depletion capacitance	F	0
VJC	B-C built-in potential	V	0.75
MJC	B-C junction exponential factor	—	0.33
XCJC	Fraction of B-C depletion capacitance connected to internal base node	—	1
TR	Ideal reverse transit time	s	0
CJS	Zero-bias collector–substrate capacitance	F	0
VJS	Substrate junction built-in potential	V	0.75
MJS	Substrate junction exponential factor	—	0
XTB	Forward and reverse beta temperature exponent	—	0
EG	Energy gap for temperature effect on IS	eV	1.11 (Si)
XTI	Temperature exponent for effect on IS	—	3
KF	Flicker noise coefficient	—	0
AF	Flicker noise exponent	—	1
FC	Coefficient for forward-bias depletion capacitance formula	—	0.5
TNOM	Parameter measurement temperature	°C	27
IRB0	Base-region recombination saturation current	A	0
IRS1	Surface recombination saturation current 1	A	0
IRS2	Surface recombination saturation current 2	A	0
ICSAT	Collector saturation current	A	0
M	Knee shape parameter	—	3

The modification to the Gummel–Poon model consists of two new contributions to the generation/recombination current and a limitation on the intrinsic collector current (see Section 4.4).

Recombination in the base region is modeled by the expression

$$I_{rb} = \text{IRB0}(e^{V_{be}/V_{\text{th}}} - 1)$$

Surface recombination is modeled by the expression

$$I_{rs} = \text{IRS1}(e^{V_{be}/V_{\text{th}}} - 1) + \text{IRS2}(e^{V_{be}/2V_{\text{th}}} - 1)$$

where IRS1 and IRS2 are proportional to the emitter perimeter.

If the model parameters M and ICSAT are given, the intrinsic collector current is modified according to the following expression:

$$I_c = \frac{I_{\text{co}}}{[1 + (I_c/\text{ICSAT})^M]^{1/M}}$$

Supported Analyses

Distortion, Noise, and Pole-Zero Analysis not supported.

O   LOSSY TRANSMISSION LINES (LTRA)

General Form

OXXXXXXX N1 N2 N3 N4 MNAME

Example

o23 1 0 2 0 lmod
ocon 10 5 20 5 interconnect

This is a two-port convolution model for single-conductor lossy transmission lines. N1 and N2 are the nodes at port 1; N3 and N4 are the nodes at port 2. It is worth mentioning that a lossy transmission line with zero loss may be more accurate than the lossless transmission.

LTRA Model

.MODEL {model name} LTRA <model parameters>

The uniform RLC/RC/LC/RG transmission line model (LTRA) models a uniform constant-parameter distributed transmission line. In case of RC and LC, the URC and TRA models may also be used. However, the newer LTRA model is usually faster and more accurate. The operation of the LTRA model is based on the convolution of the transmission line's impulse response with its inputs (see Roychowdhury and Pederson, 1991).

The LTRA model parameters are as follows:

Name	Parameter	Unit	Default
R	Resistance/length	Ω/m	0.0
L	Inductance/length	H/m	0.0
C	Capacitance/length	F/m	0.0
G	Conductance/length	Ω^{-1}/m	0.0
LEN	Length of line	m	—
REL	Breakpoint control	—	1
ABS	Breakpoint control	—	1
NOSTEPLIMIT	Do not limit time step to less than line delay	Flag	Not set

Name	Parameter	Unit	Default
NOCONTROL	Do not do complex time step control	Flag	Not set
LINEINTERP	Use linear interpolation	Flag	Not set
MIXEDINTERP	Use linear when quadratic seems bad	Flag	Not set
COMPACTREL	Special `reltol` for history compaction		RELTOL
COMPACTABS	Special `abstol` for history compaction		ABSTOL
TRUNCNR	Use Newton–Raphson method for time step control	Flag	Not set
TRUNCDONTCUT	Do not limit time step to keep impulse response errors low	Flag	Not set

The types of lines implemented so far are uniform transmission line with series loss only (RLC), uniform RC line, lossless transmission line (LC), and distributed series resistance and parallel conductance only (RG). Any other combination will yield erroneous results and should be avoided. The length (LEN) of the line must be specified.

Here follows a detailed description on some of the model parameters:

NOSTEPLIMIT is a flag that will remove the default restrictions of limiting the time step to less than the line delay in the RLC case.

NOCONTROL is a flag that prevents the default limitation on the time step based on convolution error criteria in the RLC and RC cases. This speeds up the simulation but may in some cases reduce the accuracy.

LININTERP is a flag that, when set, will use linear interpolation instead of the default quadratic interpolation for calculating delayed signals.

MIXEDINTERP is a flag that, when set, uses a metric for judging whether quadratic interpolation is applicable and, if not so, uses linear interpolation. Otherwise it uses the default quadratic interpolation.

TRUNCDONTCUT is a flag that removes the default cutting of the time step to limit errors in the actual calculation of impulse-response-related quantities.

COMPACTREL and COMPACTABS are quantities that control the compacting of the past history of values stored for convolution. Large values of these parameters result in lower accuracy but usually increase the simulation speed. These are to be used with the TRYTOCOMPACT option.

TRUNCNR is a flag that turns on the use of Newton–Raphson iterations to determine an appropriate time step in the time step control routines. The default is a trial-and-error procedure cutting the previous time step in half.

If you want to increase the speed of the simulation, follow these guidelines:

The most efficient option for increasing the speed of the simulation is REL. The default value of 1 is usually safe from the point of view of accuracy but occasionally increases the computation time. A value greater than 2 eliminates all breakpoints and may be worth trying depending on the nature of the rest of the circuit, keeping in mind that it may not be safe from the point of view of accuracy. Breakpoints can usually be entirely eliminated if the circuit is not expected to have sharp discontinuities. Values between 0 and 1 are usually not needed but may be used for setting a large number of breakpoints.

It is also possible to experiment with COMPACTREL when the option TRYTO-COMPACT is specified. The legal range is between 0 and 1. Larger values usually decrease the accuracy of the simulation but in some cases improve speed. If TRY-TOCOMPACT is not specified, history compacting is not attempted and the accuracy is high. The flags NOCONTROL, TRUNCDONTCUT, and NOSTEPLIMIT also increase speed at the expense of accuracy in some cases.

Supported Analyses

Noise and Pole-Zero Analysis not supported.

Q   BIPOLAR JUNCTION TRANSISTORS (BJTS)

General Form

```
QXXXXXXX NC NB NE <NS> MNAME <AREA> <OFF> <IC=VBE,VCE>
+ <TEMP=T>
```

Example

```
Q23 10 24 13 QMOD IC=0.6,5.0
q2 5 4 0 qnd
```

NC, NB, and NE are the collector, base, and emitter nodes, respectively. NS is the substrate node. If this is not given, ground is assumed. MNAME is the model name, AREA is the area factor, and OFF indicates an optional initial value for the element in a DC analysis. If the area factor is omitted, 1.0 is assumed. The optional initial value IC=VBE,VCE is meant to be used together with UIC in a Transient Analysis. See the description of the .IC control statement for a better way to set transient initial conditions. The optional TEMP value is the operating temperature of the device transistor. It overrides the temperature specified as an option.

BJT Model

```
.MODEL {model name} NPN <model parameters>
.MODEL {model name} PNP <model parameters>
```

The bipolar transistor model in AIM-Spice is an adaptation of the Gummel–Poon model. In AIM-Spice, the model is extended to include high-bias effects. The model autoomatically simplifies to Ebers–Moll if certain parameters are not given (VAF, IKF, VAR, IKR).

Name	Parameter	Unit	Default
IS	Transport saturation current	A	1×10^{-16}
BF	Ideal maximum forward beta	—	100
NF	Forward-current emission coefficient	—	1.0
VAF	Forward Early voltage	V	Infinite
IKF	Corner for forward-beta high-current roll-off	A	Infinite
ISE	B-E leakage saturation current	A	0
NE	B-E leakage emission coefficient	—	1.2
BR	Ideal maximum reverse beta	—	1
NR	Reverse-current emission coefficient	—	1
VAR	Reverse Early voltage	V	Infinite
IKR	Corner for reverse-beta high-current roll-off	A	Infinite
ISC	B-C leakage saturation current	A	0
NC	B-C leakage emission coefficient	—	2
RB	Zero-bias base resistance	Ω	0
IRB	Current where base resistance falls halfway to its minimum value	A	Infinite
RBM	Minimum base resistance at high currents	Ω	RB
RE	Emitter resistance	Ω	0
RC	Collector resistance	Ω	0
CJE	B-E zero-bias depletion capacitance	F	0
VJE	B-E built-in potential	V	0.75
MJE	B-E junction exponential factor	—	0.33
TF	Ideal forward transit time	s	0
XTF	Coefficient for bias dependence of TF	—	0
VTF	Voltage describing V_{BC} dependence of TF	V	Infinite
ITF	High-current parameter for effect on TF	A	0
PTF	Excess phase at $f=1.0/(\text{TF}\cdot2\pi)$ hertz	deg	
CJC	B-C zero-bias depletion capacitance	F	0
VJC	B-C built-in potential	V	0.75
MJC	B-C junction exponential factor	—	0.33
XCJC	Fraction of B-C depletion capacitance connected to internal base node	—	1
TR	Ideal reverse transit time	s	0
CJS	Zero-bias collector–substrate capacitance	F	0
VJS	Substrate junction built-in potential	V	0.75
MJS	Substrate junction exponential factor	—	0
XTB	Forward- and reverse-beta temperature exponent	—	0

Name	Parameter	Unit	Default
EG	Energy gap for temperature effect on IS	eV	1.11 (Si)
XTI	Temperature exponent for effect on IS	—	3
KF	Flicker noise coefficient	—	0
AF	Flicker noise exponent	—	1
FC	Coefficient for forward-bias depletion capacitance formula	—	0.5
TNOM	Parameter measurement temperature	°C	27

Temperature Effects

The temperature appears explicitly in the exponential terms.

The temperature dependence of the saturation current in the model is determined by

$$\mathrm{IS}(T_1) = \mathrm{IS}(T_0) \left(\frac{T_1}{T_0} \right)^{\mathrm{XTI}} \exp\left(\frac{q\mathrm{EG}(T_1 - T_0)}{kT_1T_0} \right)$$

where k is Boltzmann's constant, q is the electronic charge, EG is the energy gap parameter (in electron volts), and XTI is the saturation current temperature exponent. EG and XTI are model parameters.

The temperature dependence of the forward and reverse beta is given by

$$\beta(T_1) = \beta(T_0) \left(\frac{T_1}{T_0} \right)^{\mathrm{XTB}}$$

where XTB is a user-supplied model parameter. Temperature effects on beta are implemented by appropriate adjustment of the model parameters BF, ISE, BR, and ISC.

Supported Analyses

All.

R RESISTORS

General Form

```
RXXXXXXX N1 N2 VALUE
```

Examples

```
R1 1 2 100
RB 1 2 10K
RBIAS 4 8 10K
```

N1 and N2 are the two element nodes. VALUE is the resistance in ohms. The value can be positive or negative but not zero.

Semiconductor Resistors

General Form

```
RXXXXXXX N1 N2 <VALUE> <MNAME> <L=LENGTH> <W=WIDTH>
+ <TEMP=T>
```

Example

```
rload 2 10 10K
RMOD 3 7 RMODEL L=10U W=1U
```

This is a more general model for the resistor than the one presented above. It gives you the possibility to model temperature effects and to calculate the resistance based on geometry and processing information. VALUE, if given, defines the resistance, and information on geometry and processing will be ignored. If MNAME is specified, the resistance value is calculated based on information about the process and geometry in the model statement. If VALUE is not given, MNAME and LENGTH must be specified. If WIDTH is not given, it will be given the default value. The optional TEMP value is the temperature at which this device operates. It overrides the temperature specified in the option value.

Resistor Model

```
.MODEL {model name} R <model parameters>
```

The resistor model contains process-related parameters and the resistance value is a function of the temperature. The parameters are as follows:

Name	Parameter	Unit	Default
TC1	First-order temperature coefficient	$\Omega/°C$	0.0
TC2	Second-order temperature coefficient	$\Omega/°C^2$	0.0
RSH	Sheet resistance	$\Omega/sq.$	—
DEFW	Default width	m	1×10^{-6}
NARROW	Narrowing due to side etching	m	0.0
TNOM	Parameter measurement temperature	°C	27

The following expression is used for calculating the resistance value:

$$R = RSH \; \frac{L - NARROW}{W - NARROW}$$

Here, DEFW defines a default value of W. If either RSH or L is given, a default value of 1 kΩ is used for R.

Temperature Effects

The temperature dependence of the resistance is given by a polynomial:

$$R(T) = R(T_0)[1 + \text{TC1}(T - T_0) + \text{TC2}(T - T_0)^2]$$

where T is the operating temperature, T_0 is the nominal temperature, and TC1 and TC2 are the first- and second-order temperature coefficients, respectively.

Supported Analyses

All.

S   VOLTAGE-CONTROLLED SWITCH

General Form

```
SXXXXXXX N+ N- NC+ NC- MODEL <ON> <OFF>
```

Examples

```
s1 1 2 3 4 switch1 ON
s2 5 6 3 0 sm2 off
```

N+ and N− are the positive and negative connecting nodes of the switch, respectively. NC+ and NC− are the positive and negative controlling nodes, respectively.

Switch Model

```
.MODEL {model name} SW <model parameters>
```

The switch model allows modeling of an almost ideal switch in AIM-Spice. The switch is not quite ideal since the resistance cannot change from zero to infinity but must have a finite positive value. The on and off resistances should therefore be chosen very small and very large, respectively, compared to the other circuit elements. The model parameters are as follows:

Name	Parameter	Unit	Default
VT	Threshold voltage	V	0
VH	Hysteresis voltage	V	0
RON	On resistance	Ω	1
ROFF	Off resistance	Ω	1/GMIN

An ideal switch is highly nonlinear. The use of switches can cause large discontinuities in node voltages. A rapid change such as that associated with a switching operation can cause problems with roundoff and tolerance that may lead to erroneous results or problems in selecting proper time steps. To reduce such problems, follow these steps:

Do not set the switch impedances higher and lower than necessary.

Reduce the tolerance during a Transient Analysis. This is done by specifying the value for TRTOL less than the default value of 7.0, for example, 1.0.

When switches are placed near capacitors, reduce the size of CHGTOL to, for example, 1×10^{-16}.

Supported Analyses

All.

T   TRANSMISSION LINES (LOSSLESS)

General Form

```
TXXXXXXX N1 N2 N3 N4 Z0=VALUE <TD=VALUE>
+ <F=FREQ <NL=NRMLEN>> <IC=V1,I1,V2,I2>
```

Example

```
T1 1 0 2 0 Z0=50 TD=10NS
```

N1 and N2 are the nodes for port 1; N3 and N4 are the nodes for port 2. Z0 is the characteristic impedence of the line. The length of the line can be specified in two different ways. The transmission delay TD can be specified directly. Alternatively, a frequency F may be given together with the normalized length of the line, NL (normalized with respect to the wavelength at the frequency F). If a frequency is specified and NL is omitted, 0.25 is assumed.

The optional initial values consist of voltages and currents at each of the two ports. Note that these values are used only if the option UIC is specified in a Transient Analysis.

Supported Analyses

Noise and Pole-Zero Analysis not supported.

U   UNIFORM DISTRIBUTED RC LINES (URC)

General Form

```
UXXXXXXX N1 N2 N3 MNAME L=LENGTH <N=LUMPS>
```

Example

```
U1 1 2 0 URCMOD L=50U
URC2 1 12 2 UMODL L=1MIL N=6
```

N1 and N2 are the two nodes of the RC line itself, while N3 is the node of the capacitances. MNAME is the name of the model, LENGTH is the length of the line in meters. LUMPS, if given, is the number of segments used in modeling the RC line.

URC Model

```
.MODEL {model name} URC <model parameters>
```

The model consists of a subcircuit expansion of the URC line into a network of lumped RC segments with internally generated nodes. The RC segments are in a geometric progression, increasing toward the middle of the URC line, with K as a proportionality constant. The number of lumped segments used, if not specified on the URC line, is determined by the following expression:

$$N = \frac{\log\left(F_{MAX}\frac{R}{L}\frac{C}{L}2\pi\, l^2\frac{(K-1)^2}{K^2}\right)}{\log K}$$

The URC line is made up strictly of resistor and capacitor segments unless the ISPERL parameter is given a nonzero value, in which case the capacitors are replaced by reverse-biased diodes with a zero-bias junction capacitance equivalent to the capacitance replaced and with a saturation current of ISPERL amperes per meter of transmission line and an optional series resistance equivalent to RSPERL ohms per meter.

Name	Parameter	Unit	Default
K	Propagation constant	—	2.0
FMAX	Maximum frequency	Hz	1.0G
RPERL	Resistance per unit length	Ω/m	1000
CPERL	Capacitance per unit length	F/m	1×10^{-15}
ISPERL	Saturation current per unit length	A/m	0
RSPERL	Diode resistance per unit length	Ω/m	0

Supported Analyses

All.

V   INDEPENDENT VOLTAGE SOURCES

General Form

```
VXXXXXXX N+ N- <<DC> DC/TRAN VALUE> <AC <ACMAG <ACPHASE>>>
+ <DISTOF1 <F1MAG <F1PHASE>> <DISTOF2 <F2MAG <F2PHASE>>>
```

Examples

```
vin 21 0 pulse(0 5 1ns 1ns 1ns 5us 10us)
vcc 10 0 dc 6
vmeas 12 9
```

N+ and N- are the positive and negative nodes, respectively. Note that the voltage source does not have to be grounded. Positive current flows from the positive node through the source to the negative node. If you insert a voltage source with a zero value, it can be used as an ampere meter.

DC/TRAN is the source value in a DC or a Transient Analysis. The value can be omitted if it is zero for both the DC and the Transient Analysis. If the source is time invariant, its value can be prefixed with DC.

ACMAG is the amplitude and ACPHASE is the phase of the source during an AC Analysis. If ACMAG is omitted after the keyword AC, 1 is assumed. If ACPHASE is omitted, 0 is assumed.

All independent sources can be assigned time-varying values during a Transient Analysis. If a source is assigned a time-varying value, the value at $t=0$ is used during a DC Analysis. There are five predefined functions for time-varying sources: pulse, exponent, sine, piecewise linear, and single-frequency FM. If parameters are omitted, default values shown in the tables below will be assumed. DT and T2 are the incremental time and final time, respectively, in a Transient Analysis.

Pulse

General Form

```
PULSE(V1 V2 TD TR TF PW PER)
```

Parameter	Default Value	Unit
V1 (initial phase)	None	V
V2 (pulsed value)	None	V

Parameter	Default Value	Unit
TD (delay time)	0.0	s
TR (rise time)	DT	s
TF (fall time)	DT	s
PW (pulse width)	T2	s
PER (period)	T2	s

Example

```
VIN 3 0 PULSE(1 5 1S 0.1S 0.4S 0.5S 2S)
```

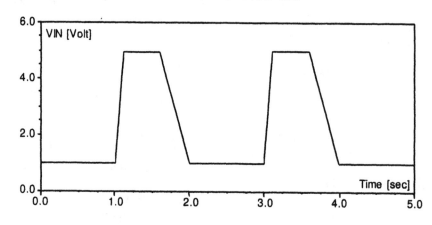

Sine

General Form

```
SIN(VO VA FREQ TD THETA)
```

Parameter	Default Value	Unit
V0 (offset)	None	V
VA (amplitude)	None	V
FREQ (frequency)	1/T2	Hz
TD (delay)	0.0	s
THETA (attenuation factor)	0.0	s^{-1}

The shape of the waveform is

$0 < time < T_D$

$\quad V = V_0$

$T_D < time < T_2$

$\quad V = V_0 + V_A \sin[2\pi\, FREQ \cdot (time + T_D]\exp[-(time - T_D) \cdot THETA]$

Example

```
VIN 3 0 SIN(2 2 5 1S 1)
```

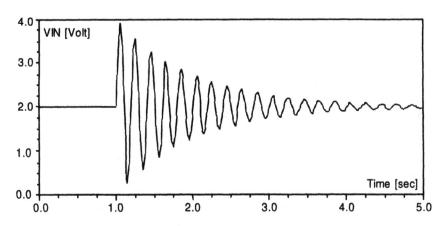

Exponent

General Form

```
EXP(V1 V2 TD1 TAU1 TD2 TAU2)
```

Parameter	Default	Unit
V1 (initial value)	None	V
VA (pulsed value)	None	V
TD1 (rise delay time)	0.0	s
TAU1 (rise time constant)	DT	s
TD2 (delay fall time)	DT1+DT	s
TAU2 (fall time constant)	DT	s

The shape of the waveform is

$0 < time < T_{DI}$

$\quad V = V1$

$T_{DI} < time < T_{D2}$

$\quad V = V_1 + (V_2 - V_1)\{1 - \exp[-(time - T_{DI})/\text{TAU1}]\}$

$T_{D2} < time < T_2$

$\quad V = V_1 + (V_2 - V_1)\{1 - \exp[-(time - T_{DI})/\text{TAU1}]\}$

$\quad\quad + (V_1 - V_2)\{1 - \exp[-(time - T_{D2})/\text{TAU2}]\}$

Example

```
VIN 3 0 EXP(1 5 1S 0.2S 2S 0.5S)
```

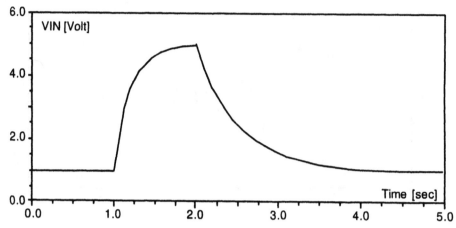

Piecewise Linear

General Form

```
PWL(T1 V1 <T2 V2 T3 V3 T4 V4 T5 V5....>)
```

Parameters and Default Values

Each pair of values (T_i, V_i) specifies the value of the source, V_i, at T_i. The values of the source in between are calculated using a linear interpolation.

Example

```
VIN 7 5 PWL(0 0 1 0 1.2 4 1.6 2.0 2.0 5.0 3.0 1.0)
```

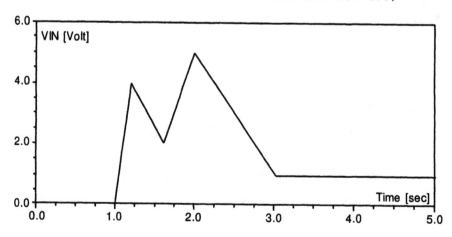

Single-Frequency FM

General Form

```
SFFM(V0 VA FC MDI FS)
```

Parameter	Default	Unit
V0 (offset)	None	V
VA (amplitude)	None	V
FC (carrier frequency)	1/T2	Hz
MDI (modulation index)	None	—
FS (signal frequency)	1/T2	Hz

The shape of the waveform is

$$V = V_0 + V_A \sin[(2\pi F_C \cdot \text{time}) + \text{MDI} \cdot \sin(2\pi F_S \cdot \text{time})]$$

Example

```
VIN 12 0 SFFM(2 1 2 5 0.2)
```

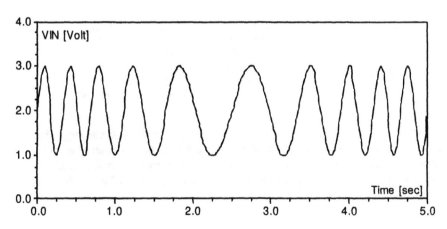

Supported Analyses

All.

W CURRENT-CONTROLLED SWITCH

General Form

```
WYYYYYYY N+ N- VNAME MODEL <ON> <OFF>
```

Examples

```
w1 1 2 vclock switchmod1
w2 3 0 vramp sm1 ON
wreset 5 6 vclk lossyswitch OFF
```

N+ and N− are positive and negative nodes of the switch, respectively. The control current is defined as the current flowing through the specified voltage source. The direction of positive control current is from the positive node through the source to the negative node.

Switch Model

```
.MODEL {model name} CSW <model parameters>
```

This switch allows the modeling of an almost ideal switch in AIM-Spice. The switch is not quite ideal since the resistance cannot change from zero to infinity but must have a finite positive value. The on and off resistances should therefore be chosen very small and very large, respectively, compared to the other circuit elements. The model parameters are as follows:

Name	Parameter	Unit	Default
IT	Threshold current	A	0
IH	Hysteresis current	A	0
RON	On resistance	Ω	1
ROFF	Off resistance	Ω	1/GMIN

An ideal switch is highly nonlinear. The use of switches can cause large discontinuities in node voltages. A rapid change such as that associated with a switching operation can cause problems with roundoff and tolerance, which may lead to erroneous results or problems in selecting proper time steps. To reduce such problems, follow these steps:

Do not set the switch impedances higher or lower than necessary.

Reduce the tolerance during a Transient analysis. This is done by specifying the value for TRTOL less than the default value of 7.0, for example, 1.0.

When switches are placed near capacitors, reduce the size of CHGTOL to, for example, 1×10^{-16}.

Supported Analyses

All.

Z   MESFETs

General Form (Level 1)

```
ZXXXXXXX ND NG NS MNAME <AREA> <OFF> <IC=VDS,VGS>
```

General Form (Levels 2 and 3)

```
ZXXXXXXX ND NG NS MNAME <L=VALUE> <W=VALUE> <OFF>
+ <IC=VDS,VGS> <TD=T> <TS=T>
```

Example

```
z1 7 2 3 zm1 off
z1 0 2 0 mesmod l=1u w=20u
```

ND, NG, and NS are the drain, gate, and source nodes, respectively. MNAME is the model name, AREA is the area factor, L is the channel length, W is the channel width, and OFF indicates an optional initial value for the element in a DC Analysis. If the area factor is omitted, 1.0 is assumed. The optional initial value IC=VDS,VGS is meant to be used together with UIC in a Transient Analysis. See the description of the .IC control statement for a better way to set transient initial conditions. The optional TD and TS values for Levels 2 and 3 are the operation drain and source temperatures, respectively, in centigrade. They override the temperature specified in the option value. If length and/or width are not specified, AIM-Spice will use default values, $L=1$ μm and $W=20$ μm for Levels 2 and 3.

MESFET Model

```
.MODEL {model name} NMF <model parameters>
.MODEL {model name} PMF <model parameters>
```

In AIM-Spice, three MESFET models are included. The difference between the models is in the formulation of the *I–V* characteristics. The parameter LEVEL selects which model to use.

LEVEL=1   Model according to Statz et al. (see Section 5.4)
LEVEL=2   Unified extrinsic model for uniformly doped channel (see
              Section 6.4)
LEVEL=3   Unified extrinsic model for delta-doped channel (see Lee et al.,
              1993)

The following are MESFET model parameters common for all levels:

Name	Parameter	Unit	Default
VTO	Pinch-off voltage	V	−2.0 (Level 1)
			−1.26 (Levels 2, 3)
RD	Drain ohmic resistance	Ω	0
RS	Source ohmic resistance	Ω	0

The following are model parameters for Level 1:

Name	Parameter	Unit	Default
IS	Junction saturation current	A	1×10^{-14}
BETA	Transconductance parameter	A/V^2	2.5×10^{-3}
B	Doping tail extending parameter	1/V	0.3
ALPHA	Saturation voltage parameter	1/V	2
LAMBDA	Channel length modulation parameter	1/V	0
CGS	Zero-bias gate–source junction capacitance	F	0
CGD	Zero-bias gate–drain junction capacitance	F	0
PB	Gate junction potential	V	1
KF	Flicker noise coefficient	—	0
AF	Flicker noise exponent	—	1
FC	Coefficient for forward-bias depletion capacitance formula	—	0.5

The model parameters for MESFET Levels 2 and 3 are listed below. Note that the default values listed correspond to the *n*-channel MESFET used as example in Section 4.4 of Lee et al. (1993).

Name	Parameter	Unit	Default
D	Depth of device (Level 2 only)	m	0.12 μm
DU	Depth of uniformly doped layer (Level 3 only)	m	0.035 μm
RG	Gate ohmic resistance	Ω	0
RDI	Internal drain ohmic resistance	Ω	0
RSI	Internal source ohmic resistance	Ω	0
RI	Resistance in series with C_{gs} (Level 2 only)	Ω	0
RF	Resistance in series with C_{gd} (Level 2 only)	Ω	0
PHIB	Effective Schottky barrier height	eV	0.5
ASTAR	Effective Richardson constant	A/m^2 K^2	4.0
GGR	Junction conductance at reverse bias	Ω^{-1} m^{-2}	40
DEL	Reverse junction conductance inverse ideality factor	—	0.04
N	Junction ideality factor	—	1

Name	Parameter	Unit	Default
LAMBDA	Output conductance parameter	V^{-1}	0.045
LAMBDAHF	Output conductance parameter at high frequencies	V^{-1}	0.045
VS	Saturation velocity (Level 2 only)	m/s	1.5×10^5
BETA	Transconductance parameter (Level 3 only)	A/V^2	0.0085
ETA	Subthreshold ideality factor	—	1.73
M	Knee shape parameter	—	2.5
ALPHA	Bulk charge parameter	V^{-1}	0
MC	Knee shape parameter	—	3.0
SIGMA0	DIBL parameter	—	0.081
VSIGMAT	DIBL parameter	V	1.01
VSIGMA	DIBL parameter	V	0.1
MU	Low-field mobility	$m^2/V\,s$	0.23
THETA	Mobility enhancement coefficient	$m^2/V^2\,s$	0
MU1	First-temperature parameter for mobility	$m^2/V\,s$	0
MU2	Second-temperature parameter for mobility	$m^2/V\,s$	0
ND	Substrate doping (Level 2 only)	m^{-3}	3.0×10^{23}
NDU	Uniform layer doping (Level 3 only)	m^{-3}	1×10^{22}
DELTA	Transition width parameter	—	5
TC	Transconductance compression factor	V^{-1}	0
ZETA	Transconductance compensation factor	—	1.0
NDELTA	Doping of delta-doped layer (Level 3 only)	m^{-3}	6×10^{24}
TH	Thickness of delta-doped layer (Level 3 only)	m	0.01 μm
TVTO	Temperature coefficient for VTO	V/K	0
TLAMBDA	Temperature coefficient for LAMBDA	°C	∞
TETA0	First-temperature coefficient for ETA	°C	∞
TETA1	Second-temperature coefficient for ETA	°C	0 K
TMU	Temperature coefficient for mobility	°C	27
XTM0	First exponent for temperature dependence of mobility	—	0
XTM1	Second exponent for temperature dependence of mobility	—	0
XTM2	Third exponent for temperature dependence of mobility	—	0
TPHIB	Temperature coefficient for PHIB	eV/K	0
TGGR	Temperature coefficient for GGR	K^{-1}	0
KS	Side-gating coefficient	—	0
VSG	Side-gating voltage	V	0
TF	Characteristic temperature determined by traps	°C	TEMP

Name	Parameter	Unit	Default
FLO	Characteristic frequency for frequency-dependent output conductance	Hz	0
DELFO	Frequency range used for frequency-dependent output conductance calculation	Hz	0
AG	Drain–source correction current gain	—	0
RTC1	First-order temperature coefficient for parasitic resistances	°C^{-1}	0
RTC2	Second-order temperature coefficient for parasitic resistances	°C^{-2}	0

For Levels 2 and 3 you can choose between intrinsic or extrinsic model by proper use of parameters RD, RS, and RDI, RSI. If you specify values for RD and RS, you select the intrinsic model with parasitic resistances applied externally. The extrinsic model is selected by specifying values for RDI and RSI.

Temperature Effects (Levels 2 and 3 Only)

The temperature appears explicitly in several exponential terms. In addition, the temperature dependence of several key parameters are modeled as follows (in terms of absolute temperatures):

Threshold voltage:

$$V_T = \text{VTO} - \text{TVTO}(\text{TS} - \text{TNOM})$$

where TNOM is the nominal temperature specified in the General Simulation Options dialog box.

Mobility:

$$\mu_{imp} = \text{MU}(\text{TS/TMU})^{\text{XTM0}}$$

$$\mu_{po} = \text{MU1}(\text{TMU/TS})^{\text{XTM1}} + \text{MU2}(\text{TMU/TS})^{\text{XTM2}}$$

$$\mu_{no}^{-1} = \mu_{imp}^{-1} + \mu_{po}^{-1}$$

where μ_{imp} is the impurity scattering limited mobility, μ_{po} is the polar optical scattering limited mobility, and μ_{no} is the effective zero-bias mobility used in calculating the drain current.

Output conductance parameter:

$$\lambda = \text{LAMBDA}(1 - \text{TS/TLAMBDA})$$

Subthreshold ideality factor:

$$\eta = \text{ETA}(1 + \text{TS/TETA0}) + \text{TETA1/TS}$$

Parasitic resistances RS, RG, RSI, RI:

$$R(\text{TS}) = R(\text{TNOM})[1 + \text{RTC1}(\text{TS} - \text{TNOM}) + \text{RTC2}(\text{TS} - \text{TNOM})^2]$$

Parasitic resistances RD, RD1, RF:

$$R(\text{TD}) = R(\text{TNOM})[1 + \text{RTC1}(\text{TD} - \text{TNOM}) + \text{RTC2}(\text{TD} - \text{TNOM})^2]$$

Supported Analyses

Level 1   All
Levels 2, 3   Distortion, Noise, and Pole-Zero Analysis not supported

---

.INCLUDE STATEMENT

General Form

```
.INCLUDE {filename}
```

Example

```
.include d:\aimspice\cmos.mod
```

filename is the name of the file, with path extension if needed, that will be included in the circuit netlist.

.SUBCKT STATEMENT

General Form

```
.SUBCKT SUBNAME N1 N2 N3 . . .
```

Example

```
.SUBCKT OPAMP 1 2 3 4 5
```

A subcircuit definition starts with the .SUBCKT statement. SUBNAME is the name of the subcircuit used when referencing the subcircuit. N1, N2, ... are external nodes, excluding 0. The group of elements that follow directly after the .SUB-CKT statement defines the topology of the subcircuit. The definition must end with the .ENDS statement. Control statements are not allowed in a subcircuit definition. A subcircuit definition can contain other subcircuit definitions, device models, and call to other subcircuits. Note that device models and subcircuit definitions within a subcircuit definition are local to that subcircuit and are not available outside. Nodes used in a subcircuit are also local, except 0 (ground), which is always global.

.ENDS STATEMENT

General Form

.ENDS <SUBNAME>

Example

.ENDS OPAMP

Each subcircuit definition must end with the .ENDS statement. The name of the subcircuit can be appended to the .ENDS statement, which indicates the end of the corresponding subcircuit. If no name is given, all subcircuit definitions are ended.

CALL TO SUBCIRCUITS

General Form

XYYYYYYY N1 <N1 N3...> SUBNAME

Example

X1 2 4 17 3 1 MULTI

.NODESET STATEMENT

General Form

.NODESET V(NODENAME)=VALUE V(NODENAME)=VALUE ...

Example

```
.NODESET V(12)=4.5 V(4)=2.23
```

This control statement helps AIM-Spice locating the dc operating point. Specified node voltages are used as a first guess of the dc operating point. This statement is useful when analyzing bistable circuits. Normally, .NODESET is not needed.

.IC STATEMENT

General Form

```
.IC V(NODENAME)=VALUE V(NODENAME)=VALUE . . .
```

Example

```
.IC V(11)=5 V(1)=2.3
```

This control statement is used for specifying initial values of a Transient Analysis. This statement is interpreted in two ways, depending on whether UIC is specified or not.

If UIC is specified, the node voltages in the .IC statement will be used to compute initial values for capacitors, diodes, and transistors. This is equivalent to specifying IC=... for each element, only more convenient. IC=... can still be specified and will override the .IC values. AIM-Spice will not perform any operating point analyses when this statement is used, and therefore, the statement should be used with care.

AIM-Spice will perform an operating point analysis before a transient analysis if UIC is not specified. Then the .IC statement has no effect.

Bugs Reported by Berkeley

Models defined within subcircuits are not handled correctly. Models should be defined outside the subcircuit definition starting with the .SUBCKT statement and ending with .ENDS.

Convergence problems can sometimes be avoided by relaxing the maximum stepsize parameter for a transient analysis.

The base node of the bipolar transistor (BJT) is incorrectly modeled and should not be used. Use instead a semiconductor capacitor to model base effects.

Charge storage in MOS devices based on the Meyer model is incorrectly calculated.

Transient simulations of strictly resistive circuits (typical for first runs or tests) allow a time step that is too large (e.g., a sinusoidal source driving a resistor). There is no integration error to restrict the time step. Use the maximum stepsize parameter or include reactive elements.

REFERENCES

M.-C. Jeng, "Design and Modeling of Deep-Submicrometer MOSFET's." ERL Memo No. ERL M90/90, Electronics Research Laboratory, University of California, Berkeley, Oct. (1990).

K. Lee, M. Shur, T. A. Fjeldly, and T. Ytterdal, *Semiconductor Device Modeling for VLSIs,* Prentice-Hall, Engléwood Cliffs, NJ (1993).

J. S. Roychowdhury and D. O. Pederson, "Efficient Transient Simulation of Lossy Interconnect," in *Proceedings of the 28th ACM/IEEE Design Automation Conference,* San Francisco, June (1991).

T. Sakurai and A. R. Newton, "A Simple MOSFET Model for Circuit Analysis and Its Applications to CMOS Gate Delay Analysis and Series-Connected MOSFET Structure," ERL Memo No. ERL M90/19, Electronics Research Laboratory, University of California, Berkeley, Mar. (1990).

M. Shur, M. Jacunski, H. Slade, and M. Hack, "Analytical Models for Amorphous and Poly-silicon Thin Film Transistors for High Definition Display Technology," *Journal of the Society for Information Display,* vol. 3, no. 4, p. 233 (1995).

H. Statz, P. Newman, I. W. Smith, R. A. Purcel, and H. A. Haus, "GaAs FET Device and Circuit Simulation in SPICE," *IEEE Transactions on Electron Devices,* vol. ED-34, pp. 160–169 (1987).

C. Turchetti, P. Prioretti, G. Masetti, E. Profumo, and M. Vanzi, "A Meyer-Like Approach for the Transient Analysis of Digital MOS IC's," *IEEE Transactions on Computer-Aided Design,* vol. 5, pp. 499–507 (1986).

D. E. Ward and R. W. Dutton, "A Charge-Oriented Model for MOS Transistor Capacitances," *IEEE Journal of Solid-State Circuits,* vol. SC-13, pp. 703–708 (1978).

APPENDIX A3

TEMPERATURE DEPENDENCE OF MOBILITIES IN SILICON

The dependence of the electron mobility on doping concentration and on temperature is given by the following expressions [see Caugley and Thomas (1967), Thornber (1982), Sze (1981), Arora et al. (1982), and Yu and Dutton (1985)]:

$$\mu_n(N_T, T) = \mu_{mn} + \frac{\mu_{on}}{1 + (N_T/N_{cn})^\nu} \tag{A3.1}$$

$$\mu_{mn} = 88\left(\frac{T}{300}\right)^{-0.57} \quad (\text{cm}^2/\text{V s}) \tag{A3.2}$$

$$\mu_{on} = 1250\left(\frac{T}{300}\right)^{-2.33} \quad (\text{cm}^2/\text{V s}) \tag{A3.3}$$

$$\nu = 0.88\left(\frac{T}{300}\right)^{-0.146} \tag{A3.4}$$

$$N_{cn} = 1.26 \times 10^{17}\left(\frac{T}{300}\right)^{2.4} \quad (\text{cm}^{-3}) \tag{A3.5}$$

The corresponding expressions for holes are

$$\mu_p(N_T, T) = \mu_{mp} + \frac{\mu_{op}}{1 + (N_T/N_{cp})^\nu} \tag{A3.6}$$

$$\mu_{mp} = 54\left(\frac{T}{300}\right)^{-0.57} \quad (\text{cm}^2/\text{V s}) \tag{A3.7}$$

$$\mu_{op} = 407\left(\frac{T}{300}\right)^{-2.23} \quad (\text{cm}^2/\text{V s}) \tag{A3.8}$$

$$N_{cp} = 2.35 \times 10^{17}\left(\frac{T}{300}\right)^{2.4} (\text{cm}^{-3}) \tag{A3.9}$$

$$\nu = 0.88\left(\frac{T}{300}\right)^{-0.146} \tag{A3.10}$$

See Figs. 2.3.2 and 2.3.4.

REFERENCES

N. D. Arora, J. R. Hauser, and D. J. Rulson, *IEEE Transactions on Electron Devices*, vol. ED-29, p. 292 (1982).

D. M. Caugley and R. E. Thomas, *Proceedings of the IEEE*, vol. 55, pp. 2192–2193 (1967).

S. M. Sze, *Physics of Semiconductor Devices*, 2nd ed., Wiley, New York (1982).

K. K. Thornber, *IEEE Electron Device Letters*, vol. EDL-3, no. 3, pp. 69–71 (1982).

Z. Yu and R. W. Dutton, *Sedan III—A General Electronic Material Device Analysis Program*, Program manual, Stanford University, July (1985).

APPENDIX A4

PARASITIC SERIES
RESISTANCES IN FETs

The drain and source parasitic series resistances R_d and R_s play an important role in limiting the device performance. These resistances may be accounted for by using the following expressions relating the "intrinsic" gate–source and drain–source voltages V_{GS} and V_{DS} to the "extrinsic" (measured) gate–source and drain–source voltages V_{gs} and V_{ds} that include voltage drops across the series resistances:

$$V_{GS} = V_{gs} - I_d R_s \tag{A4.1}$$

$$V_{DS} = V_{ds} - I_d (R_s + R_d) \tag{A4.2}$$

The measured (extrinsic) transconductance of a FET is

$$g_m = \frac{dI_d}{dV_{gs}} \bigg|_{V_{ds}} \tag{A4.3}$$

Using elementary circuit theory (see, e.g., Sze, 1985), g_m can be related to the intrinsic transconductance g_{mo} and the intrinsic drain conductance, g_{do}, of the same device as follows:

$$g_m = \frac{g_{mo}}{1 + g_{mo} R_s + g_{do}(R_s + R_d)} \tag{A4.4}$$

where

$$g_{mo} = \frac{dI_d}{dV_{GS}}\bigg|_{V_{DS}} \tag{A4.5}$$

$$g_{do} = \frac{dI_d}{dV_{DS}}\bigg|_{V_{GS}} \tag{A4.6}$$

Likewise, the measured (extrinsic) drain conductance

$$g_d = \frac{dI_d}{dV_{ds}}\bigg|_{V_{gs}} \tag{A4.7}$$

is related to the intrinsic conductances through

$$g_d = \frac{g_{do}}{1 + g_{mo}R_s + g_{do}(R_s + R_d)} \tag{A4.8}$$

In the linear region, at very small drain–source voltages, g_{mo} is small and the term $g_{mo}R_s$ can be neglected in Eqs. (A4.4) and (A4.8). In saturation, where g_{do} is very small in long-channel transistors, the term $g_{do}(R_s + R_d)$ can be neglected. However, these approximations are not valid for very short devices (see Chou and Antoniadis, 1987).

In addition to its effect on the drain–source and gate–source voltages, the potential drop across the source series resistance also modifies the substrate–source voltage. The extrinsic substrate–source voltage V_{bs} of a MOSFET is related to its intrinsic counterpart, V_{BS}, as follows:

$$V_{BS} = V_{bs} - I_d R_s \tag{A4.9}$$

In order to investigate the consequences of this shift on the extrinsic transconductance and the extrinsic drain conductance, it is convenient to define a new set of transconductances related to the substrate bias:

$$g_{mb} = \frac{dI_d}{dV_{bs}}\bigg|_{V_{ds}, V_{gs}} \tag{A4.10}$$

$$g_{mbo} = \frac{dI_d}{dV_{BS}}\bigg|_{V_{DS}, V_{GS}} \tag{A4.11}$$

where g_{mb} and g_{mbo} are called the extrinsic and intrinsic substrate transconductances, respectively. Again, from simple circuit theory, we find the following generalized relationship between extrinsic and intrinsic conductances [this can also be seen directly from Eqs. (A4.4) and (A4.8) by noting the symmetry between the definitions of gate transconductance and substrate transconductance]:

$$\frac{g_m}{g_{mo}} = \frac{g_d}{g_{do}} = \frac{g_{mb}}{g_{mbo}} = \frac{1}{1 + R_s(g_{mo} + g_{mbo}) + g_{do}(R_s + R_d)} \qquad (A4.12)$$

(see Cserveny, 1990). Equation (A4.12) shows that all extrinsic conductances are reduced by the effect on the substrate–source voltage of the voltage drop across R_s, compared to the previous results where this effect was neglected. As noted by Cserveny (1990), the error incurred in saturation by neglecting the term $R_s g_{mbo}$ in Eq. (A4.12) can be quite important. In fact, for long-channel devices in saturation, this term will normally be more important than the term $g_{do}(R_s + R_d)$.

REFERENCES

S. Y. Chou and D. A. Antoniadis, *IEEE Transactions on Electron Devices*, vol. ED-34, p. 448 (1987).

S. Cserveny, "Relationship Between Measured and Intrinsic Conductances of MOSFETs," *IEEE Transactions on Electron Devices*, vol. ED-37, no. 11, pp. 2413–2414 (1990).

S. M. Sze, *Semiconductor Devices, Physics, and Technology*, Wiley, New York (1985).

INDEX